The Political Culture of the United States

The Little, Brown Series
in Comparative Politics

Under the Editorship of

GABRIEL A. ALMOND

JAMES S. COLEMAN

LUCIAN W. PYE

AN ANALYTIC STUDY

The Political Culture of the United States

The Influence of Member Values on Regime Maintenance

Donald J. Devine

University of Maryland

Boston
LITTLE, BROWN AND COMPANY

LIBRARY OF CONGRESS CATALOG CARD NUMBER: 70-177904

FIRST PRINTING

Published simultaneously in Canada
by Little, Brown & Company (Canada) Limited

PRINTED IN THE UNITED STATES OF AMERICA

Acknowledgments and Copyrights

The author is grateful to the following publishers, polling organizations, and copyright holders for permission to use material reprinted in this book:

American Institute of Public Opinion and George Gallup, Jr. For AIPO public opinion poll data, which appear in tables throughout the book.

American Sociological Review. For Tables VI.22 and VI.23.

Catholic Digest. For Catholic Digest Poll data in Tables IV.62 and IV.64.

CBS News. For Table V.10.

Fortune Magazine. For Fortune Poll data, which appear in tables throughout the book.

Ben Gaffin and Associates. For Table V.8.

Harcourt Brace Jovanovich, Inc. For Table IV.34, reprinted with permission from Austin Ranney and Willmoore Kendall, *Democracy and the American Party System* (New York: Harcourt Brace Jovanovich, 1956), p. 503.

Louis Harris and Associates. For Table VI.21 and data in Table IV.64.

Inter-University Consortium for Political Research, University of Michigan. For Survey Research Center election study data, which appear in tables throughout the book.

The Macmillan Company. For Table V.14, reprinted with permission of The Macmillan Company from Herbert H. Hyman, "The Value Systems of Different Classes," in Reinhard Bendix and Seymour Martin Lipset, eds., *Class, Status, and Power,* 2d ed. (New York: Free Press, 1966), p. 499. Copyright © 1966 by The Free Press, a division of The Macmillan Company.

William Morrow & Company, Inc. For Table IV.36, reprinted with permission from Elmo Roper, *You and Your Leaders: Their Actions and Your Reactions, 1936–1956* (New York: Morrow, 1957), p. 33.

National Opinion Research Center. For NORC public opinion poll data, which appear in tables throughout the book.

Prentice-Hall, Inc. For Figure I.2, reprinted with permission from David Easton, *A Framework for Political Analysis* (Englewood Cliffs, N.J.: Prentice-Hall, 1965), p. 110, copyright © 1965.

Princeton University Press. For Tables IV.8 and IV.50, reprinted with permission from Gabriel A. Almond and Sidney Verba, *The Civic Culture: Political Attitudes and Democracy in Five Nations* (Princeton: Princeton University Press, 1963), pp. 102, 108. Copyright © 1963 by Princeton University Press. Copyright © 1965 by Little, Brown and Company, Inc. For public opinion poll data in *Public Opinion, 1935–1946,* Hadley Cantril, ed., prepared by Mildred Strunk, copyright 1951 by Princeton University Press, in Tables III.8, III.10, III.11, IV.2, IV.3, IV.7, IV.12, IV.17, IV.19, IV.20, IV.23, IV.28, IV.30, IV.31, IV.32, IV.35, IV.37, IV.38, IV.39, IV.41, IV.44, IV.45, IV.47, IV.48, IV.49, IV.51, IV.53, IV.54, IV.59, IV.61, IV.62, IV.64, IV.67, V.1, V.15, VI.1, VI.2, VI.3, VI.4, VI.8, VI.9, VI.11, VI.12, VI.13, VI.15, VI.18, VI.19. Reprinted by permission of Princeton University Press.

Rand McNally & Company. For Table VII.11, adapted from Daniel Goldrich, *Sons of the Establishment* (Chicago: Rand McNally, 1966), pp. 108, 113. Copyright © 1966 by Rand McNally & Company, Chicago.

Rutgers University Press. For Tables IV.4 and IV.65, reprinted with permission from Lloyd A. Free and Hadley Cantril, *The Political Beliefs of Americans: A Study of Public Opinion* (New Brunswick, N.J.: Rutgers University Press, 1968), pp. 193, 192.

Time Magazine. For Table IV.42, reprinted with permission from *Time,* the weekly newsmagazine; © 1948, by Time, Inc.

University of California Press. For Table VII.3, reprinted with permission of the Regents of the University of California from Mildred A. Schwartz, *Public Opinion and Canadian Identity* (Berkeley: University of California Press, 1967), pp. 107, 110, 116, 117.

University of Illinois Press. For Tables III.3 and III.4, reprinted with permission from William Buchanan and Hadley Cantril, *How Nations See Each Other* (Urbana: University of Illinois Press, 1953), pp. 30, 18.

Wayne State University Press. For Table IV.10, reprinted with permission of Wayne State University Press from G. B. Boynton, Samuel C. Patterson, and Ronald H. Hedlund, "The Structure of Public Support for Legislative Institutions," *Midwest Journal of Political Science,* Vol. 12, No. 2 (May, 1968), p. 166, copyright © 1968 by Wayne State University Press. For Table VII.2, reprinted with permission of Wayne State University Press, from Reid Reading, "Political Socialization in

Colombia and the United States," *Midwest Journal of Political Science*, Vol. 12, No. 3 (August, 1968), p. 361, copyright © 1968 by Wayne State University Press.

Yale University Press. For Table III.1, derived from Richard L. Merritt, *Symbols of American Community, 1735–1770* (New Haven: Yale University Press, 1966), p. 215.

Foreword

Donald J. Devine's *Political Culture of the United States* continues the interest of the Analytic Studies in the theme of political culture and socialization. It moves ahead from earlier work both theoretically and methodologically. It specifies the levels and elements of political culture in greater detail than earlier investigations, operationalizes these elements, and brings together and analyzes in terms of these levels and elements an enormous body of data from opinion surveys conducted in the United States over the past thirty-five years. From these data Devine seeks to prove that the political culture of the United States reflects a broad consensus among all groups in the population on the basic principles of Lockean liberalism. His data suggest that this consensus tends to hold up under conditions of conflict and crisis and is largely shared by minority groups. Devine's main hypothesis is that the persistence of American political institutions may in part be explained by this basic Lockean consensus.

Gabriel A. Almond
James S. Coleman
Lucian W. Pye

Preface

Recent empirical research on public opinion and democratic theory has tended to emphasize the limited influence the masses can have on politics. As a reaction to the earlier stimulus-response model of total political control by the masses, this emphasis upon limits was probably in order. Most modern political science now, however, uses a more complex system model. In this model, the members are seen neither as necessarily the greatest influence on public policy nor are they viewed as irrelevant to the political process.

The present work attempts to describe what I believe to be the major way in which the general public influences the political process. As such it is a continuation of my study of the influence of various publics on the political system. In my previous work, *The Attentive Public,*[1] I attempted to show the relative influence of the general and attentive publics upon the policy-making process in the United States. The present work focuses upon the influence of the mass public on the more basic process of regime maintenance. In the future I hope to follow Gabriel Almond's hierarchy of publics[2] and test the influence of higher elites and political leaders on both processes.

The present analysis of member value influence on regime

[1] Donald J. Devine, *The Attentive Public: Polyarchical Democracy* (Chicago: Rand McNally, 1970).

[2] Gabriel A. Almond, *The American People and Foreign Policy* (New York: Praeger, 1960; originally published 1950), pp. 137–139; Devine, pp. 31–33.

maintenance is immensely aided by the use of certain social science tools. Like any abstract science (as opposed to a practical engineering accumulation of solutions), political science can only advance with the cumulative development of conceptual and operational tools of analysis. The present study has been the fortunate inheritor of several important social science developments: (1) the analogy of system and its theoretical refinement by various authors – especially by David Easton, (2) the extremely heuristic concept of culture developed by the anthropologists, (3) the relatively recent empirical studies of political behavior in political psychology and sociology, (4) a hypothesis of an American value system developed by a historian, (5) a political science analysis of the American political process related to all branches of the discipline but especially with the behavioral-realist school,[3] (6) an important study in social psychology, and (7) over three decades of data from public opinion polls. Without the contribution of these many different perspectives, even rather primitive analyses of the social complexity of the political world like the present one would be impossible. Hopefully, with the use of these tools, some knowledge can be obtained.

The investigator is aided not only by the formal tools of the discipline but probably more importantly by his colleagues. I was extremely fortunate to have been helped in the current study by some very able political scientists. Frank Munger suggested to me that a study of the political culture of the United States could be attempted even with the limited tools presently available. Once again the conceptual skills of Gabriel Almond provide the central theme of my work. Others have also been helpful with suggestions. Whether I properly benefited from their advice is problematical. They certainly cannot be held responsible for any of the many limitations of this work.

Certain organizations were very important in the preparation of this book. The Department of Government and Politics and the College of Business and Public Administration

[3] For example, Robert A. Dahl, *Pluralist Democracy in the United States* (Chicago: Rand McNally, 1967). This "school" is analyzed and compared to two other major schools in Glendon Schubert, *The Public Interest* (Glencoe, Ill.: Free Press, 1960).

of the University of Maryland graciously provided a semester of reduced teaching and the University General Research Board an additional summer to allow the manuscript to be completed. The Computer Center at the same university also provided many hours of computing time. The Inter-University Consortium for Political Research at the Survey Research Center, University of Michigan, provided much of the data for analysis. The major poll organizations have been centrally important for the present work. The poll pioneers like Dr. George Gallup and Elmo Roper have my special thanks. I am also indebted to the people at Little, Brown, especially to Patricia Herbst, and to the editors of the series.

Finally and most importantly. I must thank my wife Ann. Her skills – secretarial, substantive, and personal – contributed substantially to this work. I would also like to thank William, Patricia, and Joseph for their understanding and patience.

Donald J. Devine

Contents

List of Figures and Tables

FIGURES

TABLES

The Political Culture of the United States

CHAPTER I

Political Culture and the Political System

CULTURE AND POLITICS

INTRODUCTION

Change of governmental form is a particular phenomenon of the second part of the twentieth century. One estimate suggests that sixty-four out of eighty-four countries of the world were less stable between 1955 and 1962 than they were between 1948 and 1954. In the years after World War II successful coups d'état occurred in seventeen of twenty countries of South America, and in many of the countries of Africa and Asia.[1] In the last three decades revolution or conquest by revolutionary force has changed the regimes of Eastern Europe, China, Cuba, most of Africa, and much of Asia. No country has been unaffected by the forces of change.

In a world of such turmoil how have some countries managed to preserve their form of government? In the face of the extreme stress characteristic of this age, traditional means of regime maintenance such as coercion and resource allocation to relevant groups have seemed inadequate.[2] It is the major argument of the present work that generalized support from

[1] Samuel P. Huntington, *Political Order in Changing Societies* (New Haven: Yale University Press, 1968), pp. 3, 4.

[2] David Easton, *A Systems Analysis of Political Life* (New York: Wiley, 1965), p. 325. Hereafter referred to as *A Systems Analysis*.

the ordinary members of the political system is one important way of explaining how some polities are able to maintain themselves in the face of severe stress. The United States seems to be one of the few countries which have been characterized by relative regime stability rather than by frequent regime change. If so, American support attitudes could be used as an empirical focus for looking at the relationship between regime maintenance and the values held by political system members.

Recent empirical findings in political science research have found that the ordinary member of the political system has limited capacity for political rationality. This has led to a feeling that one should not look to the masses to understand political phenomena. Although I do not accept the overly rationalistic, democratic-egalitarian view of the mass public as the major institutional focus for day-to-day policy formation,[3] I do believe that the member plays a crucial role in the more fundamental processes of political systems. The central aspect of this role is the extent to which the member's values influence his support or rejection of his type of government.

In order to look at the values of a people, especially the values of one's own people, it is extremely important to be able to move from the world of common sense to the realm of science. This can allow one to gain perspective. Why are terms like *class, tricolore, Internationale, Reichstag, duce, Allah,* and *Kamikaze* just words with little or no meaning to Americans, whereas *America, Stars and Stripes, Star-Spangled Banner, Congress, the president, God,* and *freedom* are expressions of great importance and significance? To Americans, the former invoke at the most mild interest, whereas the latter represent values for which many are willing to give even their lives.

To study emotion-laden values an attempt must be made to find approaches and methods which remove one from the attachments of the "real world." The theoretical framework sought must not only abstract from reality to provide perspective but it must also abstract because science cannot deal with the raw complexity of the real world. In a world of

[3] Donald J. Devine, *The Attentive Public: Polyarchical Democracy* (Chicago: Rand McNally, 1970).

almost infinite complexity, science seeks to understand. The fundamental problem of science — natural or social — is to create order out of the mass of sense perceptions which constitute reality. This can only be done, to the extent that it can be done, with the use of concepts which help us understand this "real world." [4]

Culture is one of the concepts which helps us make order out of the chaos of social interaction. One political scientist even maintains that "the discovery and elaboration of the concept of culture is one of the few genuine breakthroughs in the social sciences." [5] It is viewed as a breakthrough because it allows social science to assume a given. Science, like any other human pursuit, must start somewhere — assumptions must be made for investigation to begin.[6] Culture provides such a starting-point for any study of social life.

The concept of culture allows the analyst to forge a connection between the values of the members of a political system and the form of government in the system. The scientific perspective of culture focuses upon basic values which are widely, that is consensually, shared. The investigation of basic regime values has been generally relegated to secondary treatment in public opinion analysis. Most of the data collection done on mass attitudes is on everyday matters of public policy. Yet there is evidence that this kind of public opinion is less crucial to a political system, especially to its *type* of government, than are deep-seated, cultural values.[7] The present work attempts to empirically and theoretically analyze these culture values and the influence they have on the main-

[4] My view of science rests basically upon Karl R. Popper, *The Logic of Scientific Discovery* (New York: Harper and Row, 1965; originally published in German, 1934; in English, 1958); and Leonard K. Nash, *The Nature of the Natural Sciences* (Boston: Little, Brown, 1963). For a competent summary of the position see Arthur S. Goldberg, "Political Science as Science," in Nelson W. Polsby, Robert A. Dentler, and Paul A. Smith, eds., *Politics and Social Life* (Boston: Houghton Mifflin, 1963), pp. 26–36.

[5] Bernard C. Hennessy, *Public Opinion* (Belmont, Calif.: Wadsworth, 1965), p. 190.

[6] See note 4. For an excellent discussion of the question of assumptions see Milton Friedman, *Essays in Positive Economics* (Chicago: University of Chicago Press, 1953), ch. 1.

[7] David Easton and Robert D. Hess, "Youth and the Political System," in Seymour Martin Lipset and Leo Lowenthal, eds., *Culture and Social Character* (New York: Free Press, 1961), p. 234.

tenance and change of governmental forms. More particularly, culture is the concept chosen here to try to explain how, in a world of extreme change, essentially one regime has been maintained in the United States under deeply changing environmental conditions across almost two centuries.

CULTURE

Culture is a concept developed by anthropologists to compare member values in different social systems. When they investigated peoples with predominantly non-Western life styles and values, the differences between those styles and the European ones seemed quite large. The anthropological social scientist developed the concept of culture to identify this basic difference in life styles. On the other hand, the idea of culture has a long history in social analysis. Anthropologists, by making it an explicit *concept,* have greatly expanded its benefits for scientific analysis. Aristotle, however, used the concept of *habit* to express some of the aspects of culture. Much later, John Stuart Mill called the science of national character formation *ethology*. Use of the specific concept of human culture, however, came late in historical development, apparently after 1750 – and then in the German language only. The anthropologists Kroeber and Kluckhohn find that the term is not used in English until 1871, and it is not until the 1920's that American anthropologists use the term to any great extent.[8]

There are now a great many definitions of culture. In their monumental study, Kroeber and Kluckhohn investigated 164 of them! Kluckhohn himself defines culture as "the total life way of a people, the social legacy the individual acquires from his group. Or culture can be regarded as that part of the environment that is the creation of man." Malinowski, one of the pioneers of cultural anthropology, defined culture as "the man-made environment which must be renewed."

[8] "Ethica Eudemia," trans. J. Solomon, in *The Works of Aristotle,* W. D. Ross, ed. (London: Oxford University Press, 1925), II. 2. 1220a,b; John Stuart Mill, *A System of Logic* (London: Longmans, Green, 1896), book VI, p. 567; Alfred Kroeber and Clyde Kluckhohn, *Culture: A Critical Review of Concepts and Definitions* (New York: Random House, Vintage Books, 1952), pp. 283–302, 11.

Ralph Linton said that it was "the sum total of ideas, conditioned emotional responses, and patterns of habitual behavior" which individuals in a society have learned.[9]

Other social scientists have found that similar elements determine the meaning of the concept. A major social psychologist calls culture the "total way of life characteristic of a somewhat homogeneous society of human beings." A political scientist likewise asserts that culture "is the whole pattern of learned social values, myths, and traditions, along with the physical products of man's labors, created and shared by the members of a society." Another social scientist simply calls it the usual way organized groups of people act. A major sociologist considers culture to be the value product given in social relations. A definition from the perspective of the sociological political scientist is that culture is the "sum total of artifacts, belief systems and symbols prevalent in society which shape the behavior of its members." To a final analyst it is the "accumulation through generations of the artifacts, knowledge, beliefs, and values by which men deal with the world." [10]

Each of these definitions stresses a set of patterned values which sum to a total way of life for a people. A culture can be viewed as a value system which expresses the unique attitudes that a people passes from generation to generation. Most social scientists using the concept stress that the culture consists of supportive attitudes toward value objects such as principles, norms, rules, institutions, artifacts, and other symbols of a people.[11] Most also maintain that culture must have

[9] Clyde Kluckhohn, *Mirror for Man* (Greenwich, Conn.: Fawcett, 1963; originally published 1949), p. 24; Bronislaw Malinowski, *A Scientific Theory of Culture* (Chapel Hill: University of North Carolina Press, 1944), p. 39; Ralph Linton, *The Study of Man* (New York: Appleton-Century-Crofts, 1936), p. 288.

[10] Goodwin Watson, *Social Psychology* (Philadelphia: Lippincott, 1966), p. 306; Hennessy, p. 191; Saxon Graham, *American Culture: An Analysis of Its Development and Present Characteristics* (New York: Harper, 1957), p. 3; Talcott Parsons, *The Social System* (New York: Free Press, 1951), p. 17; William C. Mitchell, *The American Polity* (New York: Free Press, 1962), p. 22; Robin M. Williams, Jr., *American Society*, 2d ed. (New York: Knopf, 1960), p. 22.

[11] Many maintain that the physical objects are part of the culture. However, the present work will hold that the objects themselves are not behaviorally important unless the members support them attitudinally.

a relationship to behavior.[12] The value system must be related to the other social systems and it should influence the selection of what is considered socially important.[13] Most would agree that:

> culture is a product; is historical; includes ideas, patterns, and values; is selective; is learned; is based upon symbols; and is an abstraction from behavior and the products of behavior.[14]

The present study views culture basically in these social science terms. Culture is conceived as a behaviorally significant, patterned value system composed of supportive member attitudes toward principled beliefs, rule norms, artifacts, and symbols. These supportive attitudes may or may not be congruent with politically sanctioned values. In total they sum to produce the culture. To be considered cultural these values must also be transmitted, learned, and widely shared over a substantial period of time.[15]

THE CULTURAL INFLUENCE IN THE SOCIAL PROCESS

The present study is interested in the values of the ordinary member and how these affect the social process. Since culture is viewed as behaviorally significant, it is viewed as an aid in this explanation. Analysis suggests to me that culture has two behavioral dimensions important for social analysis. These relate to the influence of culture on the external social process and to the internal reaction of culture itself.

Early investigators of culture emphasized the preponderant influence of cultural factors in determining social action. For example, Linton maintained that "culture must be considered the dominant factor in establishing the basic personality types for various societies." Doob, another pioneer, observed that "by making a very few [cultural] assumptions it is possible to predict an important segment of an individual's be-

[12] Clyde Kluckhohn, "Culture and Behavior," in Gardner Lindzey, ed., *Handbook of Social Psychology* (Cambridge: Addison-Wesley, 1954), 2:923–924.

[13] See Moshe M. Czudnowski, "A Salience Dimension of Politics for the Study of Political Culture," *American Political Science Review* (September, 1968), p. 884.

[14] Kroeber and Kluckhohn, p. 308.

[15] Parsons, p. 15; also see Hennessy, pp. 191–194.

havior before his birth." He mentions such things as his language, general life style, theological ideas, clothing, and, in many cases, occupation. Albig held that "the materials of a particular culture determine the context of the mind and also, to some extent, the ways of the mind." [16]

In a general sense culture undoubtedly creates the assumptions most easily accepted in any particular social circumstance. However, to maintain a strict deterministic view (that of Leslie White, for example) [17] seems to overemphasize cultural force. The deterministic view conceives culture as an overwhelming value force which determines all social response. The individual and his society are totally determined by the past social value product of the culture. The theoretical analogy is that of stimulus-response. This model can be conceived as a reaction to two earlier analogies: one which overemphasized the pure stimulus-response *rationality* of the individual and another which overemphasized the pure stimulus-response determinism of the *non-value* environmental structure of society. The reaction, however, appears to be itself overbalanced in its emphasis on value factors. It seems to me that the more balanced analogy is one which conceives of response — social behavior — as the nondeterminate result of an *interaction process* where environmental stimuli act upon a cultural value product (which may or may not be rationally determined).

Since the analogy to be used rejects determinism, culture cannot be conceived as totally forming attitudes and behavior. Fortunately, social scientists have provided a term which describes the interaction conception of culture. Kroeber and Kluckhohn, Parsons, Watson, Mitchell, and, indeed, most of those who use the concept of culture today, use the idea of *shaping* to express the force of culture in social relations.[18]

[16] Ralph Linton, *The Cultural Background of Personality* (New York: Appleton-Century-Crofts, 1945), p. 151; Leonard W. Doob, *Public Opinion and Propaganda,* 2d ed. (Hamden, Conn.: Archon Books, 1966), p. 48; William Albig, *Modern Public Opinion* (New York: McGraw-Hill, 1956), p. 74.

[17] Leslie A. White, *The Science of Culture* (New York: Farrar, Straus and Cudahy, 1949), esp. pp. 167–189.

[18] Kroeber and Kluckhohn, p. 338; Parsons, p. 977; Mitchell, p. 22; and Watson, p. 305. Watson used the term *mold.* Actually, Linton, Doob, and Albig do not believe in total determination. Albig, for example, talks of

Culture shapes – that is, it influences – social action but it does not fully determine it.

> Human behavior cannot be understood in terms of the organism-environment model unless this be made more complex. No socialized human being views his experience freshly. His very perceptions are screened and distorted by what he has consciously and unconsciously absorbed from his culture. Between the stimulus and the response there is always interposed an intervening variable, unseen but powerful.[19]

Social action, then, is shaped by culture. But it is a shaping influence, not a fully determining force. If it were determining, there could be no change in society. This static view is rejected. Social behavior is conceived instead as the joint product of an interaction between environmental stimuli and a resisting cultural organism. Culture alone does not determine, but it is an important shaping influence.

Theorists of culture are not the only ones who provide a justification for the use of the concept of shaping when considering human behavior.[20] Psychological theory also suggests the usefulness of the concept. We turn to psychological theory because "culture is not a mystical 'force' acting at a distance. *Concretely,* it is created by individual organisms and by such organisms operating as a group." [21] Since the member is the unit, the idea of shaping could be reinforced if it was consistent with psychological theory. According to a social psychologist, each individual has a "reality world" where the outside real world is meaningfully translated for him.[22] Individuals are viewed as having reality worlds composed of views of

---

the organism's being *conditioned* by culture. Also see Graham, p. 31, and J. P. Nettl, *Political Mobilization* (New York: Basic Books, 1967), p. 27. Even a critic of the implications of the concept used the term *shape* to describe its operation; see John Dewey, *Freedom and Culture* (New York: Putnam, 1939), p. 20.

[19] Kroeber and Kluckhohn, p. 336; also see pp. 368, 337–338.

[20] Karl W. Deutsch, in *The Nerves of Government* (New York: Free Press, 1963), presents a communication model with some similarities.

[21] Kroeber and Kluckhohn, p. 367.

[22] Hadley Cantril, "Introduction," in Lloyd A. Free, *Six Allies and a Neutral* (Glencoe, Ill.: Free Press, 1959). Dawson and Prewitt call it "political self"; see Richard E. Dawson and Kenneth Prewitt, *Political Socialization* (Boston: Little, Brown, 1967), pp. 16–17; also see Williams, ch. 3.

reality, perceptions of what is good (i.e., values), and other means of interpreting environment. The reality of a reality world, however, can only be compared to the reality of a map of the world. To the same extent that a map can faithfully report only certain limited characteristics of the land it pictures (Greenland is disproportionately large on the most-used maps; distance is distorted above the equator on many maps; mountain ranges are excluded on some, railroads on others), so an individual's reality world can only picture some aspects of "reality." [23]

Since the reality world must simplify, it presumably does so through some process. Psychologists call this process *object appraisal.* By using categories of appraisal, reality worlds shape how a person views the outside world and, presumably, how he will behave.[24] One of these appraisal categories is the individual's belief set. A person's beliefs, especially his deeply held values, seem to be central in shaping how he perceives and behaves. Value appraisal may also take the form of what is called *selective perception.*

> It is essential to the psychic economy of organisms that they perceive selectively. They have a perceptual screen which passes some stimuli and blocks others. . . . An organism would be overwhelmed if it attempted to attend to every stimulus impinging on its sensory organs.[25]

This screening takes place to the extent to which commitment exists to the content of the screen. The more an individual is committed to a value, the more the screen shapes the stimuli. For strongly held values the individual will seek information congenial to his values and will interpret information in a manner congenial to the values. Indeed, research has shown that people will actually impute opposite

[23] William Buchanan and Hadley Cantril, *How Nations See Each Other* (Urbana: University of Illinois Press, 1953), pp. 2–3. Also see Walter Lippmann, *Public Opinion* (New York: Free Press, 1965; originally published 1922), esp. pp. 55, 65.

[24] M. Brewster Smith, "Opinions, Personality, and Political Behavior," *American Political Science Review* (March, 1958), pp. 1–26, and the citations that follow.

[25] Lester W. Milbrath, *Political Participation* (Chicago: Rand McNally, 1965), p. 33.

meanings to information to keep it congenial to their value system.[26]

The culture is conceived to act in an analogous manner. Indeed, this is the manner conceived by most of the analysts of culture. The value screen of culture appraises and selectively perceives stimuli which act on it. Stimuli which are congenial are accepted and allowed to act on the social process. Those which are not tend to be screened out. Other stimuli are accepted, but they are shaped by the values of the culture before they act on the social process.

The insights of psychological research also aid in describing the second important behavioral dimension of culture, the internal reaction of culture itself. In a social sense, it has been suggested that culture exists prior to the rest of social action.[27] This culture system, thus, is in a position to shape. As noted above, it shapes stimuli to conform to its values. Action is processed through this system and social behavior is influenced. As the culture system successfully deals with stress, its very success increases its shaping power and it is better able to handle stress. Psychological research indicates that man creates images which filter reality for him. Once these images satisfy him, he is very reluctant to change them.[28] Often be-

[26] *Ibid.*, p. 35; Bernard R. Berelson, Paul F. Lazarsfeld, and William N. McPhee, *Voting* (Chicago: University of Chicago Press, 1945), pp. 220–223; Herbert H. Hyman and Paul B. Sheatsley, "Some Reasons Why Information Campaigns Fail," *Public Opinion Quarterly* (Fall, 1947), pp. 412–423; and Robert E. Lane and David O. Sears, *Public Opinion* (Englewood Cliffs, N.J.: Prentice-Hall, 1964), ch. 5.

[27] Parsons, p. 17. It should be noted that Parsons has changed his view of culture somewhat. Culture is both existing pattern maintenance and part of the action system; see Talcott Parsons, "Culture and the Social System: Introduction," in Talcott Parsons *et al.*, eds., *Theories of Society* (New York: Free Press, 1961), 2: 964; and "A Rejoinder to Ogles and Levy," *American Sociological Review* (April, 1969), p. 249. Parsons' final view is closer to the one taken here, but culture is viewed as prior (see page 24).

[28] Hyman and Sheatsley, pp. 412–423; Graham, p. 37; Don D. Smith, "Cognitive Consistency and the Reception of Others' Opinions," *Public Opinion Quarterly* (Spring, 1968), pp. 1–15; Carl I. Hoveland and Walter Weiss, "The Influence of Source Credibility on Communication Effectiveness," *Public Opinion Quarterly* (Winter, 1951–1952), pp. 635–648; Charles Y. Glock, "The Comparative Study of Communication and Opinion Formation," in William Schramm, ed., *The Process and Effects of Mass Communications* (Urbana: University of Illinois Press, 1965), p. 476; and B. F. Skinner, *Science and Human Behavior* (New York: Free Press, 1965), pp. 138–140, 65, 76–77.

liefs are held under the most "irrational" or threatening circumstances. For example, in one study psychologists analyzed a religious movement which prophesied the end of the world. When the appointed day arrived, members of the movement did not change their belief patterns. Strangely, a great need to proselytize actually manifested itself. The investigators hypothesized that only by increasing the number of converts could the leaders reduce cognitive confusion. "If more people believe it, it must be true" seemed to be the reasoning.[29] Another study found that very few American prisoners were actually converted by the systematic "brainwashing" program of the Chinese Communists during the Korean War. Many individuals made accommodations, but even severe physical and psychological stress techniques did not permanently change beliefs.[30] What is true for the individual is also conceived to be true of the totality. Although culture is more complex than a simple summing operation (it must at least consider intensity of opinion holding),[31] culture is essentially a product of individual value systems. Since individuals resist change in their values, the societal culture is also seen as resistant.[32] This is the second social aspect of culture. Culture resists change in its composition. Every social system is influenced by this cultural phenomenon.[33] The relative stability of culture is to be contrasted with the dynamic nature of change within cultural limits.

A recent study of individual attitudes on candidates, religion, and party provides some empirical substantiation to the assumption that sets of values in individual reality worlds resist change. The study found that for individuals "changes in attitude are always in the direction and in amounts involving the least total disturbance to the set of attitudes." [34]

[29] Leon Festinger, Henry W. Riecken, and Stanley Schacter, "When Prophecy Fails," in Eleanor E. Maccoby, Theodore M. Newcombe, and Eugene L. Hartley, eds., *Readings in Social Psychology,* 3d ed. (New York: Holt, 1958), pp. 156–163.

[30] Edgar H. Schein, "The Chinese Indoctrination Program for Prisoners of War," *Psychiatry* (1956), pp. 149–172.

[31] Easton, p. 165; also see p. 210.

[32] Mitchell, pp. 210 ff.

[33] Parsons, "Culture and the Social System," pp. 964 ff.

[34] Dennis G. Sullivan, "Psychological Balance on Reactions to the Presidential Nominations in 1960," in M. Kent Jennings and L. Harmon Zeigler, *The Electoral Process* (Englewood Cliffs, N.J.: Prentice-Hall, 1966),

Culture value systems may also be viewed as resisting change in their composition. They can change but they resist change. One may assume that organisms seek to persist, but a value system culture additionally seeks to maintain its value composition. This is even more true for complex patterns of values. *Sets* of values tend to be maintained not only because the human unit resists change but also because value change is a crisis for complex social systems.

> There is hence a central interest in the continuity of the major cultural components of the system. For a society a basic change in values is a major crisis, because it implies restructuring everywhere.[35]

But even if there were not an empirical tendency for social systems to maintain their values, cultures could still be viewed from the perspective of value maintenance:

> Values, of course, are subject to change, but whether the empirical tendency be toward stability or not, the potentialities of disruption from this source are very great, and it is essential to look for mechanisms that tend to protect such order. . . .[36]

In social systems structures tend to be developed to maintain values so that some minimal order is maintained. That is, social systems attempt resistance to value change whether they are or are not successful in actually maintaining the values.

Two social processes are, then, assumed to be the basic ones which concern culture. The first process focuses upon the relationship between culture and social behavior. In the determination of social behavior, both culture and the social, psychological, and physical environment are viewed as important. Culture is conceived as shaping stimuli. This shaping

---

pp. 245, 257–259. In this connection one should also consider the party identification studies. The value system relating to political party seems resistant to change. See Philip E. Converse, "The Concept of the Normal Vote," and the other articles in Angus Campbell, Philip E. Converse, Warren Miller, and Donald E. Stokes, *Elections and the Political Order* (New York: Wiley, 1966). Also see *The American Voter* by these authors and many other studies which confirm their observations.

35 Parsons, "Culture and the Social Systems," p. 978.

36 Talcott Parsons, "An Outline of the Social System," in *Theories of Society*, 1: 38. Also see his *Politics and Social Structure* (New York: Free Press, 1969), pp. 444–454.

FIGURE I.1 *Stimuli-culture-social behavior process*

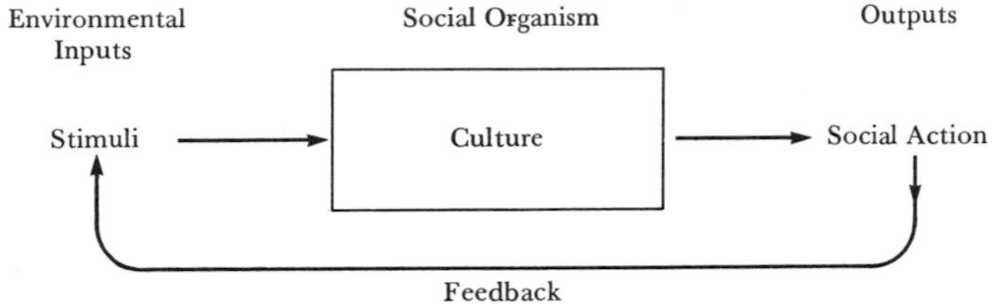

influences social action. Culture is not totally determining. Social action is the joint product of environmental stimuli acting upon a resisting cultural organism (which in turn affects both the environment and the organism). Culture thus *shapes* behavior. The other process concerns the internal operation of culture. The present view is that culture tends to *maintain* its value composition. Substantial stress can change values, and moderate influences can shift support levels. Yet culture resists this change. If culture tempers stress stimuli satisfactorily, the culture values will tend to be supported more highly. When they are more highly supported, they shape more and thus influence social output more.

The stimuli-culture interaction assumed here is diagramed in Figure I.1. It will be noted that the social process is viewed as a stimulus-organism–response-feedback, or system, analogy.[37] Environmental stimuli strike the organic culture system. Some are accepted without stress. Other stimuli are shaped by the values. In the shaping process, behavior is influenced and action is affected. In some cases the stimuli are so strong that the values are changed – but this is not the usual situation. It is important to note that the screening operation can either hinder or aid the initial stimuli. Certain environmental influences may strike the culture and be "recognized." If there is a congruence between stimulus and culture, the initial stimulus may receive added impetus. Screening then distinguishes between stimuli and expedites some and impedes others in the formation of an output. Screening, therefore, can range between the extremes of total blockage and total facilitation. The changes which are made then can

[37] For a similar analogy used in a psychological model, see Milbrath, p. 28.

affect subsequent stimuli through a process called feedback. Within the organism, culture is influenced by the interaction process — greatly or slightly — depending upon the strength and nature of the stimuli and the cultural organism. In the absence of extreme stress it tends to maintain itself.

Culture can thus, I believe, help us understand the influence of member values — which, in turn, can help us understand social processes. Culture tends to maintain its value composition while it shapes stimuli which act on its value screen.

POLITICAL CULTURE

Political culture is, simply, the political aspect of the culture of a society. John Stuart Mill noted long ago that "all questions respecting the tendency of government must stand part of the general science of society." Politics, then, must be subject to the general considerations concerning culture which have been outlined above. Kroeber and Kluckhohn advise that types of culture can be abstracted. What could be more relevant to political analysis than to abstract the political culture from the general culture? Indeed, one of the most ambitious empirical political studies of recent years was organized around this abstraction. The result was the seminal comparative work by Almond and Verba — *The Civic Culture*.[38]

One of the earliest explicit uses of the concept was by the senior author of *The Civic Culture*. In 1956 Almond stated that "every political system is embedded in a particular pattern of orientations to political actions."[39] This he referred to as a political culture. In *The Civic Culture*, political culture consists of "attitudes toward the political system and its various parts, and attitudes toward the role of the self in the system. . . . It is a set of orientations toward a special set of social objects and processes." It is a "psychological orientation toward social objects." In a book by Almond and Powell,

[38] Mill, p. 591; Kroeber and Kluckhohn, p. 367; Gabriel A. Almond and Sidney Verba, *The Civic Culture* (Boston: Little, Brown, 1965; originally published 1963).

[39] Gabriel A. Almond, "Comparative Political Systems," *Journal of Politics* (August, 1956), p. 396. For an earlier example of culture applied to politics, see Alfred McClung Lee, "Public Opinions in Relation to Culture," *Psychiatry* (February, 1945), pp. 53–57.

political culture is referred to as simply the "psychological dimension of the political system." Beer's early use of the concept defines political culture as the values, emotions, and perceptions relating to "*how* government ought to be conducted and *what* it should try to do." [40]

The present conceptualization of political culture is based upon these and the earlier insights provided by our review of the anthropological and sociological definitions. As a type of culture, political culture must have the attributes assigned to culture. It must be a patterned value system which tends to be widely supported. It must shape and tend to be maintained once supported, and it must deal with fundamental beliefs, artifacts, and symbols.[41] In addition to these cultural attributes, political culture must be limited to supportive attitudes of members as they are related to the *political* system.

Political culture will, then, be viewed as a value system which can have an influence on political and social behavior. The most important aspects of this influence are the shaping and maintaining characteristics of culture. Yet in order to study this influence, the major value objects associated with political culture as an integrated concept must be more concretely identified. These values must be fundamental, must be based on the important dimensions of culture, and they must be politically relevant. One political scientist holds that the most basic value object associated with political culture is the nation itself.[42] One's attitude toward the objectively defined

[40] Gabriel A. Almond and G. Bingham Powell, Jr., *Comparative Politics: A Developmental Approach* (Boston: Little, Brown, 1966), pp. 12, 13, 23; Samuel H. Beer, "The Analysis of Political Systems," in Samuel H. Beer and Adam B. Ulam, eds., *Patterns of Government,* 2d ed. (New York: Random House, 1962), p. 32. For a comparison of these views, see Young C. Kim, "The Concept of Political Culture in Comparative Politics," *Journal of Politics* (May, 1964), pp. 313–336. Lasswell and Kaplan use the phrase "political doctrine" to refer to similar phenomena. See Harold Lasswell and Abraham Kaplan, *Power and Society* (New Haven: Yale University Press, 1950), p. 117.

[41] Heinz Eulau, *The Behavioral Persuasion in Politics* (New York: Random House, 1963), p. 81. For an example of the incorporation of the concept in a study, see Robert A. Alford, *Party and Society* (Chicago: Rand McNally, 1963), pp. 2–6. Another work which has used culture to explain is Daniel J. Elazar, *American Federalism* (New York: Crowell, 1966).

[42] Sidney Verba, "Comparative Political Culture," in Lucian W. Pye and Sidney Verba, eds., *Political Culture and Political Development* (Princeton: Princeton University Press, 1965), p. 518.

nation would certainly appear on first thought to provide a value base which could influence future action.[43] This object can be called national identity.[44] This will be the first political culture category for value analysis. Other political scientists have noted other objects which appear to be useful to categorize fundamental political values. Indeed, there is an interesting consensus as to the important objects. Mitchell mentions beliefs, value standards, symbols, and rule norms. Deutsch mentions preferences, habits, and institutions. Mitchell, in another work, provides a system-related categorization composed of a combination of a symbol system, a belief system, and an authority system.[45]

It is the present view that these authors provide important categories for the analysis of fundamental political values. Since culture is viewed as a value system, these categories may be used to classify four subsystems of political culture: an identity system, a symbol system, a rule system, and a belief system. The identity system is the basic community element. It provides the structure for attachment to the political system as a whole and to the members of the same political system considered as co-nationals. The rule system is the repository of support for the fundamental political agreements governing a particular society: the political constitution, the norms of the regime, the "rules of the game." The symbol system is the structure for support of the basic political artifacts deemed valuable in the culture. The belief system is the repository of the fundamental principles and goals supported by the members of the political system.

For the subsystems to have the form of culture, however, they must be historical. That is, to be cultural values they must be transferred from generation to generation. Indeed the strength of a culture depends upon how well this process

[43] Lucian W. Pye, "Introduction," in *ibid.*, pp. 7–8; see also Verba, p. 523.

[44] Verba, p. 529.

[45] Mitchell, *Sociological Analysis and Politics* (Englewood Cliffs, N.J.: Prentice-Hall, 1967), pp. 28–29; Karl W. Deutsch, *Nationalism and Social Communication* (New York: Technology Press of M.I.T. and Wiley, 1953), p. 63; Mitchell, *The American Polity*, p. 23. Also see J. Roland Pennock, "Political Development, Political Systems, and Political Goods," *World Politics* (April, 1966), pp. 415–434.

is performed. The process by which these values are transferred is called political socialization.[46] The societal values must be taught to the new members and in the overwhelming majority of cases the values must be accepted if member values are to be important regime elements. As was noted earlier, values tend to be maintained — but severe stress can lead to change. One of the major ways values may not be maintained is through a break in the socialization process. The present study uses adult data in its analysis. It will hypothesize a historical value system and measure its current support. This thesis could also be tested at the point of transfer in the family and the school. However, the adult is the full member of the political system, and it is his support which is deemed of more interest here. Thus while we will be looking at value support over time, the present study will not concentrate upon the early transfer process. Fortunately, adult and child study findings are consistent for the political system under study.[47] Both stress the historical aspects of political culture.

Political culture, then, is a historical system of widespread, fundamental, behavioral, political values actually held by system members. They are classified into subsystems of identity, symbol, rule, and belief. It is a culture system and as such it tends to maintain itself while it acts to shape stimuli which

[46] Almond and Powell, pp. 24–64.

[47] For confirmation of this see especially Dawson and Prewitt, p. 27 and *passim;* David Easton and Jack Dennis, "The Child's Acquisition of Regime Norms: Political Efficacy," *American Political Science Review* (March, 1967), pp. 25–38; Easton and Dennis, "The Child's Image of Government," *The Annals* (September, 1965), pp. 40–57; Fred I. Greenstein, *Children and Politics* (New Haven: Yale University Press, 1965); Robert D. Hess and Judith V. Torney, *The Development of Basic Attitudes and Values Toward Government and Citizenship During the Elementary School Years,* Cooperative Research Project No. 1078 (U.S. Office of Education, 1965); Herbert H. Hyman, *Political Socialization* (Glencoe, Ill.: Free Press, 1959); Eleanor E. Maccoby, Richard E. Matthews, and Anton S. Morton, "Youth and Political Change," *Public Opinion Quarterly* (1954–1955), pp. 23–39; Herbert McClosky and Harold E. Dahlgren, "Primary Group Influence on Party Loyalty," *American Political Science Review* (September, 1959), pp. 757–766; Mitchell, *The American Polity,* ch. 8; Easton and Dennis, *Children and the Political System* (New York: McGraw-Hill, 1969), pp. 310–311, 413–417, and *passim;* Roberta S. Sigel, ed., *Learning About Politics* (New York: Random House, 1970); M. Kent Jennings and Richard G. Neimi, "The Transmission of Political Values from Parent to Child," *American Political Science Review* (February, 1968), pp. 85–114; Dawson and Prewitt, part 3.

act on it. Yet a knowledge of these cultural values and how they are supported

> does not always permit you to make an exact and accurate determination of the anticipated public opinion of a people toward a public issue. Such a knowledge will provide you with valuable information about the general or overall orientation of a people. However, there are many variable factors which influence the formation and expression of public opinion.[48]

Since other factors will shape the effective opinion which will affect public action, there must be some attempt to place political culture within a framework which allows for the influence of other forces and thus for the possibility of change.

A SYSTEM MODEL OF POLITICAL CULTURE

THE POLITICAL SYSTEM ANALOGY

The scientific perspective upon which the present analysis is based requires that theories be connected to empirical observation through the use of analogies.[49] Only idealized physical models or analogies can invest mathematical and logical forms with experiential relevance which make sense for these forms in the minds of the investigator and his audience.[50] Of the analogies available to the researcher one is becoming more and more accepted as most proper for the analysis of complex social entities.[51] Although there are limits to its usefulness, it appears to be the best available for political science analysis.[52]

In the present analysis, I shall generally follow the analogy (but not the full theory) of the political system developed by David Easton.[53] Easton is primarily interested in general *sys-*

[48] Clarence Schettler, *Public Opinion in American Society* (New York: Harper, 1960), p. 93; also see Hennessy, pp. 166–167, 199.

[49] Nash, p. 231.

[50] *Ibid.*, p. 232.

[51] Walter Buckley, "Society as a Complex Adaptive System," in W. Buckley, ed., *Modern Systems Research for the Behavioral Scientist* (Chicago: Aldine, 1968), pp. 490–511; also see the other articles in this fine collection.

[52] Ludwig von Bertalanffy, "General System Theory — A Critical Review," *General Systems* (1962), pp. 1–20, esp. p. 20.

[53] The works in which Easton most fully describes his analogy (as opposed to his theory) are David Easton, *A Framework for Political Analysis* (Englewood Cliffs, N.J.: Prentice-Hall, 1965; hereafter referred to as *A Framework*); "An Approach to the Analysis of Political Systems," *World*

tem persistence in the face of stress. He primarily focuses not on particular types of governments (i.e., regimes) but on political systems in general. The present work focuses upon *regimes* and how they maintain (or do not maintain) their distinctive aspects. This perspective necessitates some different emphases. Easton has dealt with a related process under the title of systems maintenance. In order to emphasize that the present focus is upon maintenance of distinctiveness in type of government, we will call our focus *regime maintenance.* However, the difference is basically just in emphasis. From the perspective used here, both generalized system persistence and regime maintenance are basic processes of political systems. Both are different ways of viewing how political systems act within a general stimulus-organism–response-feedback analogy to authoritatively allocate values for a society.[54]

The political system as a whole is too undifferentiated an idea. Easton, thus, divides it into three "levels" of community, regime, and authority. Each of these may be conceived as an alternative way of looking at how the political system acts. The community level may be viewed as the most fundamental

---

Politics (April, 1957), pp. 383–400 (hereafter referred to as "An Approach").

[54] *A Framework,* p. 50. In his earlier works Easton apparently accepted maintenance as an appropriate area for analysis. However, in a recent work with Jack Dennis, *Children in the Political System,* he states that maintenance is biased toward stability and that the persistence approach is neutral (p. 42). In my view of science (see note 5) neutrality is impossible. The very selection of major concepts implies a position. Indeed, to test a hypothesis in statistics it must be put in a positive, "biased" form. I focus upon maintenance because it is of interest and because it is so noticeable in the United States (Easton and Dennis, pp. 310–311, 413–417). It does not ignore change when it is placed within a system analogy. Socialization does not have to take place, though culture tends in that direction. The benefits of the system analogy, which is the aspect of the theory which allows for change, should not be confused with the benefits of a particular conceptual focus (*A Systems Analysis,* pp. 477–479). The present work considers systems maintenance as a type of "vital question" concerned with political systems (see *A Framework,* p. 86; *A Systems Analysis,* p. 481). Technically, it should be noted that system persistence is used as the highest level negative universal conceptualization (see Giovanni Sartori, "Concept Misformation in Comparative Politics," *American Political Science Review* [December, 1970], pp. 1033–1054) and regime maintenance as the explanatory focus. Note that this view does not lead to the logical fallacy imputed to systems analysis of affirming the consequent. See Jerome Stephens, "The Logic of Functional and Systems Analysis in Political Science," *Midwest Journal of Political Science* (August, 1969), pp. 389–391.

level in that it serves to define the member composition of the political system. The political *community* is the view of the system which concerns any group of persons who are interrelated through a common political specialization of labor. It is a focus upon *members* who are together in a common political structure.[55] The *regime* can also be viewed as fundamental but it has other objects. It

> consists of all those arrangements that regulate the way in which the demands put into the system are settled and the way in which decisions are put into effect. They are the so-called rules of the game, in the light of which actions by members of the system are legitimated and accepted by the bulk of the members as authoritative.[56]

When we deal with a specific type of nation-state, this object is of central concern. Although *authority*-level activity (that is, day-to-day policy interactions within a given regime and community) does affect regime maintenance, it does not do so directly and will not be considered fundamental. The present study will, thus, deal only with the more fundamental objects of community and regime. To emphasize that both levels refer to the essential definition of the type of government, both levels will be considered together under the process of regime maintenance.

The analogy itself is input-process–output-feedback model in which all activities influencing authoritative decision-making are interrelated within a systemic whole.[57] It is based upon the concepts *system, environment, response* (coping with stress), and *feedback*.[58] Politics is seen as an organism which acts as "an adaptive, self-regulating, and self-transforming system of behavior" where authorities allocate values for the society.[59] This system-organism responds to stimuli which may act on it either from outside its boundaries or from within. The relevant reality is divided into two components – the political and the environment of the political. The environment is all of non-political reality which may act to stress the political. The political is the structures involved in deal-

55 Easton, *A Systems Analysis*, p. 177.
56 Easton, "An Approach," p. 392.
57 *Ibid.*, p. 383.
58 Easton, *A Framework*, p. 25.
59 *Ibid.*, pp. 26, 50.

ing with environmental stress by converting it into effective demands on the political and in converting it to political decision outputs. These outputs, in turn, may feed back upon and affect the environment. In operation, environmental stimuli (wants) can stress or act upon input structure. After interaction with the structure, the affected stimuli can act upon the system. The system then reacts to the stress to try to ensure its *persistence*.[60] Figure I.2 (based on Easton's schema in *A Framework for Political Analysis*) represents the operation of the political system model with a regime emphasis.[61]

Easton conceptualizes culture as a subsystem of the environmental system. As will be seen, the present work will modify this somewhat by distinguishing political culture from the general culture. At this point, however, it is necessary to see the function of culture within the polity as conceptualized by Easton. It is external to the system. To him, as with the theorists discussed earlier, interactions are *shaped* by culture. "The general culture helps to mold the constraints within which political discussion and competition take place." [62] Of course, there are many other constraints and influences on the output of a political system, as the Easton model clearly indicates; yet the central role of culture as a limiting and shaping factor should be noted.

Easton's political system, then, is a way of looking at the complex world of politics. With the use of a few simple concepts like *system, stress, input, demand, support, output,* and *feedback,* the infinitely complex can be made manageably understandable. The political system is viewed as a self-adaptable system-organism which copes with stress through an input-conversion-output-feedback process.

POLITICAL CULTURE IN THE POLITICAL SYSTEM

It seems to me that Almond and Powell give the kind of emphasis to political culture in the political system needed

[60] *Ibid.,* pp. 86–87. The preceding description may overemphasize the input to output aspect. Authority output to environment and system are at least as important.

[61] Naturally, the reader should refer to Easton's works for a detailed explanation of his theory. I have only emphasized aspects important for my purposes. For another systemic model of the political system, see Mitchell, *The American Polity,* esp. p. 371.

[62] Easton, *A Framework,* pp. 72, 73, 113.

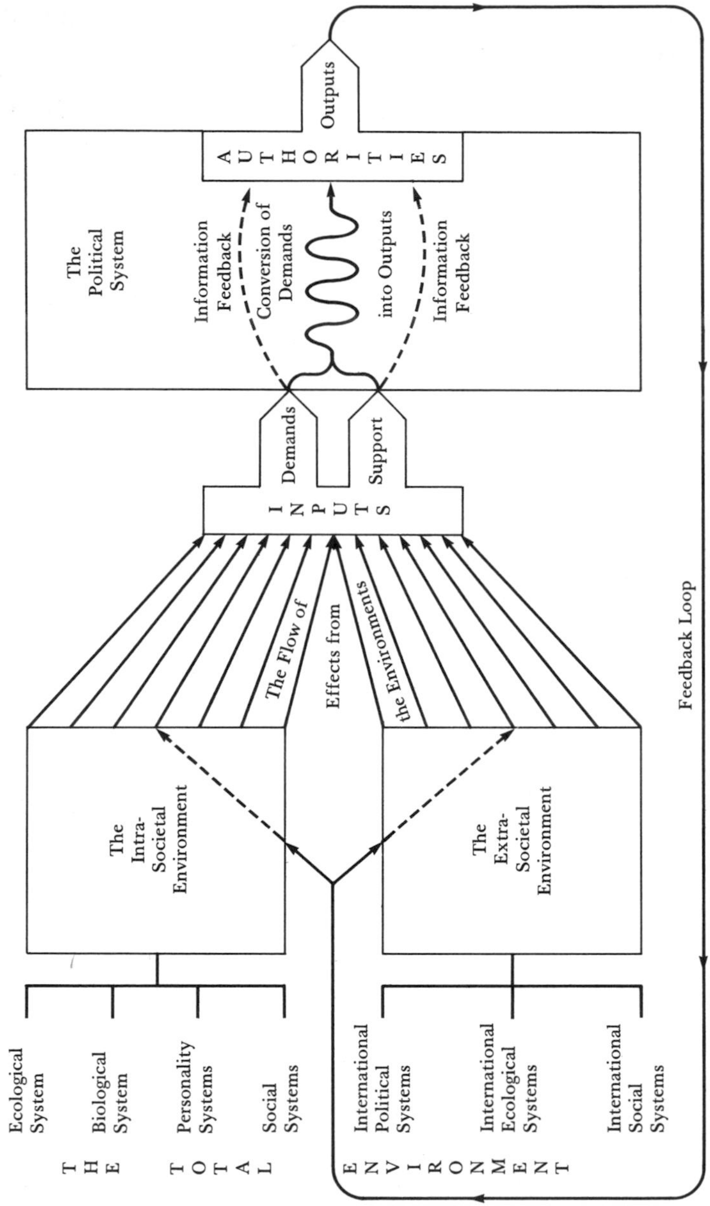

FIGURE I.2 *A dynamic response model of a political regime*

for this study. They say that "the political system consists of interacting roles, structures and subsystems, and of an *underlying* culture." [63] Political culture underlies the political system in the sense that Parsons saw culture as prior to the usual operations of the social system. In accordance with the design of this study, the system analogy must be modified to include political culture. This modification must be made in order to allow the concepts of political system and political culture to be integrated into one harmonious theoretical framework. Thus, at the same time, one can obtain the benefits of the comprehensiveness of the system and the emphasis upon individual values in the culture.[64]

It has been suggested that political culture "lies across" feedback.[65] Yet it seems to me that to harmonize with the Easton model, "culture" must be environmental, and "political" must be within the political system. It would appear that the single concept "political culture" cannot be harmoniously integrated into the model. Fortunately, Easton has anticipated the problem:

> What we choose to put inside our system, to consider within its boundaries, will depend upon what we wish to examine in detail. . . . What we leave outside, as part of its environment, will be those factors that we can accept as givens.[66]

Yet, upon reflection, political culture is both to be examined in detail and is a given. It appears to me that the apparent paradox arises because political culture actually belongs at the periphery of the inputs.[67] It is usefully conceived as an input structure prior to the political system. Concerning inputs, Easton has said:

> It will matter little whether we consider these inputs as internal or external to the political system. They stand on the border. . . . They may be equally conceived to lie within the square or outside it, as long as we recognize that they remain in the neighborhood of the boundary.[68]

[63] Almond and Powell, p. 25. Emphasis added.

[64] Pye, p. 9.

[65] Samuel C. Patterson, "The Political Cultures of the American States," *Journal of Politics* (February, 1968), p. 190.

[66] Easton, *A Framework,* p. 66.

[67] Easton, *A Systems Analysis,* p. 117.

[68] Easton, *A Framework,* p. 114.

Political culture, then, is conceived as an input subsystem structure near the boundary between the environment and the system. It acts as a *screen* on wants and is affected by them. The wants, of course, can be from any of the intersocietal or intrasocietal environments. The environmental interactions can be conceived of as stimulative impulses which act upon present cultural values. The cultural values, which tend toward maintenance, screen new impulses. The culture will not accept all want impulses and will accept some only in a radically modified form. Only stimuli which survive the screening process can have any effect upon the system. In the terminology of traditional democratic theory, only the wants that survive the screen can be considered public opinion, and only public opinion can influence further political activity.[69]

Perhaps a more important aspect of the process concerns *effective* public opinion. Environmental stimuli not only must interact with member political culture but in many cases must also be shaped by the political culture of the political system structures (e.g., parties, groups, etc.) and finally, and most importantly, by the political culture of the decision-making authorities.[70] Political culture, then, is not unidimensional but has a separate empirical identification depending upon the type of political structure concerned.[71]

Figure I.3 shows political culture integrated into the system model. Note that political culture acts as a screen. There is also a separate cultural screen for each important threshold in the political system: the member structure, the political structures, and the authority decision-makers.[72] The member screen is the first and the broadest. It generally determines

69 For the difference between changeable opinion and stable culture, see Doob, pp. 48–54; also see Samuel Stouffer, *Communism, Conformity, and Civil Liberties* (New York: Wiley, 1966), p. 13.

70 James N. Rosenau, *Public Opinion and Foreign Policy* (New York: Random House, 1961), esp. ch. 6. The reader should also be warned that *authority level* (one way of looking at the whole political system) is not the same as the *authorities* (the decision-makers).

71 I have dealt with problems of policy-making among two structures in *The Attentive Public: Polyarchical Democracy.*

72 For the theory of levels of public, see Gabriel A. Almond, *The American People and Foreign Policy* (New York: Praeger, 1960; originally published 1950), pp. 137–149. Also see V. O. Key, Jr., *Public Opinion and American Democracy* (New York: Knopf, 1961), pp. 9–10; Rosenau, ch. 4; and Devine, ch. 3 and the citations therein.

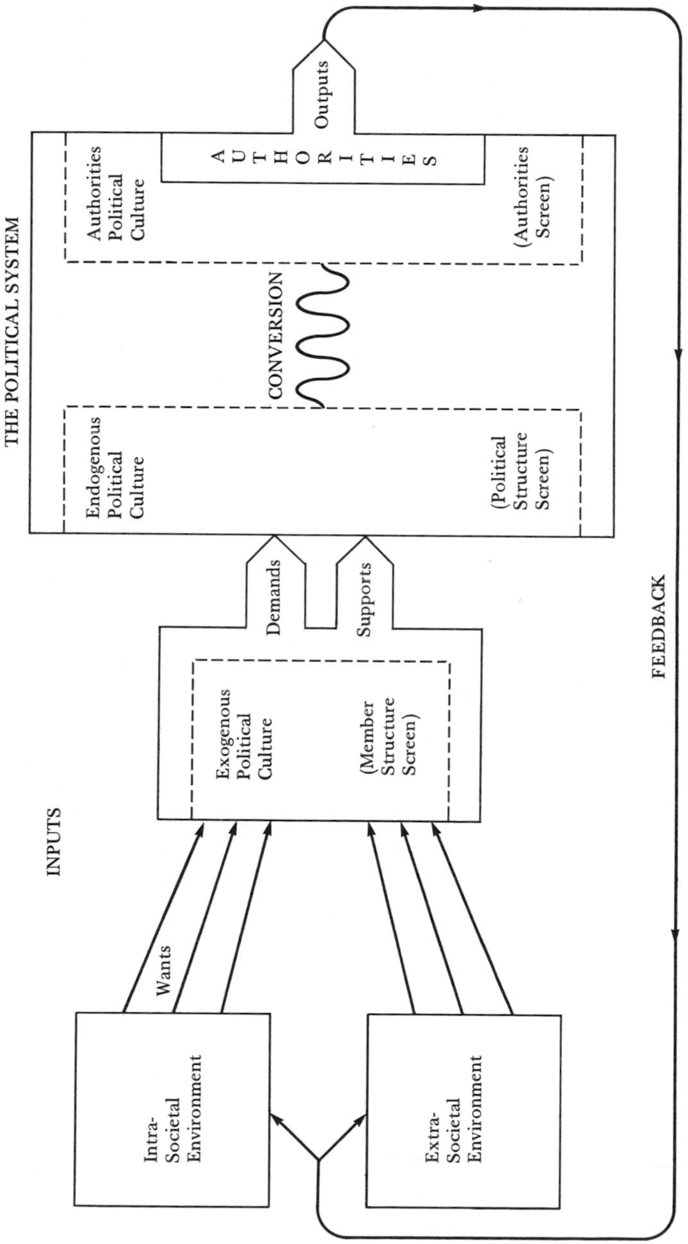

FIGURE I.3 *Political culture in the political system*

which environmental stimuli shall be received as demands and supports. (Wants, however, may reach and be processed directly by political or authority structures without mass influence.) This most extensive screen shapes the nature of the inputs which actually interact with the political system per se. The political structure screen shapes the issues which are processed by the political system.[73] The interactions between the political system structures determine how the conversion process works on these screened inputs. After the demands have been resolved by the political structures, issues are relayed to authorities, and the authorities' screen then becomes operable. It is at least theoretically arguable that the authorities' political culture can be different from that of the members. It should not be assumed, however, that there is a very different content to these cultures — this is a matter for empirical investigation. Each threshold has its own screening process and each, therefore, has an opportunity to shape the final outputs. Naturally the process may differ at each threshold; i.e., the culture may be, for example, more deterministic or relevant for the members than for the authorities. In any event, all stimuli are eventually focused upon the authorities. They resolve them and translate them into outputs (decisions or policy).

Although it is important to realize that cultures surround each important threshold of the political system, the present interest is in the *member political culture.* It is not necessarily the major determinant of input and certainly not the major influence upon specific output.[74] However, it is one of the determinants of political system output and in some types of system is a relatively important one. Member political culture is viewed as the value screen prior to the political system. This value system *shapes* the inputs into the political system, and these inputs help determine output. As these member political values shape, they may successfully deal with stress. If they do, they tend to be *maintained.* This value maintenance may then tend to maintain the regime.

73 For a discussion of similar levels see Rosenau, ch. 6, pp. 21–26. His analysis suggests that perhaps the second culture level could be subdivided. For simplicity's sake it will not be done here.

74 Easton, *A Systems Analysis,* pp. 166–167. It should also not be assumed that all outputs originate in the environment. Many are within-system generated.

DEMANDS AND SUPPORTS: MEMBER INFLUENCE ON THE SYSTEM

Political systems act to cope with stress.[75] Apparently, they have a variety of possible alternatives. Five appear to be most important for system persistence: (1) the use of coercion; (2) change in the basic structures of the regime or community; (3) the manipulation of output to achieve specific support (demand satisfaction); (4) demand regulation; and (5) the the use of diffuse support.[76] The first three responses focus upon output actions taken by political authorities. Although these responses are critical for every political system, the present emphasis is upon the individual member and his influence in coping with stress, and consequently, on input screening of demands and supports rather than on output manipulation. In terms of relative importance in system stress regulation, moreover, the use of coercion and structural change are thought to be less important than the other three possible responses for the political system which will be studied here.[77] The use of outputs to increase specific support, however, is a crucial means of stress reduction for the political system of the United States.[78] Yet present interests are upon analysis of more fundamental system processes.

The concept of political culture within a political system focuses upon the fundamental values of system members and the effect they may have on system processes – especially the major one of coping with stress. From this perspective the present analysis emphasizes the input screen shaping of member values as they regulate the number of demands made on the political system and as they influence the amount of support given the regime.

Demand overload is viewed as one of the central problems for a political system. No system could handle an infinite number of member demands.[79] The political culture is one of the main means by which demands are limited. The norms

[75] Easton, *A Framework,* p. 87.

[76] Easton, *A Systems Analysis,* pp. 249, 276.

[77] Indeed, in so far as structural change in the regime is concerned, the present view is that there has been little of any substance since 1789 for the American regime; see Chapters II and VII.

[78] See Key, ch. 21.

[79] Easton, *A Systems Analysis,* p. 61.

and values of members serve to limit "the number of desires that members will even presume to convert to demands." [80]

> More generally speaking, the types of wants converted to political demands will depend upon the image prevailing in the political culture of the purposes for which political processes may be used. The conception a member holds of the nature of the political instrumentalities, of his role in political life and the role of others will condition the kind of claims that he feels justified in politicizing.[81]

For example, Easton notes that although many Americans feel strongly about the style of clothing that should be worn — the length of a hemline, perhaps — the expectation of the political culture of the United States is such that these wants are not considered proper areas for governmental decision-making.[82] Under other political systems — that of Communist China comes to mind — standards of dress are deemed appropriate for political regulation.

Member values are important in screening stimuli. Screening is such that many wants are never translated into demands that must consume the limited resources of the political conversion process. System members thus use their own values to restrain themselves. Their values shape the type and number of political demands they make.

Although the concept of political culture is important for the idea of demands, it is central for what Easton calls diffuse support. Specific support (i.e., *quid pro quo* support) for a political system can be attained by coercion, structure change, or demand satisfaction. However, Easton holds that

> no system could endure for very long if it did not seek to build up a reservoir of support frequently described as patriotism, love of country, loyalty and the like — upon which it could count regardless of the particular trials and tribulations or frustration of desires that the members might experience at the moment.[83]

Diffuse support is this reserve of member good will toward the basic levels of the political system. It is support beyond fear

[80] Easton, *A Framework,* p. 122.
[81] Easton, *A Systems Analysis,* p. 104.
[82] *Ibid.*
[83] Easton, *A Framework,* p. 125; and *A Systems Analysis,* pp. 267–272.

or a *quid pro quo* for actions done by authority. It is member support for the community and the regime without regard to its pragmatic performance (though performance undoubtedly affects it in the long run). This broad support given by the political system member allows the day-to-day policy process to take place within some given rules and understandings. Diffuse support in a system allows authorities to make decisions which may not (especially in the short run) benefit some groups. Members accept these decisions in the belief that the regime as a whole is legitimate and just. Without this broad member support, regimes could not maintain themselves without extensive reliance upon coercion.

Political culture is conceived as a major structural component of diffuse support. It is one of the input structures which screen environmental stimuli to the political system. The whole input structure is viewed as composed of political culture values and other relevant social structure (e.g., class and status structure, education structure, roles).[84] Political culture is the repository of the supportive attitudinal values actually held by members as they relate to the regime and community. Interaction between environmental stimuli, political culture, and other input structure determines the diffuse support which acts on the political system. The screening and shaping influence of political culture is paramount among these determinants of diffuse support.

Political culture, as repository of diffuse support, may or may not be functional for system persistence or maintenance. Strong support for the political culture aids coping with stress only if the political culture produces positive diffuse support for the regime. If the member values are not congruent with the existing regime or community, strong culture support could lead to negative diffuse support and to regime instability. For reasons which will be more apparent in the next chapter, I shall assume that American political culture values are congruent with its regime. This will allow both culture support and regime support to be considered as operating in the same direction. That is, for the American system if political culture is supported so is the regime.

[84] Frank C. Nall, II, "Role Expectations: A Cross Cultural Study," *Rural Sociology* (March, 1962), pp. 28–41.

Easton identifies two major forms of diffuse support. *Overt support* relates to supportive *behavior* or *actions* directed toward the political system. This type of support is obviously of central importance. Unless attitudes are translated into behavior they do not directly influence output.[85] Again, however, our interests lead us in another direction. Although the simple observation that there has been one regime in the United States for almost two centuries is an indication that covert support is behaviorally significant, actual behavior cannot be investigated here.

The design and interests of the present study lead to the analysis of covert diffuse support. *Covert support* is, simply, "a supportive frame of mind with respect to others or some object." [86] Covert diffuse support concerns fundamental supportive *attitudes* directed toward the political regime. Covert support is measured in terms of direction and strength of member attitudinal attachment to values.[87] The direction of support toward a political system is measured as positive, negative, or neutral.[88] Strength can be measured in terms of the simple net number supporting, the degree to which support is widespread among groups (especially among powerful groups), and the intensity of support.[89] These aspects, except for intensity, will be investigated in the present study. Intensity is a crucial aspect of support but not much data is available. The little which is available is reported, but intensity usually can be estimated only from the other measures of support.

Let us review the interrelationship between member values, political culture, and demands and support. Member support for values defines the political culture. Political culture, in turn, is the repository of major determinants of demands and supports. Environmental stimuli act upon the political culture screen. Demands are part of the residual of the interaction between stimuli and culture and other input structures. Diffuse supports are the other major residuals after interaction

[85] Easton, *A Systems Analysis*, pp. 159–160.
[86] *Ibid.*, p. 160.
[87] *Ibid.*, p. 161–164.
[88] *Ibid.*, p. 163.
[89] *Ibid.*, p. 164.

with input structure—especially political culture. These demands and supports may be converted to issues by political institutions and to outputs (policies and decisions) by authorities. It is through these means that members may influence regime maintenance and other political processes.

CULTURE SYSTEMS AND REGIME MAINTENANCE

The political system is viewed as an interactive, adaptive system. The political culture is likewise viewed as an adaptive, self-regulating, and self-transforming subsystem of the political system. As a system it seeks to persist. Political culture, however, also acts as a *culture* system. As a system it acts to cope with stress and merely to persist. As a culture system, political culture seeks to maintain the content of its value system. That is, it seeks to preserve itself as a patterned value system.[90] Of course, if the environmental stress is extreme, the content can change. Any system can change. However, as the literature on culture suggests, member values held for long periods of time are highly resistant to change. A value system like political culture will tend to maintain itself.

A political culture system acts to reduce value stress by shaping stimuli to conform with the present content of its value screen. If change has to be made in the political system, culture's shaping strength is used to shift change elsewhere before it affects values.[91] Since a political system is conceived as having a political culture, it is both a pure system and a value system. The system seeks simple persistence, the value system seeks maintenance. Political system stress may reduce support for values. If the political system has a strong political culture, however, stimuli are shaped, values are retained, and the crisis is solved by other structures. Values which survive this process over extended periods of time tend to be supported even more strongly and are, thus, even more able to shape future perceptions. The more values are accepted, the more they shape and are retained, and the more they can reduce stress for the whole political system.

90 Parsons, "Culture and the Social System," pp. 964 ff.; Parsons, *The Social System*, p. 379; and Parsons, "A Rejoinder," p. 249.

91 Talcott Parsons and Winston White, "The Link Between Character and Society," in Seymour Martin Lipset and Leo Lowenthal, eds., *Culture and Social Character* (New York: Free Press, 1961), p. 103.

It should be noted, again, that a strong member culture does not necessarily lead to the maintenance of a particular regime. The political culture values could be inconsistent with the regime and its values (perhaps it is the political culture from the previous regime) and this could act to reduce support for the regime. Even if the political culture is regime consistent, the social system of the country in question may be too structurally rigid to deflect change to other structures (either in the political system or in the environment). For the present analysis, however, these are not viewed as important limiting conditions. For the American political system in general, political culture is viewed as supportive of its regime. It is the purpose of this book to inquire into the nature and role of the influence of the political culture of the United States as it affects the American regime.

How have political systems been able to persist in a world of stress and change? Many have chosen regime structure change. Others have been able to deal with stress and to maintain their political regime. Some have been maintained by authority responses such as coercion or community change. All have relied to some degree upon the manipulation of outputs to satisfy members – especially those in powerful groups. In return for specific rewards (and perhaps among members in less powerful groups even limited rewards such as preservation of order may suffice), the regime is suported. This support, however, is tenuous. Although regimes may be maintained for substantial periods of time by these means, ordinary member support is of enough consequence in all political systems that it must be reckoned with. Most at least attempt to manipulate it so that members may be controlled. On the other hand, a very few political systems have been able to rely upon prevailing value support from their members to help maintain their governmental form. The major thesis of the present work is that the United States is among this latter group.

The goal of the remainder of this book is to describe and explain the political culture values of the United States. To do this, we must postulate a value system which can be tested as a potential American political culture.

CHAPTER II

The American Political Culture System

CONSENSUS, FUNDAMENTALS, RATIONALITY, AND AMERICAN DEMOCRACY

DEMOCRACY AND FUNDAMENTALS

The thesis of the present work is that one can identify an American value system which acts as a political culture. It is argued that the United States has had a high degree of consensus on fundamental political values at the community and regime levels. This political consensus is conceived as having had an important shaping influence upon the content of its public policy, upon the maintenance of its democratic regime, and upon the persistence of its community. Indeed it has been suggested that no regime could be maintained for long without coercion, in the absence of member support. Yet competent political scientists have held that not only are beliefs in political fundamentals not necessary for regime maintenance but that consensual belief in fundamentals is incompatible with the existence of a democratic regime.

The eminent modern political philosopher Carl Friedrich has said that there is no necessity for fundamental values to be agreed to by all and, specifically, that in the Western democracies – including the United States – this type of agreement has not existed. In his *The New Image of the Common Man* Friedrich closely reviews world history and carefully concludes that basic disagreement on such "fundamentals" as nationality and economic class does not threaten democratic government. Rather, these differences are crucially *necessary* for the "give-and-take" of the democratic process. Indeed, the morale of groups presently unsatisfied with the process demands that these issues be kept open. To Friedrich, consensus is not needed but, rather, people must agree to solve *specific* problems without considering basic beliefs. It is only under aristocratic or totalitarian government that fundamental beliefs are necessary to sustain weak and nonpopularly based regimes.[1]

V. O. Key, Jr., another well-regarded political scientist, in his monumental study of public opinion examined much empirical data and concluded that Friedrich was correct.[2] However, if we look more closely at Key's careful reporting of the results of his research, we begin to detect a purely semantic problem.

> Whatever the characteristics of popular attitude that permit governments to operate as if a basic consensus existed, they do not seem to consist of ideas that amount to a consensus on political fundamentals unless we mean by that phrase nothing more than a popular recognition of the legitimacy of the regime.[3]

In terms of the present study, the identity, symbol, and rule subsystems relate directly to the "legitimacy of the regime," and the belief system is a regime-level support of legitimacy. Perhaps agreement on fundamentals does mean "nothing more" than acceptance of regime legitimacy.

Friedrich himself implicitly accepts the importance of agreement on "fundamentals." He twice mentions the necessity for the belief in the value of "agreeing to disagree" – only he does not want to call this agreement on fundamentals. In fact, he admits that if pushed far enough one can show an agreement on fundamentals but that the agreements "become increasingly vague, until finally they no longer bear any rela-

[1] Carl J. Friedrich, *The New Image of the Common Man* (Boston: Beacon Press, 1950; originally published 1942), pp. 155–156, 157, 160–171, 185.

[2] V. O. Key, Jr., *Public Opinion and American Democracy* (New York: Knopf, 1961), p. 41.

[3] *Ibid.*, p. 50.

tionship to practical affairs."[4] Basically, Friedrich (and Key also) is opposed to agreement on fundamentals because aristocrats and totalitarians seem to be the ones advocating the concept, because it seems to assume superior knowledge on the part of authority as to what should be believed, and because it would deny change.[5]

Political culture as a concept, though, does not suffer from the objections raised by Friedrich and Key. Advocacy is not a proper objection.[6] The concept is assumed to be descriptive. Second, the question of imposing beliefs is not relevant. The concept of political culture merely implies that, empirically, most systems attempt to impose beliefs (e.g., through socialization). Third, the system model used here does not deny change. Indeed, the model explicitly allows for change through the environmental inputs and feedback effects. Finally, I maintain that though Friedrich and Key may not think that "vague" beliefs affect "practical affairs," they would not disagree that if "pushed far enough" there is something that an analyst could call political culture. The Easton distinction between regime and authority level is directly relevant here. Political culture is hypothesized to be controlling at the regime and community levels and specifically *not* directly at the authority level — the level of "practical affairs." Of course, culture affects authority-level decisions, but the *agreement* is at the regime or community level. In any event, affecting practical affairs does not mean determining them; i.e., culture shapes, not determines.

Consider, further, another of Key's meticulous statements of his findings:

> While the data suggest that certain attitudes widely distributed within the population may bear on how the political system works, to call these beliefs a "consensus on fundamentals" im-

[4] Friedrich, pp. 172, 177, 185. Also see his *Man and His Government* (New York: McGraw-Hill, 1963), esp. p. 345.

[5] Friedrich, *The New Image of the Common Man,* pp. 156, 185, 173, 177; Key, pp. 41, 548. Also see a recent article by Friedrich on the necessity of moral beliefs: Carl J. Friedrich, assisted by Morton Horowitz, "Some Thoughts on the Relation of Political Theory to Anthropology," *American Political Science Review* (June, 1968), pp. 536–545.

[6] For a traditional defense of the need for consensus see Murray Clark Havens, *The Challenges to Democracy* (Austin: University of Texas Press, 1965).

putes to them a more explicit and ordered content than they seem to possess.[7]

In other words, Key specifically advises that he has empirically identified "widely distributed" beliefs that "may bear on the way the political system works." Yet the widely distributed beliefs which affect the system are not fundamentals. Key only seems to disagree with the concept of fundamental because it implies "explicit and ordered content." Since Key accepts so many of the empirical indications of the existence of fundamentals, while he rejects the concept, this may again be due to ethical opinions regarding the antidemocratic implications of the concept. By using the concept of political culture instead, and by placing it within a system model, perhaps the objections can be made less relevant. It is interesting that Key does note the "screening function" of national identity and, indeed, he suggests that this may explain its relevance.[8] It should be further emphasized that the concept of political culture does not necessarily imply a very "explicit and ordered" system. The elements in the cultural system may be quite stable but the resolution of *conflicting* elements within the system may produce very different outputs – depending upon the weighing of the particular value at the time. Robin Williams observes that "American society does not have a completely consistent and integrated value-structure"; yet he finds basic value-systems important molders of the social process.[9] It is here hypothesized that such cultural values actually shape the political process whether they are called fundamentals or political culture.[10]

IGNORANCE AND CONSENSUS

It seems, then, the argument that belief in fundamental values is unnecessary and undemocratic is not strong enough to justify the abandonment of the concept of political culture. A more important objection to the use of the concept, however, may be the question of psychological capacity. Are the

[7] Key, p. 49.

[8] *Ibid.*, p. 43. Also see his discussion of consensus in *Politics, Parties and Pressure Groups*, 4th ed. (New York: Crowell, 1958), pp. 243–244.

[9] *American Society*, 2d ed. rev. (New York: Knopf, 1960), p. 413.

[10] This discussion is continued in Chapters V, VI, and VII.

TABLE II.1 *Familiarity with Bill of Rights*

Question: What do you know of the Bill of Rights? (NORC)

	1944	*1946*
Never heard of it, not sure if had	23%	31%
Had heard, but could not identify it	39	36
Confused, unsatisfactory, or incorrect	15	12
Reasonably accurate idea of content	23	21

Source: *Public Opinion Quarterly* (Summer, 1944), p. 283; *ibid.* (Winter, 1946–1947), p. 604.

masses sophisticated enough to hold a consistent value system? Much recent empirical research on public opinion would suggest a negative response. The data from the two most important voting studies certainly stress the limited rationality of the American citizen. *Voting* stated that people did not make individual rational decisions but that each voted the way "trusted people around him" voted. *The American Voter* found that only 2.5 percent of the people (3.5 percent of voters) could be classified as having *any* sophisticated view of ideology in politics.[11] Certainly, knowledge of some of the basic political values in the United States is limited (shocking, most intellectuals would say). Consider the information in Table II.1, which stunned educators and politicians in the 1940's.[12] Only one person in five could give a coherent explanation of the Bill of Rights.[13]

[11] Bernard R. Berelson, Paul F. Lazarsfeld, and William N. McPhee, *Voting* (Chicago: University of Chicago Press, 1954), p. 309; Angus Campbell, Philip E. Converse, Warren E. Miller, and Donald E. Stokes, *The American Voter* (New York: Wiley, 1964; originally published 1960), p. 131.

[12] In the tables, the following abbreviations will be used for the main polling organizations: AIPO: American Institute of Public Opinion (Gallup Poll); AV: Almond-Verba five-nation study; FOR: *Fortune* magazine poll; ICPR: Inter-University Consortium for Political Research; NORC: National Opinion Research Center; OPOR: Office of Public Opinion Research; SRC: Survey Research Center at the University of Michigan.

[13] This may be a widespread phenomenon. In Canada only 18 percent could give a correct definition of the "Magna Charta"; see Hadley Cantril, *Public Opinion, 1935–1946* (Princeton: Princeton University Press, 1951), p. 428.

TABLE II.2 *Knowledge of representative*

Question: Do you know the name of the present representative in Congress from your district?

	1942 (AIPO)	*1947 (AIPO)*[a]	*1965 (AIPO)*	*1966 (AIPO)*	*1966 (SRC)*	*1970 (AIPO)*
Yes	46%	38%	43%	46%	39%	53%
No	54	62	57	54	61	47

Source: *Public Opinion Quarterly* (Fall, 1947), p. 485; *Gallup Opinion Index* (August, 1966), p. 10; ICPR, 1966; *Gallup Opinion Index* (April, 1970), p. 20.

[a]In 1947 the wording was slightly altered. For the exact wording, consult the source.

Polls have shown that only about 55 percent of the adult population knows how many United States senators each state has.[14] Table II.2 shows that less than 50 percent knows the name of their congressman. Only 35 percent can give a tolerable definition of the electoral college, less than 20 percent can name the three branches of the federal government, and only slightly over half of the population can give even a tolerable definition of the American economic system.[15] Yet the lack of knowledge should not be exaggerated. All the questions above are purely informational, rather than contextual. They probably are not important in problem resolution – if the meaning of the principles involved (rather than the form) is known. In other words, symbolism in these cases is weak but behavioral "screening" is not necessarily so. For example, a slightly higher percentage of people can *name* a United States senator from their state than can tell how *many* senators there are.[16] Although less than half could tell an interviewer his representative's name in 1958, 77 percent could correctly state which party controlled the House – and this was a year when the Congress was controlled by Dem-

[14] Hazel Erskine, "The Polls: Textbook Knowledge," *Public Opinion Quarterly* (Spring, 1963), pp. 136–141. One authority estimates that one-third of the American people "live in a world that psychologically does not include foreign affairs." See Leonard S. Cottrell, Jr., and Sylvia Eberhart, *American Opinion on World Affairs* (Princeton: Princeton University Press, 1948), p. 14.

[15] Erskine.

[16] Cantril, p. 790.

TABLE II.3 *Complication of government*

Question: Some people say politics and government are so complicated that the average man cannot really understand what is going on.

	1952 (SRC)	*1956 (SRC)*	*1960 (SRC)*	*1960 (AV)*	*1964 (SRC)*	*1966 (SRC)*	*1968 (SRC)*
Agree	69%	62%	56%	62%	66%	69%	71%
Disagree	29	35	39	32	31	27	19
Depends, DK, other	2	3	5	6	3	4	0
N	799	1,762	1,954	970	1,571	1,291	1,341

Source: ICPR.

ocrats and the more visible presidency by the Republicans.[17] And although few could explain some of the more informational questions noted above, on some questions there is great familiarity: 80 percent know what a veto is, 93 percent know the number of years of a presidential term, and 71 percent can identify a party platform.[18]

LIMITED RATIONALITY AND POLITICAL CULTURE MOOD

Almond and Verba maintain that views of democratic citizenship should not be placed "unreasonably high." [19] After all, the citizen plays many other roles – spouse, parent, working man – which demand much of his time and effort. The expertise demanded is just not worth the effort to the average American.[20] In fact, he is well aware of his limitations as shown in Table II.3. About two-thirds of the citizens of the United States feel that government is too complicated to be understood. Table II.4 gives a further idea of the range of

17 Survey Research Center, *1958 Election Codebook* (Ann Arbor: Inter-University Consortium for Political Research, 1958). However, perception is lower when one party gains seats but not control; see *1966 Election Codebook* (Ann Arbor: Inter-University Consortium for Political Research, 1966).

18 Erskine.

19 Gabriel A. Almond and Sidney Verba, *The Civic Culture* (Boston: Little, Brown, 1965; originally published 1963), p. 340.

20 On the whole question of effort, capacity, and interest, see Joseph A. Schumpeter, *Capitalism, Socialism and Democracy* (New York: Harper and Row, 1942), ch. 22, esp. pp. 258–259.

TABLE II.4 *Issue comprehension*

Question (in part): How well do you understand these issues? (AV, 1960)

	National issues		Local issues	
Very well	7%	45%	21%	64%
Moderately well	38		43	
Not so well	37		22	
Not at all	15		10	
Depends, DK, other	3		4	
N	970		970	

Source: ICPR.

competence people believe they have. It should be observed that few think they understand issues *very* well, and few do not understand *at all.* It should also be noted that understanding is much greater for the more familiar and less complex local issues.

Yet people in the United States do not feel that they are ignored in the political process. On the contrary, as Table II.5 shows very clearly, over two-thirds of the population feel that they have some control over their government. But if most people realize that they cannot fully understand, yet feel that they can control, how do they think they influence government? The obvious answer is through elections. Americans feel that they cannot act on each detailed piece of legislation but they can influence broad policy by the simple act

TABLE II.5 *Subjective citizen competence*

Question: People like me do not have any say about what the government does.

	1952 (SRC)	*1956 (SRC)*	*1960 (SRC)*	*1960 (AV)*	*1964 (SRC)*	*1966 (SRC)*
Agree	31%	29%	21%	37%	29%	35%
Disagree	66	70	71	61	69	60
DK, other	3	1	2	2	2	5
N	1,799	1,762	1,954	970	1,571	1,291

Source: ICPR.

of voting.[21] Naturally, the idea of broad influence would be especially relevant within a theory of political culture.

On the other hand, there is no question that it is difficult to translate the complexity of the real world with its manifold issues into the simplicity of voting. Yet Key presents empirical evidence that "voters, or at least a large number of them, are moved by their perceptions and appraisals of policy and performance." [22] He suggests that rational decisions can be made if the decision is reduced to approval or disapproval of the authorities in power. In this way the *choice is reduced to the simplicity of the voting act,* and a simple, rational, meaningful decision can be made, within structural limits, for or against the authorities. A limited rationality is thus achieved through elections, which can control — if only very broadly — those with political power.

Although elections are the major way public mood can influence policy, it also acts in more indirect fashion.

> The "masses" participate in policy-making in indirect and primarily passive ways. Their moods, interests, and expectations set limits on the discretion of their representatives — those exercising influence within private organizations and in the media of communications, as well as those holding public office. The public also affects the selection of elites through the public electoral process, the private electoral process of interest groups, or through "buying" or refusing to "buy" the policy recommendations of the communications elite.[23]

The mass public thus has "a rather limited though important

[21] For the scholarly presentation of this view, see Robert A. Dahl, *Who Governs?* (New Haven: Yale University Press, 1961), ch. 28.

[22] Robert Axelrod, "The Structure of Public Opinion on Policy Issues," *Public Opinion Quarterly* (Spring, 1967), pp. 51–60; Gordon Tullock, "Theoretical Forerunners," *The Calculus of Consent* (Ann Arbor: University of Michigan Press, 1965), pp. 323–340 (a NORC survey in 1944 showed that only one-quarter of those who voted for candidates agreed with all the candidates said [ICPR]); V. O. Key, Jr., *The Responsible Electorate* (Cambridge: Harvard University Press, Belknap Press, 1966), p. 150.

[23] Gabriel A. Almond, *The American People and Foreign Policy* (New York: Praeger, 1960; originally published 1950), p. 139. Also see Key, *Public Opinion and American Democracy,* ch. 21; and James M. Rosenau, *Public Opinion and Foreign Policy* (New York: Random House, 1961), ch. 4.

TABLE II.6 *Voting: citizen affect*

Question: Voting is the only way that people like me can have a say about how the government runs things.

	1952 (SRC)	*1956 (SRC)*	*1960 (SRC)*	*1960 (AV)*[a]	*1964 (SRC)*	*1966 (SRC)*[a]	*1968 (SRC)*
Agree	80%	72%	76%	71%	73%	68%	57%
Disagree	17	26	14	25	25	27	42
Other, DK	3	2	10	4	2	5	1
N	1,799	1,762	1,954	970	1,571	1,291	1,339

Source: ICPR. Also see the Almond-Verba data for the United Kingdom, Germany, Italy, and Mexico, which suggests that this view of the centrality of voting is widespread across culture.
[a]In 1960 and 1966, the wording was slightly altered. For the exact wording, consult the source.

function" in setting *limits* on political authorities.[24] Whether through elections or more indirect means, public mood appears to be able to have a broadly rational effect on political output.

Certainly the public feels that voting is the direct way they influence decisions, as is shown in Table II.6. About 70 percent feel that they are effective only through the electoral process. The public, then, believes that it has an influence on how the government acts. Indeed, more than seven in ten believe that the way people vote decides "how things are run." [25] Although voting is probably not very important in the totality of life attractions to most people, within the political sphere it is considered the most important activity.

In summary, it appears that, in the United States, people have a broad influence upon authorities through voting and other expressions of mood. The limited rationality of the American citizen and perhaps the inherent complexity of government do not allow detailed control of specific policies but do allow an influence. Elections, if they offer a choice between competing elites (e.g., through political parties), do allow the system member to evaluate the performance of the

24 Dahl, p. 324, and his *Congress and Foreign Policy* (New York: Harcourt, Brace, 1950), pp. 42–44.
25 SRC, 1952, 1956; AV, 1960. Ninety percent thinks elections make government pay some attention (SRC, 1964).

authorities in power and provide a means by which he may rationally reject the present authorities in favor of others. This is a significant means of control.

The present view is that political culture can act through elections, as do other expressions of mood, to influence output. This is a major way that member values can influence what authorities can do. But culture can influence in as many ways as do other expressions of political mood. Political culture can be made politically effective through the relative formality of a letter to an official, through informal conversations with authorities,[26] and through the more subtle means of feelings relayed to officials by intermediaries who have contact with normally nonpolitical members.[27] Political culture influence of this type does not necessitate high levels of rationality.

THE AMERICAN POLITICAL CULTURE

ENVIRONMENTAL INFLUENCES

Apparently, then, the concept of political culture can be used for the United States. It does not seem necessarily to be an abandonment of American democratic principles to consider the idea of fundamentals. Neither does the concept seem to demand an unachievable level of rationality for Americans. It may, therefore, be appropriate to now proceed to the systematic analysis of political culture values in accordance with the theoretical framework outlined above. In accordance with this perspective both the values and the influences upon them are important. Therefore, economic, ecological, psychological, and social factors in the environment of the political system must be considered. They will help explain the value content of the system they act upon.

It has become commonplace to note the substantial changes,

[26] Lewis Anthony Dexter, "What Do Congressmen Hear?" in Nelson W. Polsby *et al.*, *Politics and Social Life* (Boston: Houghton Mifflin, 1963), pp. 485–495.

[27] Paul Lazarsfeld, Bernard Berelson, and Hazel Gaudet, *The People's Choice* (New York: Duell, Sloan and Pearce, 1944), ch. 16; Elihu Katz and Paul Lazarsfeld, *Personal Influence* (Glencoe, Ill.: Free Press, 1955), p. 330; and Elmo C. Wilson, "The Press and Public Opinion in Erie County, Ohio," *Journalism Quarterly* (March, 1941), pp. 14–17.

TABLE II.7 *What do you consider to be the most urgent problem facing you and your family today? (AIPO, 1967)*

Financial	60%	Old age	3%
Sickness	8	Miscellaneous	10
Vietnam	5	No problems	16
Race	4	No opinion	2
Employment	3		

Source: *Gallup Opinion Index* (November, 1967), p. 30. For similar findings, see Gerald Gurin, Joseph Veroff, and Sheila Field, *Americans View Their Mental Health* (New York: Basic Books, 1960), p. 24.

usually economic, which have taken place since the settlement of the United States. Indeed, economically and in comparison with growth in other time-spans (though a good yardstick has yet to be conceived) change does seem to be dramatic. The United States has grown from an underdeveloped nation in the seventeenth century to the wealthiest world power in the twentieth. In fact it is four times more economically powerful than its nearest competitor and its GNP is approximately 40 percent of the combined world wealth total. The nation has changed from a small agricultural to a large industrial power. Where in the early years of development over 90 percent of the people were in agriculture, today less than 9 percent are so occupied. Population grew from about 4 million at the time of the adoption of the Constitution to more than 200 million today.[28] Even the size of the country has been materially expanded – from thirteen colonies to fifty states.

With changes of this magnitude, it would be strange if economic influences did not affect regime maintenance. Indeed, people are generally most attuned to economic problems, as is shown in Table II.7. The undoubted centrality of economics, in fact, has led certain analysts beyond values to posit a strong relationship between economic development and regime maintenance in a democracy.[29] Perhaps some threshold must

[28] Bruce M. Russett *et al., World Handbook of Political and Social Indicators* (New Haven: Yale University Press, 1964), pp. 152–154; Bureau of the Census, *Historical Statistics of the United States* (1961), pp. 7, 72; also see William H. Young, *Ogg and Ray's Introduction to American Government,* 12th ed. (New York: Appleton-Century-Crofts, 1962), pp. 5–6.

[29] Seymour Martin Lipset, "Some Social Requisites of Democracy,"

be reached before democracy can be achieved, but a strong developmental relationship between the two is not evident—certainly not one strong enough to be considered totally deterministic.[30] This is not to deny the role that abundant material and economic success could play in maintaining a regime. Indeed it may even be, as one authority concludes, that "economic abundance is conducive to political democracy."[31] Certainly most democracies of the world have wealthy economies. Whatever the exact relationship to a particular type of regime, it appears that economic satisfaction tends to reduce stress for any regime. For the United States this has meant that for most of its history its economic system has provided a major base of support for its political regime—whatever form it might have taken.

The physical environment, as it is first met by new settlers, also seems to be an important influence upon political values. If political culture transmits experience it would seem reasonable to assume that the early impressions are the ones which shape the later ones. An economic historian emphasizes the effects this initial experience can have:

> Physical environment determines to a large extent where man shall live, what kind of work he shall do, what he may produce, and the routes over which he must travel and transport his products. Because of its influence upon his economic life, physical environment goes far to determine man's social and political point of view, his habits and desires, and even his physical frame.[32]

The richness of America's natural resources presented at least an opportunity for regime maintenance. The immediate

---

American Political Science Review (March, 1969), pp. 69–105; Phillips Cutright, "National Political Development," in Nelson W. Polsby, Robert A. Denter, and Paul A. Smith, eds., *Politics and Social Life* (Boston: Houghton Mifflin, 1963), pp. 569–582; and David M. Potter, *People of Plenty* (Chicago: University of Chicago Press, 1954), pp. 112–127.

30 Deane E. Neubauer, "Some Conditions of Democracy," *American Political Science Review* (December, 1967), pp. 1002–1009.

31 Potter, p. 112. For an interesting three variable model, see Martin C. Needler, "Political Development and Socioeconomic Development: The Case of Latin America," *ibid.* (September, 1968), pp. 889–897.

32 Harold Underwood Faulkner, *American Economic History,* 6th ed. (New York: Harper, 1949), p. 1.

countryside was habitable and exploitable. Climate was temperate. Land was abundant and could be secured cheaply. Nature had provided the materials for development and had also created an ocean of sufficient size to isolate the United States from the wars and intrigues of Europe.[33]

Besides its direct influences, the physical environment produced a psychological effect. Many of the physical aspects of America seemed to be related to the values of John Locke – the political theorist whose ideas were most influential in the mother country early in the formative colonial period. This English philosopher provided a rational explanation for the founding and development of the colonies. To describe the state of nature, Locke noted that "in the beginning, all the world was America." [34] To Americans their forests provided convincing proof of an original state of nature.[35] The colonists found themselves in circumstances where there was no law, so they made compacts. The Mayflower Compact was "proof" that natural man could contract for a civil society. The frontier provided many examples of Locke's emigration right.[36] The frontiersman, indeed the voyage to the new world itself, gave perfect examples of the republican Locke's free individual, subject only to his God. He was the epitome of the natural man who used his reason to shape the environment. Had Locke not given the ideal?

> The natural liberty of man is to be free from any superior power on earth, and not to be under the will or legislative authority of man, but to have only the law of Nature for his rule.[37]

[33] Williams, ch. 1 and p. 398; also see William C. Mitchell, *The American Polity* (New York: Free Press, 1962), pp. 120–121; Harvey M. Karlen, *The Pattern of American Government* (Beverly Hills, Calif.: Glencoe Press, 1968), ch. 2.

[34] John Locke, *Second Treatise on Government,* in Ernest Barker, ed., *Social Contract* (New York: Oxford University Press, 1962), sec. 49, p. 29.

[35] Louis Hartz, *The Liberal Tradition in America* (New York: Harcourt, Brace and World, 1955), p. 60.

[36] *Ibid.,* p. 61.

[37] Locke, sec. 22, p. 15. Note that Locke was involved in the colonies in more than a theoretical manner. He was involved in the settlement and had even developed a constitution for the colony of Carolina (Barker, "Introduction," p. xix).

Economic abundance and an environment susceptible to human manipulation provided the basis for a type of political experience not easily explainable with traditional theories of government. The appearance and wide acceptance of the Whig liberal theory of John Locke in the intellectual circles of the day provided a handy explanation. Indeed as time passed the theory seemed more and more explanatory. Locke explained the foundations and the development of the country. His theory early became accepted and became more so as Locke provided the explanation successively not only for the founding of the colonies but for the breaking of the contract with Britain and the formation of the new independent regime. As Hartz has put it, "History was on a lark, out to tease men not by shattering their dreams, but by fulfilling them with satiric accuracy."[38] The American experience seemed incredibly well explained by the Lockean thesis.

Economic, physical, and psychological environments were undoubtedly important predisposing factors in the development of American political values. Yet there is a limit to what they can explain. The environmental factors may provide the setting and society may not be understandable without knowledge of them; yet

> they are far from explaining even the most elementary sociological problems. From them we cannot predict very specifically about such matters as the American language, the forms of family life or economic organization, the nature of religious beliefs or political ideology. . . .[39]

To investigate political values, attention must be focused upon social phenomena.

THE LIBERAL TRADITION IN AMERICA

It is the contention of this work that a consensual political culture has existed in the United States, essentially unchanged during its entire history. Its roots are in the republican principles of Locke and were reinforced and adapted to American circumstances by Madison and the other "Founding

[38] Hartz, p. 60.
[39] Williams, p. 19.

Fathers" of the Constitution. The historian Louis Hartz presents the most comprehensive argument for this thesis. He calls the American political culture the "liberal tradition." It is not the ideology of a single political faction in the United States but rather a tradition which covers both American "liberals" and "conservatives." This tradition is based upon the values of John Locke, the philosopher of the liberal stage of sociological history. American historians have tended to emphasize American history as *sui generis.* They have not tended to place American social phenomena within universal social patterns. This approach is useful, but it does not lead to comparative insight and analysis. It also tends to overemphasize conflict. When broad patterns are not clearly defined, standards of similarity are unclear and all difference appears large.

From a comparative sociological perspective, the most generally accepted schema for viewing more recent social phenomena is the classification of feudal, liberal, and socialistic — with change taking place in the direction indicated by the order of listing. Of course since the system analogy directs our study, development is viewed as nondeterministic, i.e., change does not have to take place, nor does each stage have to be followed by the one suggested in the classification.[40] Yet empirical data do show a general tendency in the direction indicated. It also must be noted that our theory recognizes that change can take place within these categories. Our perspective, however, focuses upon change *between* the categories as the most significant. For example, in Europe major change can be seen to have taken place during the change from feudal to liberal and from liberal to socialistic stages. Other change seems comparatively slight.[41]

[40] The classification schema is thus Marxian, but a nondeterministic Marxian schema. It should also be noted that the categories are not in some sense "real" — they are analytic. It can even be argued that the liberal stage is not a separate stage but a continuation of the feudal. On both points, see Schumpeter, pp. 13, 82, 416, 61, 202–203, 259, 214, and *passim.* Also see Karl A. Wittfogel, *Oriental Depotism* (New Haven: Yale University Press, 1957), ch. 9.

[41] Of course, it is only *comparatively* slight. The Revolution, for example, was violent; R. R. Palmer, *The Age of Democratic Revolution: The Challenge* (Princeton: Princeton University Press, 1959), pp. 188–189. For its comparative mildness, however, see pp. 234–235.

In transitions between major periods of social change, old and new structures come in contact. As perspectives and interests differ, conflict takes place between institutions representing each major social stage. In Europe, to establish itself the liberal bourgeoisie had to struggle against feudal institutions and privileges. In return the feudal and mercantile institutions resisted. In a transition between major periods, social and political stress is high. This degree of stress can affect values. A country that did not experience this period of transition would not experience this stress. One would expect a rather different value tradition.

> One of the central characteristics of a nonfeudal society is that it lacks a genuine revolutionary tradition, the tradition which in Europe has been linked with the Puritan and French revolutions: that it is "born equal" as Tocqueville said. And this being the case, it lacks also a tradition of reaction: lacking Robespierre it lacks Maistre. . . .[42]

It shall be the position of the present study that the maintenance of political values in the United States is peculiarly related to the major social phenomenon of avoiding this period of major change. Sociologically, the "outstanding thing about the American community in Western history" was its lack of a feudal or traditional social structure.[43] In America, imposition of the liberal order did not mean the destruction of a previous social order. The United States avoided the divisions within the community, so noticeable in Europe, that seriously threatened stable social and political systems. The ability to establish a more modern regime without the conflict necessary to overthrow an established social order provided a basis for the establishment of a commonly accepted regime. The political system would naturally have to endure the stress of the present, but it could escape being burdened with the problems and divisions of the past. It could profit from the intellectual legacy of the past without incurring the animosities of those who represented the previ-

[42] Hartz, p. 5. Also see Samuel P. Huntington, *Political Order in Changing Societies* (New Haven: Yale University Press, 1968), pp. 125–126.
[43] Hartz, p. 3.

ous institutional fulfillment of that legacy. The lack of a feudal past and its burdens allowed the possibility of the development of a very widely accepted political regime.[44]

One major insight derived from using our sociological schema, then, is the emphasis upon the comparative uniqueness of the American situation (especially as compared to the European) due to the lack of the weight of feudal structure.

> Of great significance in facilitating America's development as a nation, both politically and economically, was the fact that the weight of ancient tradition . . . was largely absent. It was not only a new nation, it was a new society, much less bound to the customs and values of the past than any nation of Europe. Crevecoeur well described the American as a "new man," the likes of which had never been seen before.[45]

But was there a completely new beginning? Was there no continuity within the newness? Seymour Martin Lipset holds that the "new man" is a complete break with the past and that he is the result of the American Revolution. It is the values of the Revolution which provide the identity necessary to create the new American man.[46] However, the use of our nondeterministic, nondiscrete classification opens the possibility of a combination of noncontinuity and continuity. The second major insight provided by our schema is the possibility of social structure noncontinuity and value continuity.

One of the earliest and most influential systematic studies of the norms and values of the American people was performed over 130 years ago by Alexis de Tocqueville. In his analysis, Tocqueville stressed the unique lack of inhibiting social structure at the founding of the American community.

> The emigrants who fixed themselves on the shores of America in the beginning of the seventeenth century, severed the democratic principle from all the principles which repressed it in the old communities of Europe and transplanted it unalloyed to the New World. It has there been allowed to spread in perfect

44 *Ibid.*, p. 35. Also see Robert A. Dahl, "The American Oppositions: Affirmation and Denial," in Robert A. Dahl, ed., *Political Oppositions in Western Democracies* (New Haven: Yale University Press, 1966), pp. 43–46.

45 Seymour Martin Lipset, *The First New Nation* (Garden City, N.Y.: Anchor Books, Doubleday, 1967), p. 107.

46 *Ibid.*, pp. 75–102.

> freedom, and to put forth its consequences in the laws by influencing the manners of the country.[47]

In another place he mentions that aristocracy was not transferred to New England and that it would never take root in the rest of the country. He seems impressed, then, with the uniqueness of a country without the legacy of feudal social institutions. However, Tocqueville sees a continuity within the uniqueness. To him, "not an opinion, not a custom, not a law, I may even say an event, is upon record which the origin of that people will not explain." [48]

Tocqueville maintained that the American social and political situation was the result of *both* the initial uniqueness of the land and the people who inhabited it, and the political traditions of the country of origin.[49] Although the social structure was newly created, the political values were transferred from the English mother country.

> Born in a country which had been agitated for centuries by the struggles of faction, and in which all parties had been obliged in their turn to place themselves under the protection of the laws, their political education had been perfected in this rude school, and they were more conversant with the notions of right, and the principles of true freedom, than the greater part of their European contemporaries. At the period of the first emigrations, the parish system, that fruitful germ of free institutions, was deeply rooted in the habits of the English; and with it the doctrine of the sovereignty of the people had been introduced into the bosom of the monarchy of the House of Tudor.[50]

To Tocqueville, the "new man" did not spring whole from the Revolution but arrived as the first colonist. This earliest English settler was the joint product of the environment in which he found himself and the political heritage which he brought with him. The "new man" arrived in the new world a new social man but an old political man. Only the symbiosis was a new man.

[47] Alexis de Tocqueville, *Democracy in America,* trans. Henry Reeve (New Rochelle, N.Y.: Arlington House, Heirloom Edition), 1:xlvi; also see p. 11.

[48] *Ibid.,* pp. 31–37, 11.

[49] *Ibid.,* p. 11; also see 2:v.

[50] *Ibid.,* 1:12.

The American situation was, thus, both the result of a favorable institutional newness and a political tradition. Although the community was later violently divided and a revolution successfully consummated, a value continuity persisted with this division. The American revolutionaries "did not reject their English patrimony totally, but rather built on some parts of it (such as English law and English philosophy), while they sloughed off others (such as monarchy, aristocracy, and the idea of royal prerogative).[51] The patrimony which persisted was based upon a very old tradition.

> The English colonists took these late medieval and Tudor political ideas, practices, and institutions across the Atlantic with them during the great migrations in the first half of the seventeenth century. The patterns of thought and behavior established in the New World developed and grew but did not substantially change during the century and a half of colonihood.[52]

Transferred through the colonial period these values became the basis for every American regime and were finally transferred to the Constitution. Indeed, this document is best viewed as basically a product of the values of

> the common law, the principles of Magna Carta and the Bill of Rights, the writings of Locke and Blackstone, and other characteristic products of their English motherland. In the main . . . this monumental heritage had passed to America far back in colonial days and at the time when the national Constitution took form was already deeply embedded in the constitutions, laws, and usages of the states.[53]

It is this value tradition which the present work maintains has been transferred from generation to generation virtually intact both as the American political culture and as the values consistent with the constitutional regime.

The whole value tradition is summed up in the "English

51 William Nisbet Chambers, *Political Parties in a New Nation* (New York: Oxford University Press, 1963), p. 9.

52 Huntington, pp. 96–97. Indeed, Huntington emphasizes an even older foundation for the tradition. Locke was certainly a product of the earlier tradition; see Sterling P. Lamprecht, *The Moral and Political Philosophy of John Locke* (New York: Russell and Russell, 1962), esp. pp. 5–22.

53 Young, p. 30. For the continuity of the values from the Mayflower Compact to the Constitution, see Willmoore Kendall and George W. Carey, *The Basic Symbols of the American Political Tradition* (Baton Rouge: Louisiana State University Press, 1970).

liberal philosophy" of John Locke.[54] The value essences of this philosophy as they relate to politics will be investigated below. Basically they are viewed as the central aspects of the American value system. Even Locke's epistemology, however, has an interesting dualism not unlike the dualism de Tocqueville found for America.[55] Locke saw man as rational and free, and his consent is needed for government to be legitimate. He is unrestricted in that he begins life with a mind like a white paper — not unlike America before settlement. But this man also has a tradition within which his reason operates. This is especially so for his values, which are based upon a natural law.[56] Like America man is born free but is limited by a value tradition.

Locke viewed government as contracted by this complex but essentially free man to preserve himself from the insecurity of the state of nature.[57] Government was thus somewhat unnatural — limited to protecting the individual's life, liberty, and property through popular consent, established laws, impartial judges, and limited but effective executives. A government that exceeds these bounds is illegitimate, and the people have the right to break the contract.

> It was this set of beliefs that Locke expounded at the time when this nation was in its early years. And it was this ideology that was transported to America via colonists and books, where it found a favorable environment and reception. As a result it became the orientation of Americans, and formed not only the basis of their claim for freedom from England but subsequently the basis of the Constitution itself. To this day, indeed, we are largely Lockeans; there are few Platonists, Hegelians, Marxists, followers of Rousseau, or any of the other European philosophers in America. Our own contribution to philosophy, pragmatism, is a descendant and variation of the basic tenents of Locke's liberalism. In the sense of adherence to the central doctrines of Locke, we are all "little liberals." [58]

[54] Mitchell, p. 107.

[55] On the dualism, see Leo Strauss, *Natural Right and History* (Chicago: University of Chicago Press, 1953), pp. 8, 202–251; and F. A. Hayek, *The Constitution of Liberty* (Chicago: University of Chicago Press, 1960), p. 406.

[56] See Lamprecht, *passim,* but esp. pp. 9, 10–15, 22.

[57] Locke, Sec. 9.

[58] Mitchell, p. 108.

The descriptive power of Locke's thought as the center of the American value system has been noted from many different perspectives. The most widely used political science textbooks focus upon Locke and the tradition associated with him as the most important early value influences upon American political values.[59] Anthropologists have almost universally documented Lockean values as characteristic of the United States (see Chapter IV). Kohn, a major scholar of comparative nationalism, in his study of American nationalism specifically names Locke as the philosopher of the United States. Hayek, a political philosopher, even holds that the English liberal tradition was the only major influence upon American political philosophy. Other studies more interested in political culture per se refer to the liberal tradition as being the controlling influence on American political culture.[60] Even some of those who reject (or ignore) the influence of the liberal tradition prior to 1776 seem to implicitly accept it since the Revolution.[61] Almost all analysts believe that there has been a long period of value continuity in the United States.

[59] Dan Nimmo and Thomas D. Ungs, *American Political Patterns,* 2d ed. (Boston: Little, Brown, 1969), pp. 62–68; Marian D. Irish and James W. Prothro, *The Politics of American Democracy,* 4th ed. (Englewood Cliffs, N.J.: Prentice-Hall, 1968), p. 73; James Macgregor Burns and Jack Walter Peltason, *Government by the People,* 6th ed. (Englewood Cliffs, N.J.: Prentice-Hall, 1966), p. 39; and Young, pp. 11–14. There are other examples. See Harvey M. Karlen, *The Pattern of American Government* (Beverly Hills, Calif.: Glencoe Press, 1968), pp. 53, 85; and Carl J. Friedrich and Robert G. McClosky, eds., *From the Declaration of Independence to the Constitution* (Indianapolis: Bobbs-Merrill, 1954), p. xxxii.

[60] Hans Kohn, *American Nationalism* (New York: Collier Books, 1961), pp. 20–21, 31–32; Hayek, pp. 176–186; Walter Dean Burnham, "Party Systems and the Political Process," in William Nisbet Chambers and Walter Dean Burnham, eds., *The American Party Systems* (New York: Oxford University Press, 1967), p. 281; Samuel C. Patterson, "The Political Culture of the American States," *Journal of Politics* (February, 1968), p. 193; Robert R. Alford, *Party and Society* (Chicago: Rand McNally, 1963), p. 4; Erwin C. Hargrove, "On Canadian and American Political Culture," *Canadian Journal of Economics and Political Science* (February, 1967), pp. 107–111; G. Horowitz, "Conservatism, Liberalism, and Socialism in Canada: An Interpretation," *Ibid.* (May, 1966), pp. 143–171; and Roger D. Masters, "The Lockean Tradition in American Foreign Policy," *Journal of International Affairs,* No. 2 (1967), pp. 253–277.

[61] Lipset, *The First New Nation,* p. 102. The same could be said for Hofstadter — Lipset quotes him at this point. Also see "A Changing American Character?" in Seymour Martin Lipset and Leo Lowenthal, eds. *Culture and Social Character* (New York: Free Press, 1961), p. 166.

But what about the tradition today? Has it survived the socialist stage? What about the New Deal? Did it not forsake liberalism? Hartz suggests that this may seem to be the case because Roosevelt did not have to talk to socialists or communists. If he could *not* have ignored them – as the liberal nature of the society allowed him to – Roosevelt would have had to temper all of his appeals. That is, he would have had to make qualifications in his statements to indicate his support of liberal ideals. However, as long as it was only an effective argument between whig ("conservative") and petit-bourgeois ("liberal") strains *within* liberalism, he could talk quite radically and be accepted by his audience as liberal. Hartz points out that this lack of left-wing opposition allowed Roosevelt to sound decisive where British Liberals and French Radicals became lost in qualifications. Undoubtedly, there was to some degree a problem of "liberal convictions held in the face of nonliberal innovations," but the New Deal innovations were always modified to the liberal tradition.[62]

The continuity of the tradition is interestingly outlined in an exhaustive study done in 1937, as shown in Table II.8. McDiarmid found that certain symbols used by presidents of the United States in their first inaugural addresses show a surprising stability. Of course, some have varied widely, especially identity symbols. For example, "union" symbols used extensively before the Civil War have essentially disappeared since 1881. "America" was used extensively during the period of socialist challenge in the 1930's and less frequently in other periods. However, symbols of the rules of the game, the Constitution, self-government, and the basic beliefs in God, freedom, the past, and confidence in the future have been used throughout the history of the United States. Of course, the data presented is only through Franklin Roosevelt's first inauguration, but a reading of the addresses of Truman, Eisenhower, Kennedy, Johnson, and Nixon surely shows few changes. Even if it is allowed that the stimuli of the 1930 depression strained the relevance of the liberal tradition to the American political leadership, it was a deviation rather than a repudiation. Certainly, the return to the tradition be-

[62] Hartz, p. 263.

TABLE II.8 *Continuity of the liberal tradition: frequency of symbol use in presidential inaugural addresses to 1932*

President	*Estimated number of words*	*Our form of government Our institutions, Our system, etc.*	*America(n)*	*United States*	*Union*	*Our country, My country*	*Our glorious past, Our traditions, Our memories, etc.*	*Our forefathers, Founders, sages, Framers, heroes, etc.*
Washington	1,300	3	2	2	1	5	1	
J. Adams	2,300	5	8	3	2	3	2	1
Jefferson	1,400	2			2	3	2	3
Madison	1,100	1		2	2	6	1	2
Monroe	3,100	10	1	12	10	3		3
J. Q. Adams	2,700	3		1	20	5	1	6
Jackson	1,100	1		1	1	1		4
Van Buren	3,700	21	2	1	2	5	3	5
W. H. Harrison	7,600	11	7	11	10	8		8
Tyler	1,600	6	2	3	4	1	2	2
Polk	4,500	6		12	31	10	1	3
Taylor	1,100	4	2	1	2	2	1	4
Pierce	3,100	6	2	2	5	4	12	9
Buchanan	2,600	5	2	8	10	4	3	3
Lincoln	3,300	3	2	4	20	5	5	1
Johnson	350		1					
Grant	1,100			1	2	3		
Hayes	2,400	2	1	2	3	6	2	1
Garfield	2,900	4	1	7	11	2	5	2
Arthur	500	2	1			2	1	1
Cleveland	1,600	7	4	1		2	5	2
B. Harrison	4,100	4	5		5	3	3	3
McKinley	3,800	9	9	12		4	5	3
T. Roosevelt	900	3					4	4
Taft	5,000	1	7	6		5	2	
Wilson	1,400	2					1	
Harding	3,200	5	24	1	1	5	4	4
Coolidge	3,800	4	11			8	6	
Hoover	3,700	8	12	2		8	3	1
F. D. Roosevelt	1,700	3	3	1		1	3	2

God, Almighty, Divine Providence, etc	*Constitution(al)*	*Free(dom)*	*Liberty(ies)*	*Independent(ce)*	*Self-government, Gov't by the people, etc.*	*Economy (in gov't)*	*Virtues of American people, Their intelligence Their honesty, etc.*	*Present prosperity, Bounteous blessings, Happy conditions, etc.*	*Confidence in future*
8	2	3	4	1	3				2
6	8	3	5	2	1		11	2	2
2	3	7	3		7	2		3	1
1	1	3	2	1	1	1	2	5	1
4	6	3	3	2	4	1	7	22	1
5	12	5	4	2	4	1		5	2
3	3	4	2	2	1	1	2		1
3	6	3	1	1	2		10	11	4
7	38	19	20	4	8		5		2
3	6	2			2	1	2		
5	18	13	5	4	4	2	1	4	1
1	5	1	1		1	1	2	2	1
4	11	4	2	1	1	1	1	6	3
4	16	9	4	4	6	2	2	4	1
4	32	3	1	1	6		5	1	2
1		3			1				2
2	3	1			2	2			2
1	10	3			4	1			2
2	19	10	5	1	7	1	1	3	4
1	4	1			1	1	2	2	1
3	7	2	2		8	2	3	2	1
1	14	6	2	2	3	2	4	5	3
5	2	11	1		7	2	5		2
1		3			2		2	7	2
1	4	3	1		1	1			1
2			2				1	1	1
5	2	5	5	2	7	1	5		11
1	7	9	3	11	5	6	4	5	2
3	2	6	4	3	14	1	3	7	3
4	4				1				3

Source: John McDiarmid, "Presidential Inaugural Addresses – A Study in Verbal Symbols," *Public Opinion Quarterly* (July, 1937), pp. 80–81; see also the discussion on pp. 79–82.

gan soon thereafter. As Hartz poignantly notes, "Were not Truman's speeches during his second administration an odd blend of the fighting Roosevelt and the complacent capitalist oratory of Herbert Hoover?" [63]

The liberal tradition is, thus, offered as the core of the beliefs which comprise the American political culture. Its main representative is John Locke. It is maintained that Locke's republican principles have provided the basis for the behaviorally significantly political beliefs of the United States from their founding until the present.[64] This argument is derived essentially from Tocqueville and more explicitly from Louis Hartz, but also from the many other sources cited here. The argument is not based upon authority, however, but upon empirical investigation. The most recent empirical works do suggest that values have been maintained. One work summarizes the present value situation as strongly supportive of the traditional values – Williams says, "the same major values – only more so." [65] The present work will try systematically and empirically to test the extent to which the Lockean liberal tradition is actually held as the present value system of the United States.

GIVENNESS AND CHANGE

It is not the intention of this work to maintain that American public policy outputs have been the same over the years but rather that they have been shaped by the same general value system. Daniel Boorstin has identified an important

[63] Hartz, p. 272.

[64] This description is even accepted by some of the American regime's most radical critics. Indeed, the liberal tradition is precisely what is wrong with the United States. See Marc Pilisuk and Thomas Hayden, "Is There a Military-Industrial Complex Which Prevents Peace?" *Journal of Social Issues* (July, 1963), pp. 70–113. Of course, not all accept the thesis. The present test is especially to be compared with these. See particularly Michael Mann, "The Social Cohesion of Liberal Democracy," *American Sociological Review* (June, 1970), pp. 423–439; and William H. Form and Joan Rytina, "Ideological Beliefs on the Distribution of Power in the United States," *ibid.* (February, 1969), pp. 19–31. In my opinion, in neither case does the data support the conclusions reached by the authors.

[65] Robin M. Williams, Jr., "Individual and Group Values," *The Annals* (May, 1967), p. 36; see also Lloyd A. Free and Hadley Cantril, *The Political Beliefs of Americans: A Study of Public Opinion* (New Brunswick, N.J.: Rutgers University Press, 1967), pp. 174–176.

property of this system – its "givenness." It is "the belief that values in America are in some way or other automatically defined: *given* by certain facts of geography or history peculiar to us." [66] They seem given because they have been accepted for so long. The values of the liberal tradition are "given" because they are unquestioned, they seem natural; they are the only values felt to be compatible with the national environment. The tradition is, in this sense, non-philosophical; it is accepted as part of the given of the political culture. As Boorstin explains it, the givenness is perceived first as a gift from the Founding Fathers, who "equipped our nation at its birth with a perfect and complete political theory, adequate to all our future needs." Givenness is, second, a gift from the functioning authority institutions of the society. Finally, it is a combination of the past and the institutions which gives a high sense of legitimacy to the ideas, the procedures, and the other outputs processed through the institutions. In sum, it is a culture where values are viewed as empirically and traditionally given in a sense more profound and deep-rooted than one constructed through simple rationality.[67]

In essence, Boorstin suggests a shaping function for givenness. We would say that a high rate of acceptance of the political culture and satisfaction with its operation lead to a belief that its values are so good, so obviously good, that they are the only ones compatible with the survival of the system. When political culture reaches this point, givenness can be said to be operational. Givenness is simply a measure of the strength of the screening influence of political culture. Givenness is relayed to the political system in the form of high levels of diffuse support. The importance of givenness is that at that point the political system is highly able to cope with stress through the protection available from its political value screen. Even relatively severe want stress can be ignored as not consistent with the regime values which are supported. Even in the face of war, extreme depression, or other strong stresses,

[66] Daniel J. Boorstin, *The Genius of American Politics* (Chicago: University of Chicago Press, 1953), p. 9.

[67] *Ibid.*, pp. 9–10. In this sense the American culture is empirical rather than rationalistic. See Giovanni Sartori, "Politics, Ideology and Belief Systems," *American Political Science Review* (June, 1969), p. 402.

givenness cultures can maintain themselves. Regimes with congruent givenness political cultures produce positive diffuse support except under the most stressful conditions. The present hypothesis is that the American liberal tradition has generated a political culture that does display givenness, does produce positive diffuse support, and does shape demands so as to maintain both its own value composition and its regime.

The liberal tradition, thus, is conceived as an important shaping factor for the political system. Since its influence is a shaping one, however, it is not totally deterministic. Even where givenness is apparent, change takes place. Indeed, it appears that Americans have a (non-political) cultural belief in the importance of change.[68] To understand American change values, though, a distinction between political and social change must be made. In the United States some change is made through political means – yet it is small in comparison with the total amount of change in the society. American political culture limits the amount of political change because of the nature of its content. Lockean liberalism values limited government. Change is expected to take place mainly in the environment of the political system – not through the governmental process. In a regime which values limited government, one should expect to look outside politics for major societal change.

American belief in the importance of change is expressed primarily through private commercial and associational institutions. The political scientist Samuel Huntington not only notes that change in the United States centers in the nongovernmental structures, but he also expresses how change in both sectors is related.

> Change in some spheres renders unnecessary or impossible change in others. In America the continuity and stability of government has permitted the rapid change of society, and the rapid change in society has encouraged continuity and stability in government.[69]

[68] Charles W. Morris, *Varieties of Human Values* (Chicago: University of Chicago Press, 1956), p. 24.

[69] Huntington, p. 132; also see Chambers, *Political Parties in a New Nation,* pp. 7–8. From a similar perspective Parsons calls the United States a "nonpolitical" society, see Talcott Parsons, *Structure and Process in Modern Society* (New York: Free Press, 1965), p. 206.

Major change thus takes place outside of the political institutions. The limited amount of political change the institutions do allow is sufficient since most change is external to the politcal system. A study of general social culture in the United States would thus be very different from the current study. There the screening process would be much weaker. The liberal tradition, by allowing the market to control much social activity, escaped the necessity for centralization to achieve modernization which occurred in practically every other country.[70] Lockeanism could triumph in both social and political spheres by allowing change in the one and stability in the other.

Givenness, thus, is seen as a characteristic of a particular type of political culture. This type of culture is one which relays especially high levels of diffuse support toward the regime and the community. The American environmental circumstances, the anti-governmental bias of America's liberal tradition, and the ability of its social system to cope with stress outside of the political system have all led to the creation and maintenance of just such a givenness culture in the United States. This type of culture, in turn, provided the basis for high levels of diffuse support which tend the regime toward maintenance.

THE LIBERAL TRADITION POLITICAL CULTURE

The political culture of the United States is conceived as being composed basically of Lockean values, which can be called the liberal tradition. It is maintained that widespread agreement on these values exists in the American system. Environmental stimuli are processed through the screen of this agreement. Some want stimuli are easily translated into demands if they are in accordance with the political culture. Other wants are substantially changed before they become demands and still others which are antagonistic to the tradition can be ignored. Through its influence upon the conversion process, political culture influences the outputs arrived at by the system. Because of the Lockean nature of the American political culture, most wants are channeled to other social systems. The remaining wants are routed to the political sys-

[70] Huntington, pp. 135–139.

tem where they too are converted and shaped by the liberal tradition.

Stimuli are, thus, shaped in the direction of the cultural tradition. This is one of the major means a regime has of maintaining itself. A givenness political culture such as that assumed to be operative in the United States is especially able to provide the political system with a means of stress control. In a regime which de-emphasizes the use of coercion, maintenance could prove difficult. How has the American regime been able to maintain itself?

> Why has it survived so long? Why did it not go the way, for example, of the Restoration Charter in France? Has this not been precisely because fundamental value struggles have not been characteristic of the United States? Surely the Constitution did not fare very well in the only time any such struggle did appear in the United States, the time of the slavery controversy in the middle of the nineteenth century. Civil war broke out.[71]

As long as values are widely accepted, political regimes can cope with stress. The more widely accepted they are, presumably the more functional they are to regime maintenance. However, there is no expectation of unanimity. For political culture to be considered existent it will be enough to demonstrate that a consensus exists.[72]

The view that the Lockean liberal tradition is the value consensus in the United States has not been widely disputed. However, some have dissented on some aspects, though close analysis suggests that their objections are not really applicable to the present conceptualization. For example, Edward Handler charges Boorstin with excluding change from his givenness concept. When givenness is converted to nondeterministic political culture as it is here, though, change is explicitly introduced into consideration. A second critic has attacked

[71] Hartz, p. 85.

[72] For an interesting discussion of unanimity as the basis of constitutions, see James M. Buchanan and Gordon Tullock, *The Calculus of Consent* (Ann Arbor: University of Michigan Press, 1962), ch. 6; also see chs. 7 and 8; on consensus, see Key, *Public Opinion and American Democracy,* ch. 2. This concept will be treated more extensively in Chapter V.

Boorstin for his use of the concept of consensus. John Higham maintains that consensus was actually a "cult" created to inculcate conservatism. If true, this criticism would also be applicable to political culture. Yet a few years later Higham and others concluded that Hartz and Boorstin actually emphasize consensus because they analyze the United States from a comparative perspective.[73] In comparison with other political systems the United States is consensual. This perspective is the justification for the use of the organizing concept of culture for the present study of American values (as it is also the justification for the inclusion of this book in a comparative politics series). The comparative perspective provides the rationale for stressing the lack of an American feudal period and the resulting favorable environment for developing a highly supported regime. Certainly, there are value differences in the United States. Yet, from a world perspective, they are small.

A third criticism of the liberal tradition hypothesis is directly related to the second. Harry Jaffa disputes Boorstin because he interprets him as saying that America has not had profound disputes. Jaffa maintains that, indeed, the United States had the most basic kinds of disagreement – they simply have not been European conflicts. He maintains that equality and liberty are the context within which American conflict takes place. But is Jaffa's contention really incompatible with the present thesis? The assertion that conflict takes place within the context of equality and liberty actually seems to confirm it. Liberty and equality, as liberal tradition values, provide a context within which conflict operates in the United States. In Europe, by contrast, the liberal must not only resolve internal liberal conflicts (such as those between equality and liberty) but he must additionally contest with tory conservatism and socialism. This third critic, indeed, does note that the American focus upon equality and liberty

[73] Edward Handler, "Introduction," in Edward Handler, ed., *The American Political Experience* (Lexington, Mass.: Heath, 1968), pp. 2–5, also see pp. 147–148; John Higham, "The Cult of Consensus," in Handler, pp. 40–48; John Higham with Leonard Krieger and Felix Gilbert, *History* (Englewood Cliffs, N.J.: Prentice-Hall, 1965), p. 223.

is unique.[74] It is unique because acceptance of the liberal tradition has allowed Americans to argue exclusively over liberal values.

A final critic also disputes one aspect of Boorstin's givenness conception. Neal Riemer maintains that the United States has had a "theory" and that its development is "more indebted to past theory and less a product of the present American 'landscape' than Boorstin appreciates." In contrast to Boorstin, however, the present thesis explicitly recognizes that there *has* been a "theory" – a liberal (or republican) theory. Riemer, however, does not attribute this American theory to Locke. To him, American theory is traceable only to James Madison. He does not mention Locke at all but suggests that Madison had to create the peculiar American theory of federalism without much aid from republican values. He had to choose only the values which seemed appropriate to the American circumstances. Yet, at another point, Riemer *does* acknowledge that liberal values shaped Madison's theory.

> Republican values by themselves do not constitute a theory. They are only part of a theory in the sense that they establish the purpose and standards which may direct men, shape institutions, and influence public policies. In addition to values, any political theory must embody an appraisal of facts and factual generalizations as it must also work out programs of action designed in the light of the facts to advance the values of a given society.[75]

In this sense the liberal tradition looks back to Locke and other early liberals in terms of the *values,* and is newly American in terms of the *appraisal* of facts and their *application* to a particular environment. The American contribution is summed up in the Constitution and *The Federalist Papers.* Riemer thus importantly modifies Hartz's single liberal tradi-

[74] Harry V. Jaffa, *Equality and Liberty: Theory and Practice in American Politics* (New York: Oxford University Press, 1965), pp. 127, 128. Also see *Mississippi Valley Historical Review* (September, 1962), pp. 292–293.

[75] Neal Riemer, "Two Conceptions of the Genius of American Politics," *Journal of Politics* (November, 1958), p. 701; see also pp. 697 and 699. Note that Riemer uses the term *shape.* For an indication of Madison's regard for Locke, see James Madison, *The Forging of American Federalism,* Saul K. Padover, ed. (New York: Harper and Row, 1965), p. 35.

tion. It rather consists of two streams, the older from the early liberals and relating to the values (i.e., from John Locke) and the newer from the Founding Fathers and relating to appraisals of the values (i.e., from James Madison). This modification will be accepted for the present study.

In summary, the American political culture can be referred to as the liberal tradition. It is maintained that the values of Locke, Madison, and the other liberals of the seventeenth and eighteenth centuries represent essentially the same values which comprise the American political culture at the present. Although it has changed, the remarkable phenomenon is that American political culture has survived so unchanged under substantial environmental pressure. It is further hypothesized that these same values shape current political behavior. In the analysis which follows, the Hartz thesis will be used as a working hypothesis. Each element of the system model of the political culture will be investigated to test the extent to which the United States fulfills adherence to the liberal tradition. The data to be investigated will, hopefully, reveal the degree to which the tradition is shared as the American political culture. For it is on empirical grounds that the concept of political culture, and its presumed relationship in the American case to the liberal tradition, must be accepted or rejected. To be accepted, a convincing case must be made that there is enough agreement to constitute culture and enough "shaping" to constitute behaviorally significant political culture.

OPERATIONALIZATION OF THE STUDY

Boorstin has advised us that one of the cultural affinities of the American people is their empirical tradition. Americans want to "let the facts speak for themselves." [76] In a scholarly study, especially, one must be aware of cultural biases. Do facts speak for themselves? Let us take the simple reporting of the results of the 1960 election by the highly respected and objective Congressional Quarterly Service:

> On election day, Nov. 8, the largest number in history—68,838,979 Americans—cast ballots for President. Kennedy

[76] Boorstin, pp. 107–118. The value of empiricism is viewed as a general rather than a political culture value.

> emerged the victor with a solid majority in the electoral college. But his popular-vote plurality over Nixon was only 112,803 votes – the smallest vote margin of the 20th Century.[77]

Certainly we have a simple statement of facts which could not be disputed. The figures were based upon returns made to Congress. Americans to the number of 68,838,979 cast ballots for president, and a plurality voted for Kennedy.

Let us take another look at the facts. (1) Of course, Americans do not cast ballots for president. They cast ballots for electors who cast ballots for president. (2) William Andrews, in an interesting article, notes that about 1 million persons voted in 1960 but their votes were invalidated.[78] The number of Americans who voted in 1960 may have been as many as 70 million. (3) Was Kennedy elected, then, by a plurality?

> It is customary to list the popular vote in such a way that Kennedy appears a narrow winner over Nixon in 1960. Actually in order to reach this distribution of the total vote it is necessary to exaggerate the Kennedy vote from Alabama since on the slate of Democratic electors in Alabama there were uncommitted electors. Eventually six of the uncommitted electors voted for Harry Byrd and five electors voted for Kennedy. If the Kennedy popular vote in Alabama is reduced to a proportion, say 5/11 in this case, of the vote for Democratic electors and if only this reduced popular vote is added to his national total, Nixon, not Kennedy, has the larger popular vote total nationally in 1960.[79]

The "facts" of the 1960 election or any other facts do not speak for themselves or if they do it is in the enigmatic language of the oracle of Delphi.

If facts do not speak for themselves, how can political culture be empiricized? The full truth is that it cannot be conclusively determined. All the analyst can do is set forth his

[77] *Congress and the Nation* (Washington, D.C.: Congressional Quarterly Service, 1964), p. 39.

[78] William G. Andrews, "American Voting Participation," *Western Political Quarterly* (December, 1966), p. 650.

[79] William H. Flanigan, *Political Behavior of the American Electorate* (Boston: Allyn and Bacon, 1968), p. 27; also see *Congressional Quarterly Weekly Report* (February 17, 1961), pp. 285–288.

operational measures in an explicit manner and define his theoretical structure and concepts so that the reader can evaluate their adequacy. The final test of the theory is how well it will stand up to tests of falsification and how heuristic it proves in the long run. Unfortunately, with the resources available, it has not been possible to test the hypotheses as rigorously as I would have liked. Operationalizations (operational definitions) especially are a problem. However, operationalizations are always a problem for scientific analysis. They can neither be as analytically precise as concepts nor can they be as complex as the reality they try to represent. Thus, in the present study, the operational measures of liberal tradition values will not do justice to the subtleties of these values. The operationalizations can only attempt to measure what are thought to be central aspects of very complex concepts. This problem is inherent in scientific analysis and cannot be avoided. Yet this approach is useful. By utilizing the conceptual richness of the concept of culture, by being guided by the biological system analogy, and by using the best data available, it is maintained that the study design and its operationalization can explain enough "fact" to justify the loss of a certain richness of description. Abstraction sacrifices complexity so that we may understand. However, to deal with concepts scientifically the abstractions themselves cannot be used. They must be operationalized in a form useful for empirical testing.

The concept of political culture naturally has two component elements. The first of these is the political. The primary data restriction, then, is that the data be political. From the foregoing analysis it follows that the political data will be related to the liberal tradition. However, at the operational level more detailed criteria must be set. In the following empirical sections of this book most data will be gleaned from politically oriented national probability samples (opinion "polls") taken between 1935 and 1970. Three operational criteria were used for selection of the poll questions to be used. First, any question relating to public attachment (negative or positive) to an obvious liberal tradition value is included. Second, any question dealing with individual assessment of any

American political institution (widely defined) or process is included.[80] Finally, to have as wide a data base as possible, questions relating to the widespread politically related attitudes noted by the national-character literature are included. Much non-political, environmental data is also collected but it is collected as subsidiary to the political data. The incidental collection of this data fortunately allows the testing of some environmental aspects of the system model. However, unless data collection is primarily limited to the political there could be no criteria for data inclusion.

The second element of our main concept is the cultural. Culture suggests generally accepted values. Operationally, the data collected must relate to widespread attitudes held over a long period of time. The idea of culture as an agreement on fundamentals also demands that substantial majorities support the agreement. What the exact percentage of support should be is difficult to conceptualize and it is essentially arbitrary. One study required that 75 percent agreement had to be secured for a consensus.[81] The present study will not be as restrictive. I do not believe that this is a realistic level for a consensus for conceptual and operational reasons. In the real world there usually are substantial numbers of people who do not have an opinion on values. Others simply do not care.[82] Often these people account for 10 to 20 percent of the population. The study which used the three-quarters criterion did so by forcing these people to choose between two pro-con alternatives. In my view this procedure represents an unrealistic, forced manipulation of the population to be analyzed. In the real world these people are allowed not to have an opinion and a realistic reconstruction of that world should allow the

[80] Data are limited, however to structural aspects, not to personality factors, which may be institutionally related. For example, questions on the presidency will not be used if the question mentions the president's name. They are excluded because of the psychological tendency to evaluate the institution from a partisan perspective rather than on the basis of the value given to the institution itself; see Hadley Cantril, *Gauging Public Opinion* (Princeton: Princeton University Press, 1944), p. 39.

[81] Herbert McClosky, "Consensus and Ideology in American Politics," *American Political Science Review* (June, 1964), p. 363.

[82] The authors of *Voting* note that often the "don't know" responses are often "don't care" replies; see Berelson *et al.*, p. 309.

same.[83] Indeed, the concept of acquiescence requires that this group be treated separately—and, perhaps, that in most cases of maintenance they should be counted together with those who positively support the regime.[84] For these reasons I believe that it is necessary to separate and include the no-opinion responses in any conceptualization of mass values.

There is also a data restriction which suggests that high operational criteria for a consensus is inappropriate. All the data used here are derived from other peoples' investigations.[85] Naturally, then, I must use data which may not be in the exact form most appropriate for my purposes. Specifically, there are few questions which directly ask if people support basic regime values. Most of the questions refer to values as they relate to some other object. Often there is a clash of values within a single question. Any reference to another object will influence support for a value—especially reference to another value. Operationally this will appear to be reduced support for the value under study, whereas actually support for the value itself is consensual. Considered alone, the value would be highly supported.[86]

Because of these considerations, I believe that a consensus must be defined more realistically. Three-fourths appears even conceptually too restrictive because of the no-opinion phenomenon. *Conceptually,* perhaps two-thirds support would be more appropriate. With a large number of no opinions, however, even that figure would be too high. It may be that a relative measure is needed. *Operationally,* both the problem

[83] It appears that forcing out the "don't know" or "no opinion" can seriously distort findings. Consider the datum that in 1948 10 percent of the people could not name the two major party candidates for president of the United States. To *force* this kind of person to give an opinion is to confuse the findings; see Hyman and Sheatsley, p. 41.

[84] Daniel Goldrich, *Sons of the Establishment* (Chicago: Rand McNally, 1966), pp. 5–15. Also see the definition of a policy majority in Austin Ranney and Willmoore Kendall, *Democracy and the American Party System* (New York: Harcourt, Brace and World, 1956), p. 55.

[85] For the problems of secondary analyses, see Herbert Hyman, *Survey Design and Analysis* (Glencoe, Ill.: Free Press, 1955), p. 214.

[86] For example, consider Tables IV.47 and IV.48. In the first table, a simple question on belief in a liberal tradition value is asked, and 95 percent support the value. This type of direct question is rare, however. Table IV.48 presents the more usual form. The value of free speech is considered together with the value of unlimited free speech.

FIGURE II.1 *The policy system and the cultural system: percent support for attitude*

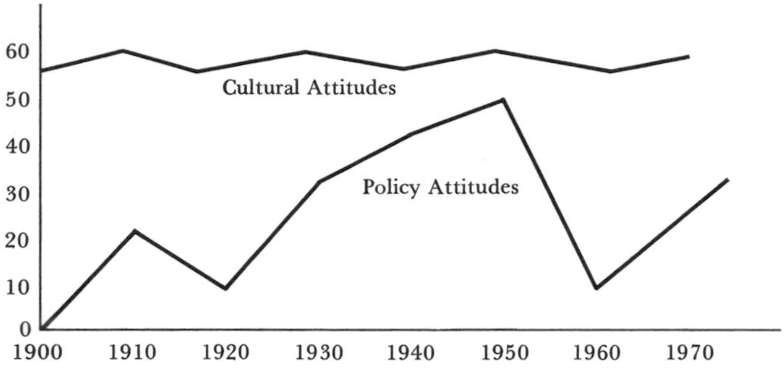

of no-opinion responses and the problems of secondary analysis suggest that for political culture to be operational, there must be (1) at least a *majority for* the value and (2) a *substantial difference* (at least twenty percentage points) between the majority and minority positions. This operational definition is admittedly subject to objection, but considering the above-noted limits to operationalization, it seems the best that can be formulated. Interestingly, even with these operational problems, most value support reported here exceeds the two-thirds requirement and usually exceeds the three-fourths criterion.

Parsons notes that to become a structural part of a social system, interactions must become institutionalized.[87] Since values are viewed as structural, they must meet this criterion. If political culture is to have meaning it must persist so that it can shape the political process. Much opinion exists on the level of policy-making. After being subject to the cultural screen, it fluctuates rather widely through time. It is opinion which is not central to the cultural system. It is the arena within which the differences between political parties and "ideologies" at the authority level are fought to resolution. Figure II.1 indicates the simple empirical difference between the two systems. Policy system attitudes fluctuate widely over

[87] Talcott Parsons, "Culture and the Social System: Introduction," in Talcott Parsons *et al.*, eds., *Theories of Society* (New York: Free Press, 1961), 2:977.

time while the cultural system attitudes are relatively stable.[88] Not all stable attitudes are cultural (the other criteria must also be met) but only the stable attitudes can be conceived as part of the political culture.

The study also attempts to use more than one criterion of judgment where possible.[89] While the methods used will be generally similar (usually opinion polls), as many *measures* of the phenomenon as possible will be used. Confirmation will also be attempted through the use of as many *time periods* as data availability will allow, and of as many *independent surveys* as exist.

The study of values is by no means original to the present work. A large body of scholarly research has used an approach called national character study. This method was used extensively between 1942 and 1953 but has declined since.[90] National character study was developed for small societies. The scale problem for large heterogenous states has forced most analysts to abandon the approach.[91] Indeed, Margaret Mead's description of the method points out its limitations (and, of course, its benefits):

> Anthropological sampling is not a poor and inadequate version of sociological or socio-psychological sampling, a version where *N* equals too few cases. *It is simply a different kind of sampling,* in which the validity of the sample depends not so much upon the number of cases as upon the proper specification of the informant. . . .[92]

Certainly this method can be usefully applied to small societies and can provide helpful insights into the values of a large state. On the other hand, the problems of representative sam-

[88] I would suggest that individual consistency would be even greater than for party identification; see Philip E. Converse, "The Nature of Belief Systems in Mass Publics," in David Apter ed., *Ideology and Discontent* (New York: Free Press, 1964), p. 240.

[89] Karl W. Deutsch, *Nationalism and Social Communication* (New York: Technology Press of M.I.T. and Wiley, 1953), p. 81.

[90] E. Adamson Hoebel, "Anthropological Perspectives on National Character," *The Annals* (March, 1967), p. 2.

[91] *Ibid.*, p. 5. Also see Potter, chs. 1 and 2.

[92] Margaret Mead, "National Character," in Alfred Kroeber, ed., *Anthropology Today* (Chicago: University of Chicago Press, 1953), pp. 645–655.

pling undoubtedly become more difficult as the size of the population increases.[93] This view emphasizes, however, that if one looks for the regularities that are *common to all,* then, naturally, no matter who is chosen he will be able to represent the common characteristics of all.[94]

However, John Stuart Mill noted almost three-quarters of a century ago that

> What is obtained even after the most extensive and accurate observation is merely a comparative result. . . .
>
> . . . Since, therefore, the comparison is not one of kinds but of ratio and degrees and since in proportion as the differences are slight it requires a greater number of instances to eliminate chance; it cannot often happen to any one to know a sufficient number of cases with the accuracy requisite. . . .[95]

Mill raises the question as to how alike a people can be. Obviously, if there are patterns held by all, the anthropological technique is adequate. But what if the common characteristics are only tendencies?

Inkeles provides a helpful clarification: "National character refers to relatively enduring personality characteristics and patterns that are modal among the adult members of a society." Once the concept of *modal* patterns is understood the limitations of direct observation are more apparent. The appreciation of these limits has led Inkeles to give credence to the observation now frequently made that national character cannot be studied scientifically in large states.[96] However, the modern random sample polling operation guided by the concept of mode can overcome the limitation of individual direct observation even in large heterogeneous nations.

[93] Any book on random sampling could be consulted. Three good ones are Charles H. Backstrom and Gerald D. Hursh, *Survey Research* (Evanston, Ill.: Northwestern University Press, 1963), ch. 2; Herbert M. Blalock, *Social Statistics* (New York: McGraw-Hill, 1960), ch. 22; and Fred N. Kerlinger, *Foundations of Behavioral Research* (New York: Holt, Rinehart and Winston, 1964), ch. 4.

[94] Margaret Mead, "National Character and the Science of Anthropology," in Seymour Martin Lipset and Leo Lowenthal, eds., *Culture and Social Character* (New York: Free Press, 1961), p. 19.

[95] Mill, pp. 565–566.

[96] Alex Inkeles, "National Character and Modern Political Systems," in Francis L. K. Hsu, ed., *Psychological Anthropology* (Homewood, Ill.: Dorsey Press, 1961), p. 173. Also see Kroeber and Kluckhohn, pp. 330, 176.

Another social scientist, Don D. Smith, calls national character studies subjective and unscientific. Yet he does not totally reject the approach; he does argue, however, that major changes must be made both in the approach to attitudes and in methodology. Methodologically, he suggests that this can be done with a good data base which allows examination of modal frequencies.[97]

If Inkeles' and Smith's view of national character is accepted, then national character can be a useful method of social analysis. The current work could then be called a study in national character. By the addition of the survey and the idea of modal opinion, as well as by the incorporation of the guidelines suggested by Smith, the study of national character in large states is made possible. Yet it appears to me that political culture is more useful as a concept than national character for the following reasons. First, it more directly relates to the substantial literature on the behavioral characteristics of culture. Second, as an aggregate concept political culture becomes an analytic social structure rather than a psychological attribute of individuals. Third, national character has some overtones of ethnocentrism and blood-related, "inborn" characteristics.[98] Fourth, political culture can more easily be identified with the modal operational method, and the use of advanced techniques.[99] Finally, political culture seems more amenable to inclusion in the system model.

The data for the present study of mass political culture has been taken mainly from the most reliable public opinion polls. Fortunately, more than three decades of useful results are now available to the opinion analyst. Not too many years ago an investigation like the current one would have been impossible. The American Institute of Public Opinion (Gallup Poll) has been taking polls continuously since 1935. More recently, the Survey Research Center has undertaken detailed, highly technical, and competent studies of the biennial Amer-

[97] Don D. Smith, "Modal Attitude Clusters: A Supplement for the Study of National Character," *Social Forces* (June, 1966), pp. 526–527, 529.

[98] Inkeles, pp. 174–175; Kluckhohn, ch. 5.

[99] Moshe M. Czudnowski, p. 878. Edward H. Shills, "Authoritarianism 'Right' and 'Left,' " in Richard Christy and Marie Jehoda, eds., *Studies in the Scope and Method of the Authoritarian Personality* (Glencoe, Ill.: Free Press, 1954), pp. 24–49.

ican elections. Many other polls – taken at less frequent intervals or done only once – are also useful to the student of culture and opinion. On the other hand, few polls have been directly devoted to cultural problems (the Almond-Verba study being the most conspicuous exception). It is not surprising, therefore, to find that much data is not available. The commercial polls are naturally more interested in matters of current policy. However, the poll takers have been generous enough to provide sufficient data to allow at least a preliminary survey of American political culture.

Even with the best data, however, the task is extremely ambitious.

> For it is evident that American life and American energy have expressed themselves in a great diversity of ways, and any effort to define the American as if nearly two hundred million persons all correspond to a single type would certainly reduce complex data to a blunt, crude, and oversimplified form.[100]

The analyst who made this observation is worried lest an investigation like the current one be led in the direction of superficially stereotyping nations. This is certainly a danger, but it can be averted.

The approach to the study of values presented here attempts to avoid simplistic stereotyping through the use of a rather sophisticated scientific conceptual framework. The concept of political culture directs the whole analysis. Its heuristic implications provide a valuable scientific tool for value research. Hopefully, the use of modal distributions, the employment of a complex system model, the analytical separation of the concept of political culture into four component systems, the use of all reliable data available, and the presentation of a wide range of replies to survey questions in tabular form across a substantial time period – all these methods may provide a revealing description of that complex which has been called the American political culture.

One of the first to view social and political problems in the United States from the perspective of social science, Alexis

[100] David M. Potter, "The Quest for the National Character," in John Higham, ed., *The Reconstruction of American History* (New York: Harper and Row, 1962), p. 213.

de Tocqueville, viewed them from an implicit model of political culture.

> The manners of the Americans of the United States are, then, the real cause which renders that people able to support a democratic government. . . . I should say that the physical circumstances are less efficient than the laws, and the laws very subordinate to the manners of the people.[101]

Of course, "physical circumstances" correspond to the environmental in our model, the "laws" refer to the authority structure and "manners" can be interpreted as political culture. In a like manner Madison emphasizes the importance of cultural influences. He asks, what restrains abuses in government?

> I answer: the genius of the whole system; the nature of just and constitutional laws; and, above all, the vigilant and manly spirit which actuates the people of America. . . .[102]

Both Tocqueville and Madison suggest that the basic values of Americans provide a most important explanation of the maintenance of the American democratic regime. The present study is an empirical test of this hypothesis.

What specifically will be tested? The analysis to this point suggests that there are four hypotheses. The first is descriptive. Since a descriptive hypothesis is one which seeks to define empirically the major concept used in a study,[103] a crucial goal of this research is the empirical description of the American political culture. (1) The descriptive hypothesis of the present work is that the supported American political culture is Lockean. (2) For support to be cultural, tradition support must be shown to be widely distributed through major social groupings in the United States. The remaining hypotheses are explanatory. An explanatory hypothesis seeks to explain one variable through the use of another.[104] (3) In this study, the values of the liberal tradition are used to explain the types of demands and supports processed by the political system. The resulting hypothesis is that the high levels of political cul-

[101] Tocqueville, 1:313; also see 2:7.

[102] *The Federalist Papers* (New York: New American Library, 1961), No. 57, p. 353; also see No. 62, p. 382; No. 39, p. 240.

[103] Hyman, p. 78.

[104] *Ibid.*, p. 178.

ture support in the United States shape wants in the direction of the liberal tradition and that this shaping leads to the maintenance of the culture values. The final hypothesis is (4) that the maintained political culture helps to determine whether the regime itself will be maintained. Political culture is used to explain the stability of political regimes. American data is related to other available diffuse support data and it is hypothesized that the higher the level of support for the political culture in a system, other things being equal, the greater the chance the regime will be maintained.

Although there are four hypotheses, they are not equally central to the present study. As has been noted above, the book is primarily descriptive. The major goal is the empirical description of the political culture of the United States. To do this we must now turn to the empirical measurement of the four subsystems of political culture.

CHAPTER III

The Basic Requisites of Political Culture

THE AMERICAN IDENTITY SYSTEM

The community is conceived as the most basic level at which a political system acts.[1] A political community is "a group of persons bound together by a political division of labor." [2] The concept of community as used here in its most general sense carries little of the sociological implications of the quality of relationship existing between individuals. Political community in the present sense is simply the political object which is the people of a nation-state who are bound together politically. This object could even be kept together solely by force or output manipulation.[3] In a study of values, however, one is not primarily interested in studying the manipulability of political community but in measuring the degree of support for the value object among the system members. The value perspective leads to a focused study of member orientations to the object of the community rather than to a study of either the object itself or to authority influence on the object. The identity system analysis describes the extent to which members support, or do not support, the values associated with the political community.

1 David Easton, *A Systems Analysis of Political Life* (New York: Wiley, 1965), pp. 176, 177–179. Hereafter referred to as *A Systems Analysis*.
2 *Ibid.*, p. 177.
3 *Ibid.*, pp. 185–189.

According to the available literature, members of a political system may be related to the political community in two major ways. The first and more primitive relationship is called the sense of national identity.[4] The *sense of national identity* is the *personal identification between the member himself and the political system.* It is concerned with the degree to which an individual identifies with his community. Through the sense of national identity the individual can orient himself politically and can identify with a group of others. The nation with which the individual identifies does not have to be the nation-state with which he is objectively associated, although for the great majority of members, it is. For many national minorities within political systems, however, the nation with which the member identifies is not the same as the one which authoritatively allocates values for him. For the present study, though, national identity is mainly concerned with the extent to which the member does or does not vertically relate to the nation-state within which he is domiciled.

The second important way members relate to the political community is through what can be called the sense of community identity. The *sense of community identity* is the *mutual identification of individuals with each other in a political system.*[5] As used here, sense of community identity is limited to a consideration of how strongly members identify with others who share the same political system. It helps to explore the cohesiveness within the national group. Sense of community identity is distinguished from sense of national identity in that the former focuses upon the horizontal member-to-member interactions rather than upon vertical member-to-political-system relationships. It is the realization of togetherness within a web of relationships, the belief that the web should *embrace others* besides me, as opposed to my awareness of my own location in the web itself. Both of these ways of

[4] Sidney Verba defines two dimensions of national identity. Although I will use both of these dimensions, only the first will be referred to by that term. His second dimension will be related to Easton's sense of community. See Sidney Verba, "Comparative Political Culture," in Lucian W. Pye and Sidney Verba, eds., *Political Culture and Political Development* (Princeton: Princeton University Press, 1965), pp. 534–536.

[5] Easton, p. 325.

relating to the political community will be conceived as essential components or subsystems of the identity system of political culture. The identity system, thus, comprises both vertical and horizontal attachments or orientations toward the political community.

SENSE OF NATIONAL IDENTITY

In most political systems, at some point in time members begin to develop attachments to some groups beyond the most parochial. One of the relationships which may develop is that between the individual and the nation-state. Because of the importance of the nation-state in the modern world, the present study will focus upon identification with this important stucture. From the perspective of the study of values, the interest is centered upon the vertical affective orientation of the member toward the abstract symbol of his political system.[6] The support of the member for this abstraction, other things being equal, provides screening capacity for the political culture. As such, sense of national identity presents the first of the elements of the culture which must be investigated.

One authority on national identity maintains that identification with the nation-state is the most crucial aspect of political culture.

> It is the sense of identity with the nation that legitimizes the activities of national elites and makes it possible for them to mobilize the commitment and support of their followers. The most potent kind of commitment that political elites can arouse is to the political system per se — that is, a commitment to it over and above its actual performance. It is only such a rain-or-shine commitment which will allow a system to survive the many kinds of crisis that are likely to arise during processes of rapid social change.[7]

It is the present view that sense of national identity is at least *one* of the crucial aspects of political culture. It certainly

[6] Of course, this means that vertical identity could just as easily be considered within the second subsystem of political culture. It was not, because the concept of an identity system seeks to stress identification with the idea of nation-state as an important *preliminary* subsystem. The concept of symbol attempts instead to emphasize general artifact attachment.

[7] Verba, pp. 529–530.

seems to be one of the most basic requisites for a shaping culture. Presumably, a member must conceive that a separate nation exists before he can have an affective orientation to it and before nationality can have an influence upon his other attitudes. There are two levels to sense of national identity, then. In the first place there must be a threshold of realization that a nation exists and in the second some affective orientation to it. Both of these aspects will be considered below.

It was mentioned above that a political system could be created and maintained without identity support through the use of force. In the Lockean liberal tradition, however, outside force is not a legitimate means for system creation and maintenance.[8] In a liberal polity, therefore, one might expect that the value system, which is here viewed as an alternative to force, would be developed very early in the process of creating a political community. One view, mentioned above, maintains that American identity post-dated the Revolution. This identity was first expressed in the person of Washington as a charismatic unifying leader and in the Revolution intellectuals as system-creating elites.[9] The present analysis will attempt to show that national identity actually developed prior to the adoption of the Constitution and even before the Revolutionary War. By indicating that identity is a continuous process, we make the possibility of the existence of any tradition more defensible. For if identity arose exclusively from the personalities, symbols, and values of a revolution, the existence of a tradition extending back to the late seventeenth century is certainly very questionable. At this point, then, the hypothesis minimally requires that American national identity be a developmental process, reaching back to a tradition in Colonial America. However, this is only the minimal descriptive hypothesis for American national identity. The concept of political culture also requires that there be, at some point in time, a consensual support for national

[8] John Locke, *Second Treatise on Civil Government,* in Sir Ernest Barker, ed., *Social Contract* (New York: Oxford University Press, 1962), sec. 176, pp. 103–104. Hereafter referred to as *Second Treatise.*

[9] Seymour Martin Lipset, *The First New Nation* (Garden City, N.Y.: Doubleday, 1967), ch. 2.

identity. In a liberal polity based upon government by consent [10] the hypothesis would lead one to predict that sense of national identity would be especially strong in the United States. For the current attitude data available, therefore, Americans must have a wide commitment to the political culture values of the nation.

THE CONCEPT OF NATIONAL IDENTITY

Karl Deutsch identifies what he maintains are the agreed-upon elements of nationality: the physical country, a shared history, a system of interlocking values, interlocking habits and roles which link to social institutions, and a common attachment to symbols.[11] Vertical identity is simply one's identification with this entity. But there are different explanations as to how national identity is achieved. The choice of the most explanatory theory can help us answer the minimal hypothesis expectation for the identity system. Richard Merritt suggests that the important theories can be placed in three categories: (1) common political structures create a national identity or, similarly, common experiences in other formal institutions serve the awareness function; (2) single important, dramatic events (such as the Stamp Act or the Revolution in the American case) lead to nationalism; or, (3) national identity is the product of long-term increases in social communication.[12]

Merritt rejects the first explanation because to him there is too much evidence to indicate that identification exists prior to the establishment of formal governmental or other formal society-wide social institutions. The second explanation is more difficult for him to dismiss. Most prominent historians have used this type of explanation. The greatest problem with this explanation is that the historians do not choose the same historical event as crucial. Merritt notes that Albert Harkness, Jr., chooses the War of Jenkins' Ear (1739); Max Savelle opts for the time between King George's War

[10] *Second Treatise,* sec. 106, p. 62.

[11] Karl W. Deutsch, *Nationalism and Social Communication* (New York: Technology Press of M.I.T. and Wiley, 1953), p. 14.

[12] Richard L. Merritt, *Symbols of American Community 1735–1775* (New Haven: Yale University Press, 1966), pp. 4–14.

(1748) and the French and Indian War (1754); Carl Lotus Becker decides in favor of the French and Indian War itself; while Carl Bridenbaugh, Bernard Knollenberg, John C. Miller, Edmund S. Morgan, and Allan Nevins argue for the Stamp Act (1765) as the critical historical act in national identity.[13]

It is this lack of a single, precise event which makes the present work opt for the third approach. Deutsch suggests that the best way to measure this long-term growth is to study the communication interactions in a society. To him, "Processes of communication are the basis of the coherence of societies, cultures, and personality." [14] Where communication interactions are concentrated, a nation exists; where they fall off rapidly in intensity, the nation ends. Members of a single people are thus identified by more intensive social communication and economic intercourse. An important part of this communication is devoted to the common beliefs, symbols, and habits of culture.[15] The culture is transmitted through communication. It is therefore suggested that the history of national identity can be studied through the *development* of patterns of social interaction and communication. Over time, then,

> we can trace the gradual assembling of the conditions for national consciousness in one population after another. At an early stage, we find a population with at least the minimum of compatibility in the culture patterns, if not in the speech, of the smaller groups — tribes, villages, or families — of which it may consist. Its area of habitation is delimited largely by . . . uninhabited areas and by natural obstacles. . . . Common experience of gaining a living in the area may strengthen that compatibility which in turn had made it possible for that experience to be made out and remembered as common. Experience and complementarity may then continue to reproduce each other . . . in a syndrome of . . . social learning in which individuals, usually over several generations, learn to become a people.[16]

[13] *Ibid.*, pp. 8–10.
[14] Deutsch, p. 61.
[15] *Ibid.*, pp. 75, 61–63; also see Merritt, pp. 17–18.
[16] Deutsch, p. 148.

Merritt provides a framework with which the development of national identity in a country can be investigated. Total community (which includes national identity) is achieved when there is a willingness or conscious consensus within some social group "to create common structures and processes of government and to comply with their injunctions."[17] Merritt's idea of community amalgamation, however, seems to be a much more final process than is implied by Verba's "sense of identity with the nation." In Merritt's sense amalgamation is the whole process leading up to and including the creation of the regime. Sense of national identity, however, will be viewed as a process leading to amalgamation.[18] But it will not necessitate creation of the regime to be considered existent. Identity will, further, not be defined to mean "the same as." Merritt defines identity as the "degree of . . . identification with the new group."[19] Verba refers to a psychological membership or "positive identification with" the nation.[20] In the present work national identity will measure the degree of affective identification with a common national entity, but it does not require community amalgamation. A sense of identity can precede the formation of the regime.

Merritt presents six developmental criteria which can be used as indicators for the existence of an integrated community: (1) when a common name for the country is accepted by outsiders and members; (2) when political parties or leadership groups transcend purely regional boundaries; (3) when people feel a common identification with each other — a "We, the people" feeling; (4) when communication is more effective within the system than it is outside of it; (5) when symbols such as flags, anthems, and historic myths are shared; and, (6) when the constituent institutions of government have been amalgamated.[21] Based on the distinctions already made, only the first four criteria will be deemed necessary for national

[17] Merritt, pp. 21–22.

[18] See the discussion of the definitional problem in Mildred A. Schwartz, *Public Opinion and Canadian Identity* (Berkeley: University of California Press, 1967), pp. 8–15.

[19] Merritt, pp. 112–113.

[20] Verba, pp. 529–531.

[21] Merritt, pp. 22–23.

identity to be considered existent. The other two criteria ask for conditions which can usually be met only after a community exists as a separate political regime.

THE ORIGINS OF AMERICAN NATIONAL IDENTITY

In his excellent study of the development of American community between 1735 and 1775, Merritt has provided an empirical description of the growth of colonial American system identity. His basic identity question is, "When did the British colonials begin to identify themselves as Americans?" That it was early seems certain,[22] but when? The methodology employed by Merritt assumes a developmental process influenced by formative events. Merritt does not dismiss the historians of the "crucial event" school. Rather, he places these events along a continuum, where they mark phases of identity. Time, then, is the horizontal dimension. At each time period, Merritt measures awareness of American identity through rises in the level of symbols of attachment to things American. That is, what is the ratio of British symbols of identity used by the population to American symbols? Since the direct evidence is not available, he systematically analyzes the place-name symbols employed in a sample of newspapers published in five major colonial settlements. In every settlement — Boston, New York, Philadelphia, Williamsburg, and Charleston, South Carolina — each symbol was counted in four randomly selected issues per year. All symbols were characterized as American, British, and other.[23]

Fourteen major events (and six other subcategories not listed here) were of special interest: the War of Jenkins' Ear, the Albany Conference (1754), the French and Indian War, Pontiac's Conspiracy (1763), the Royal Proclamation of Reorganization (1763), the Stamp Act (1765), the Townshend Acts (1767), the Boston Massacre (1770), the Tea Act, the "Intolerable Acts" (1774), the Quebec Act (1774), the First Continental Congress (1774), the Battle of Lexington (1775), and the convening of the Second Continental Congress in

[22] On this point, see Robert A. Dahl, *Pluralist Democracy in the United States* (Chicago: Rand McNally, 1967), p. 73.
[23] Merritt, pp. 24–25.

TABLE III.1 *Median distribution of identity place-name symbols, America, 1735–1775*

	1735–1739	*1740–1744*	*1745–1749*	*1750–1754*	*1755–1759*	*1760–1764*	*1765–1769*	*1770–1775*
American symbols	16%	11%	11%	17%	20%	16%	28%	32%
British symbols	22	17	17	22	20	19	25	24
Other symbols	65	72	70	62	61	65	45	41

Source: Derived from Richard L. Merritt, *Symbols of American Community, 1735–1770* (New Haven: Yale University Press, 1966), p. 215. Columns do not always sum to 100 percent due to rounding.

1775.[24] Using these events as orienting points, Merritt charts the usage of symbols of identity. A derived distribution of the symbols through time appears in Table III.1. Although the development is not uniform, the general direction is toward an increase across time. The use of American symbols of identity doubled from 1735–39 to 1770–75.

Unfortunately, the use of five-year intervals in Table III.1 does not indicate the full identity process. Within the five-year periods there are wide variations in identifications. Years of conflict show great shifts toward the use of American symbols. Peace brings a shift back to British symbols. On the other hand, the time between the high identity periods progressively shortened as the break with Britain approached. The table also does not indicate the relative shift between British-American and purely American symbols of identity. In 1774 and 1775, on the eve of revolution, American symbols were double the British ones.[25]

Not only did the colonists become aware of this separation but the British became more aware of it earlier. The idea of "separateness" actually grew more slowly in America than in the mother country. It was not until about 1760 that Americans began viewing their land as truly distinct from their British homeland.

> By 1763 the colonists had crossed an important threshold in symbol usage and in perceptions of the American community. The linear trend line . . . indicates that it was in 1763 that the

[24] Merritt, pp. 29–31.
[25] *Ibid.*, ch. 5.

> trend in colonial symbol usage crossed the 50 per cent line; after that date symbols of American origin more often identified the colonies as American than British in every year save two, 1765 and 1766.[26]

This was also true in conceptions of the distinctiveness of the American people, though this trend was not fully evident until 1773. Symbolic ties between American states began to become more important than mother-country–colony ones. Although the whole complex development was anything but smooth, identity cannot easily be solely attributed to the historian's single, dramatic event. In fact, there is evidence in Merritt's study that the French and Indian War actually prolonged the process of finding a common American identity. Other events like the Stamp Act undoubtedly reinforced identity but they were not the first nor single cause.[27] Symbol usage, rather, sloped toward a long-term development of national identity.

In summary, there was a high degree of identification with an American community by the late 1760's – well before the creation of the new political system. All four of Merritt's criteria for community-level identification of a political system were met perhaps a decade before the Revolution. At the latest, it preceded the meeting of the Second Continental Congress and effective American control of the royal colonial governments by two years,[28] and the Declaration of Independence by three. Once the governmental institutions were formed, however, the identity system was reinforced by identification with regime symbols.

POST-REGIME NATIONAL IDENTITY

If a political system is to be able to survive the severe stress to which systems are on occasion subjected, support must be higher than the recognition levels evident at the establishment of the political community.[29] The feeling of "separate-

[26] *Ibid.*, p. 125.

[27] *Ibid.*, pp. 131, 147–181.

[28] Ralph Volney Harlow, *The United States* (New York: Holt, 1949), p. 103.

[29] See Easton, p. 187, regarding the development of a strong sense of community in periods of stress.

ness" is only a starting point. For authorities to operate effectively with a legitimacy based on consent, they must themselves be identified with valued community objects.[30] One of the means which can be used is the development of a consensual support for national identity.

The building of a sense of national identity was especially evident in the United States during its formative period. An appreciation of the place of George Washington as a central national symbol is necessary for an understanding of this type of identity. As noted, Washington was a central identity symbol for the new nation.[31] He was truly regarded by his contemporaries as "first in war, first in peace and first in the hearts of his countrymen." [32] In the transition period from the beginning of the Revolution to adoption of the Articles of Confederation, Washington, the Continental Congress (a much inferior identity symbol), and the Declaration of Independence (which was specifically intended as a unifying symbol) [33] were the *only* symbols of national identity. The states served an ambivalent identity purpose. While they vied for identity appeal with the national regime (and probably were the winners),[34] they also helped create the peculiar identity necessary for a federal system, i.e., identification with a provincial and national regime. That the states were not so much rivals as partners in the identity process was one of the implicit justifications for the establishment of a federal system.

The Articles of Confederation also served an identity function by creating a transition regime prior to the more permanent 1789 one. The Lockean tradition of distrust of centralized

[30] Verba; also see Lucian W. Pye, *Politics, Personality and Nation Building* (New Haven: Yale University Press, 1962), pp. 38–42; and Seymour Martin Lipset, *Political Man* (Garden City, N.Y.: Doubleday, 1960), pp. 64–70.

[31] Lipset, *The First New Nation,* pp. 21–26.

[32] Harlow, p. 162. Harlow, not a nationalist historiographer, summarizes the Washington qualities as high prestige, respected judgment, acknowledged courage and determination, wide experience, administrative competence, and recognized honesty; also see William Nisbet Chambers, *Political Parties in a New Nation* (New York: Oxford University Press, 1963), pp. 35–36.

[33] Harlow, p. 108.

[34] *Ibid.,* p. 124: "Local pride was strong, and confidence in the local legislatures even stronger."

government was strong, and "in the colonies there had never been any genuine, widespread, deep rooted desire for union."[35] This value was overcome to some extent by substantial environmental pressures demanding union. Without the Articles, however, the Constitution might never have been accepted. It was probably a necessary transition to overcome the fear of a central government. It was the Constitution itself which provided the most important symbol for national regime identity. The place of the Constitution in the United States is probably unmatched in any other society. As Kohn has said, "For the American Constitution is unlike any other: it represents the lifeblood of the American nation, its supreme symbol and manifestation." [36] It represents — and to some extent is — the regime. As the Constitution became accepted it provided an especially strong affective orientation for the nation.

Other identification devices have also been important in the development of an American identity. Turner suggested that the frontier was the prime cause of the development of a separate identity.

> Up to our own day American history has been in a large degree the history of the colonization of the Great West. The existence of an area of free land, its continuous recession, and the advance of American settlement westward, explain American development.[37]

Recently, many historians have questioned the influence of the frontier in explaining the uniqueness of the United States. As Hartz correctly notes, "Frontiers are to be found in other lands." [38] However, it is not in terms of its uniqueness that the frontier has been important in American development. The frontier spirit interacted with the social structure already existing and influenced the political culture; but it was not the primary influence.

[35] *Ibid.*, p. 124.

[36] Hans Kohn, *American Nationalism* (New York: Collier Books, 1961), p. 20. This important American symbol will be considered in more detail below.

[37] Frederick Jackson Turner, *The Frontier in American History* (New York: Holt, 1921), p. 1.

[38] Louis Hartz, *The Liberal Tradition in America* (New York: Harcourt, Brace and World, 1955), p. 95.

Deutsch helps us determine the significance of the frontier when he notes that national identity will be the more effective as consciousness of differences with others are exaggerated. The psychological awareness of a difference is more important, in the long run, than the strictly physical effects.[39] The significance of the frontier was that it made Americans *think* they were different and helped them identify with the nation.

> The expanding Western empire and the interpretations of its function in American life helped the United States in its newly awakened nationalism to turn away from, and against, England and Europe. The influence of the ever moving frontier tended to lessen the consciousness of common roots and origins binding the two shores of the North Atlantic.[40]

The *idea* of the frontier, thus, helped to further exaggerate already existing differences between the new United States and its older ties with Europe and especially Great Britain. The frontier was not, as Turner claimed, the reason for American differences but an emphasis which served to further justify the separateness which had been previously argued by the Declaration of Independence.

The frontier was not the only identity symbol in which interpretation was as important as or more important than the environmental factor itself. The "American School" of history presented a description of American origins and development which emphasized separateness and uniqueness. George Bancroft, whose monumental *History of the United States* appeared between 1834 and 1874, epitomized this school.

> Like all nationalist historians, Bancroft wished not so much to tell what had happened as to see it as it should have happened in a logical historical development, the goal of which was the justification of the nation, its deeds and its glories. Nationalist historiography desires not only to describe a people's life but to help form it and to make its history appear as the fulfillment of a supposed national destiny. Such a historiography is less important for the knowledge of history itself than for the under-

[39] Deutsch, pp. 147–151. Also see Henry Nash Smith, *The Virgin Land: The American West as Symbol and Myth* (New York: Vintage Books, 1957; originally published 1950), p. v.

[40] Kohn, p. 33; also see pp. 36–39, 56, for examples from other countries of similar phenomena.

> standing of the image which a nation forms of itself and of its own nationalism.[41]

Bancroft and his school, like Turner, sought to create a powerful image of a unique and self-sufficient nation which could be easily identified – not only by common roots but, as in the case of Bancroft and the American School, by a common destiny.

Awareness of "separateness," the adoption of the Constitution as a community-related symbol, and the reinforcement techniques of the nationalist historiographers all served to create a strong sense of national identity. Yet identity must be constantly renewed or diffuse support is reduced. Identity has been developed as a continuous process. As such we may be able to specify certain critical points in the process. Can a point of threshold (where identity is high enough to support some regime) be identified? Is there a point where the system is relatively well entrenched? If we can accept the political party system as being a central measure of the development of a political system in a democracy, perhaps a tentative conclusion on this point could be reached.[42] Chambers finds that the main characteristics of the American party system were fully developed by 1860 and that the process began roughly with the election of 1828.[43] A high level of identity was probably reached by the 1830's and, certainly by midcentury. Strangely, the system was almost destroyed at the time of its maturation. Yet it was a strange type of threat to a system – both sides to the controversy claimed the defense of the major regime symbol even through the most intense fighting of the Civil War. After the fighting ended identity was probably regained rather quickly.

Figure III.1 charts the development of American identity as hypothesized by this analysis. Merritt suggests the early development with the threshold of identity being crossed in the

[41] *Ibid.*, p. 40.

[42] A fuller discussion of the justification for this assumption can be found in Austin Ranney and Willmoore Kendall, *Democracy and the American Party System* (New York: Harcourt, Brace and World, 1956), ch. 5, esp. pp. 114–115; and Herman Finer, *Theory and Practice of Modern Government*, rev. ed., (New York: Holt, 1949), ch. 13.

[43] Chambers, pp. 11–14.

FIGURE III.1 *The development of a sense of American national identity*

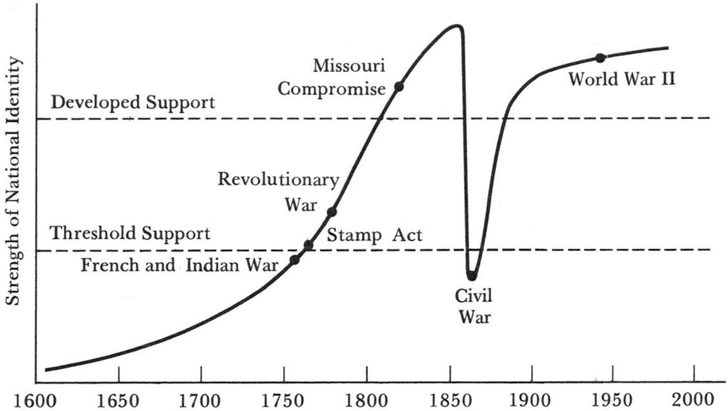

1760's. It is suggested that the sense of identity increased until the collapse at the time of the Civil War. Full development takes place about 1820 and the maximum at about 1850.[44] After this period, identity became threatened. Even then, a higher sense of national identity existed just before the War than had existed before 1750. After the war, the sense of identity probably recovered at a rapid rate.

Of course, this analysis is to a great extent speculative. However, the Merritt analysis, the Chambers research, and the Civil War and present attitudinal data do allow us some reference points for hypothesizing consensus and cleavage points.[45] Between these reference points very little can be relied upon. Despite the limitation of data and resources, however, the analysis appears to be a reasonable historical description of the American identity process.

MAINTENANCE OF NATIONAL IDENTITY

What of the current status of the identity system? Much of the previous discussion has had to proceed with inferential data. Fortunately, recent attitudinal data is more directly related to attachment to the nation. The subtleness of identity as a concept, however, does not allow for a simple measure-

[44] Ranney and Kendall, pp. 109–110.

[45] See Chapter V for a discussion of cleavages. Also see Figure III.2 and the related discussion.

TABLE III.2 *Behavioral measurement of national identity*

Question: If you were free to do so, would you like to go settle in another country? (AIPO)

	1948	*1971*
United States	94% (No)	88% (No)
Sweden	85	82
Finland	—	81
West Germany	—	73
France	72	—
Netherlands	56	84
Norway	66	—
Italy	64	—
Great Britain	53	59

Source: *Public Opinion Quarterly* (Fall, 1948), p. 544; and *Gallup Opinion Index* (May, 1971), p. 25.

ment. The minimal requirement must be that the operationalization measure the degree to which the individual has an affective orientation to the vertical object of the nation. If possible the measure should be comparative so that American orientation can be judged in relation to similar attachment in other political systems.

Tables III.2 and III.3 attempt to measure this vertical orientation to various political systems. National identity is measured, first, in terms of tenacity of attachment in a behavioral manner. To what degree would the members of the system be willing to desert the national community in a physical sense? The second measure is more directly related to emotional commitment to the nation. The more behavioral sense of national identity is derived from two comparative studies done by Gallup International. The people in nine Western countries were asked if they would emigrate from their political system. Table III.2 shows that Americans were least willing to do so. In the other countries commitment varied widely. The first measure of sense of national identity, therefore, seems to meet the descriptive hypothesis expectation of strong identification for the United States as a liberal tradition country. Of course, other dimensions than strictly national identity (among others, the political ones of state and local identity) [46] are included in

[46] In 1946 Gallup found that 72 percent of the American people did

TABLE III.3 *Attitudinal measurement of national identity*

Question: Which country in the world gives the best chance of leading the kind of life you would like to lead? (1948)

	Percentage identifying with their own country
United States	96%
Australia	83
Britain	51
Norway	50
Mexico	45
France	43
Italy	36
Netherlands	31
Germany	30

Source: William Buchanan and Hadley Cantril, *How Nations See Each Other* (Urbana: University of Illinois Press, 1953), p. 30.

the measure but these are viewed as complementary to the national identity.

Table III.3 presents the data for the more attitudinal measure of national identity. This poll, conducted by Benson and Benson, Inc., for UNESCO, was taken in 1948. Americans were found to identify their native country with their own preferred way of living at a relatively high rate. As with the first measure, more than nine out of ten Americans identified with their country, again the greatest percentage so identifying of the countries tested. Similar questions were also asked by the Gallup organization in 1946. When given a choice of place to live, 98 percent of the Americans chose their own country as compared to 80 percent of Frenchmen.[47] By either

not want to move even within the nation; see Hadley Cantril, *Public Opinion, 1935–1946* (Princeton: Princeton University Press, 1951), p. 449.

[47] *Public Opinion Quarterly* (Winter, 1946–1947), p. 629. It should be noted that the time element may be important. After all, 1946 was immediately following World War II and 1948 was not too much later. Unfortunately, not much data is available for any later years. If it were, the other western European countries would undoubtedly score much higher. In this regard note the differences in identity for Germany in 1948 (Table III.3) and 1971 (Table III.2). It might also be noted here that although a country could be weak in support for one measure (or even for a whole subsystem), this lack could be offset by high support levels elsewhere.

criterion the United States appears to have a rather widespread sense of national identity.

Another indication of the sense of national identity is to compare identifications for groups within the country as possible competing influences for national identity. Within a liberal tradition culture parochial identities should complement national identity.[48] In their five-nation study, Almond and Verba use the example of marrying across party lines. In the United States 90 percent of those interviewed would not care if such cross-party marriage took place. The percentages are slightly less in Britain. Yet the strength of patterns of identity cleavage that can exist within a country is apparent when the data on Germany, Italy, and Mexico are examined. In these countries party divisions seem to be important challenges to national identity. In the most extreme case, for Italy, almost half of the party identifiers do not believe in marriage across party lines.[49]

If there is a strong sense of national identity, one might also expect that national identity would be stronger than cross-national identities. Marxists have suggested that class may provide a possible source for a competitive supra-national group identification. Does class provide such competition? The UNESCO study previously referred to presents some data which can test this possibility. In each of the seven countries studied the respondent was asked whether he identified with people of his *own class* abroad. The subject was then asked if he identified with people of his *own nationality* who were not of his own class. The difference between the percentage identifying with their country and the percentage identifying with their class can be used as a crude index of relative national identity. As will be seen from Table III.4 the scores range between − 5 and + 35 with the United States again the highest in terms of national identity.

A review of the available data on the vertical dimension of the identity system suggests that the United States has a highly supported sense of national identity. From both a behavioral

[48] Verba, pp. 533–534.

[49] Gabriel A. Almond and Sidney Verba, *The Civic Culture* (Boston: Little, Brown, 1965; originally published 1963), pp. 97–101 and data from ICPR. Also see Chapter V for other possible cleavages.

TABLE III.4 *Relative influence of national and class identity, 1948*

	Identify with country not of own class	*Identify with class not of own country*	*Index*[a]
United States	77% (Yes)	42% (Yes)	35
Germany	64	30	34
Norway	64	41	23
Mexico	56	40	16
France	63	48	15
Australia	78	67	11
Britain	67	58	9
Italy	50	41	9
Netherlands	56	61	−5

Source: Derived from Buchanan and Cantril, p. 18.
[a]Index equals the difference between columns 1 and 2.

and a more attitudinal perspective, the United States ranked first among the countries studied. Even allowing for cross-cultural problems of question interpretation and instrument design, American support appears high. This first dimension of political culture, thus, seems to fulfill the expectations of the liberal tradition hypothesis.

SENSE OF COMMUNITY IDENTITY

Until this point, our analysis concentrated upon the vertical individual-to-system-symbol relationships within the identity system. However, as Verba tells us, there are two main dimensions:

> The sense of national identity deals with a vertical form of identification — the sense of attachment that individuals have to some superordinate political unit and to the symbols of the political unit. A second significant dimension of identification is the horizontal identification with one's fellow ctizens — the sense of integration individuals have with other people who inhabit the political system. . . . In particular the extent to which members of a political system have trust and confidence in their fellow political actors is a crucial aspect of the horizontal integration of a political culture.[50]

Easton calls this second dimension the sense of community. Sense of community identity is defined here as the amount of

[50] Verba, p. 535.

affective orientation the member of a political system has toward the other members of the community. Support is measured by the cohesiveness of all of the members.[51] For the system, then, the interest is in how widely the sense of community is diffused among its members. For a sense of community to be high, the cooperative attitude of *trust* must be widely distributed in the society.

The horizontal individual-to-community relationship is conceived as resting upon what, for democratic regimes, Tocqueville called the "mother of science" — the feeling of association. "The progress of all the rest depends upon the progress it has made." [52] Pye links this "associational sentiment" with the ability of a nation to develop.

> The feeling of basic distrust leaves people unsure of their control over the world and hence fearful that the world is either against them or indifferent to them. Distrusting others, they must distrust their own capacity to influence others, and hence they have feelings of impotence. . . . When basic trust is replaced with cynicism, a people will suspect that behind the screen of political promises their leaders are really "out to get everything for themselves."[53]

The liberal regime, especially, depends upon the widespread diffusion of this identity value. The community relationships necessary to a political culture of contracts of the Lockean type would be impossible without trust. Locke is quite clear in emphasizing the necessity for trust in the development of a liberal political system.

> Since, then, those who like one another so well as to join into society cannot but be supposed to have some acquaintance and friendship together, and some trust one in another; they could not but have greater apprehension of others than of one another.[54]

[51] Easton, pp. 325, 185–189.

[52] Alexis de Tocqueville, *Democracy in America,* trans. Henry Reeve (New Rochelle, N.Y.: Arlington House, Heirloom Edition), 2:118.

[53] Pye, p. 55; for an early psychological use of trust, see T. W. Adorno *et al., The Authoritarian Personality* (New York: Harper and Row, 1950), esp. p. 411.

[54] *Second Treatise,* sec. 107, p. 63.

A test of the liberal tradition hypothesis would therefore be to measure the incidence of trust throughout the community. If trust is not widespread in the United States one could conclude that there is little support for the existence of the hypothesized tradition as the American political culture. Beyond the testing of the hypothesis itself Verba wonders whether any type of society that expects to work toward common goals can exist without trust. He doubts that individuals will freely turn power over to leaders in a situation of community distrust.[55] The importance of freely given power to a Lockean regime, thus, makes trust an especially important value for the present hypothesis. Therefore I will maintain that a country with a liberal tradition political culture must have even more trust than non-tradition cultures.

In his own study of Burma, Pye presents evidence of a political system where trust is not a widely held value. Pye maintains that Burmese treat outsiders and especially strangers with suspicion. The child is socialized to make strong and basic distinctions between friends and enemies. In such a culture, contracts and cooperation are not much in evidence. A quite dramatic and at the same time an eminently scholarly example of the debilitating affects of the lack of trust in a community is presented by Edward Banfield and his wife for a village in southern Italy. They maintain that the "Montegranesi" of that area act as if they were following the rule: "Maximize the material, short-run advantage of the nuclear family; assume that all others will do likewise." The consequences of following the rule, which they call "amoral familism," are: no civic associations, no private welfare, and no charitable work. In short, there is no spontaneous sense of community. Huntington summarizes other research which suggests that lack of trust is also characteristic of the cultures of the Arab world, Latin America, Ethiopia, and Iran.[56]

[55] Verba, p. 536.

[56] Pye, pp. 205–207 (also see his "Introduction" in Pye and Verba, p. 22); Edward C. Banfield, with the assistance of Laura Fasano Banfield, *The Moral Basis of a Backward Society* (Glencoe, Ill.: Free Press, 1958), p. 85, ch. 5 and 6 *passim*, and pp. 16–19; Samuel P. Huntington, *Political Order in Changing Societies* (New Haven: Yale University Press, 1968), pp. 28–29. For a country which does have high trust see Harry Eckstein, *Division and Cohesion in a Democracy: A Study of Norway* (Princeton: Princeton University Press, 1966).

Contrast this with what Tocqueville reported of the United States when he visited in the 1830's:

> The Americans make associations. . . . Whenever at the head of some new undertaking, you see the Government in France, or a man of rank in England, in the United States you will be sure to find an association.[57]

One historical work would indicate that associational trust long preceded Tocqueville's visit. One hundred years earlier, Americans had started almshouses, charitable societies, and even volunteer fire departments.[58] Indeed, a large-scale study of books on American traits found that associational activity was mentioned as an important descriptive characteristic throughout all of American history.[59] Even the large-scale immigration of "alien" culture members did not inhibit the maintenance of this value. Apparently, the immigrants quickly accepted the cultural value of associational trust.

> The peasants, in their coming, did not bring with them the social patterns of the Old World. These could not be imposed on the activities of the New. . . .
>
> . . . The crowded tenement neighborhoods spontaneously generated associations. Here people could not help meeting one another, their lives were so much in the open, so much shared. . . . Within the ghetto could grow understanding, then sympathy and in time, cooperation.[60]

Burial associations, mutual aid societies for savings, charitable and philanthropic institutions, newspapers, and theaters were created within the culture of trust — even in the urban tenement areas.[61] At the beginning of the twentieth century, Lord Bryce noted the persistence of these associative characteristics in the United States.[62]

[57] Tocqueville, p. 114.

[58] Carl Bridenbaugh, *Citizens in the Wilderness* (New York: Knopf, 1964), p. 299.

[59] Lee Coleman, "What Is American? A Study of Alleged American Traits," *Social Forces* (May, 1941), p. 497.

[60] Oscar Handlin, *The Uprooted: The Epic Story of the Great Migrations that Made the American People* (Boston: Little, Brown, 1952), p. 171; also see Nels Anderson, *The Urban Community* (New York: Holt, Rinehart and Winston, 1959), p. 5.

[61] Handlin, pp. 174–185.

[62] James Bryce, *The American Commonwealth* (New York: Putnam, Capricorn Books, 1959), vol. 2, part 4, ch. 4.

TABLE III.5 *Sense of community identity*

Question: Do you think most people can be trusted? (NORC)

	1948	*1964*
Yes	66%	77%
No	30	21
Don't know	4	2

Source: Hazel Erskine, "The Polls," *Public Opinion Quarterly* (Spring, 1969), p. 153; *ibid.* (Winter, 1948–1949), p. 756.

How strong is the sentiment today? It has been estimated that the United States now has 100,000 private welfare groups, 6,000 private foundations, 36 million members in fraternal and service organizations, 13 million labor union members, 320,000 churches and synagogues with 118 million members, and 5 million business firms.[63] But what of member support of associational sentiment? In 1960 the Almond-Verba study asked if Americans thought that human nature was fundamentally cooperative. Eighty percent agreed that it was.[64] Table III.5 shows opinion data more directly related to trust. More than two-thirds support the trust value. Sense of community identity appears to be supported by a consensus majority of Americans.

Table III.6 presents the trust value in a comparative perspective. In all the reported years, a majority of the Americans feels that most people can be trusted. However, a consensus is not operationally demonstrated by this trust question. In my opinion, the second choice for these questions confounds two Lockean societal dimensions. Although there should be trust in civil society, trust in nature would be foolish.[65] In any

[63] Richard C. Cornuelle, *Reclaiming the American Dream* (New York: Random House, 1965), pp. 38–40. Also see Eli Ginzberg, Dale L. Heistand, and Beatrice G. Reubens, *The Pluralistic Economy* (New York: McGraw-Hill, 1965), chs. 4–5.

[64] ICPR, 1960.

[65] Of course, there is no reason to suppose that the typical person answering this question makes the Lockean distinction between human nature in the natural condition and under civil government. This degree of sophistication does not appear possible, for reasons enumerated in Chapter II. One can probably assume that, judging upon the society with which they are familiar, most people feel that others are cooperative—i.e., within civil society—but that the other view of man reduces the commitment to trust under some operational circumstances.

TABLE III.6 *Community trust in the United States and other selected Western nations*

Question: Some people say that most people can be trusted. Others say you can't be too careful in your dealings with people. How do you feel about it?

	U.S. 1968	*U.S. 1966*	*U.S. 1964*	*U.S. 1960*	*U.K. 1960*	*Mexico 1960*
Most people can be trusted	55%	52%	53%	55%	49%	30%
You can't be too careful	43	46	45	40	39	65
It depends, other, etc.	2	2	2	5	12	5
N	1,343	1,291	1,450	970	970	970

	West Germany 1960	*Italy 1960*	*U.S. 1954*[a]	*U.S. 1948*[a]	*West Germany 1948*[a]
Most people can be trusted	19%	7%	62%	66%	6%
You can't be too careful	59	83	NA	30	93
It depends, other, etc.	22	10	NA	4	1
N	970	970	NA	NA	NA

Source: ICPR: SRC, 1968, 1966, 1964; AV, 1960; Samuel A. Stouffer, *Communism, Conformity and Civil Liberties* (New York: Wiley, 1955), p. 86 (data for 1954); *Public Opinion Quarterly* (Winter, 1948–1949), p. 756 (NORC, 1948).

[a]The wording was slightly altered (see Table III.5). The 1948 German Poll was taken by the occupying powers after Germany's defeat in World War II and a decade of rule by Hitler. This could explain the very low figures in that year as compared with those for 1960.

event, American trust is comparatively higher than for the other countries tested (except for Britain, where the difference is slight). Comparatively, the United States has a rather high proportion of people who trust their fellow members of the political community. Table III.5 even suggests that a better operationalization of trust shows consensual levels of identity within the community.

If Americans follow the Lockean tradition, again, they cannot be overly trusting. There should be a suspicion of power in the state of nature for the vertical trust relationship.[66] We should expect, therefore, that there would not be a naive trust

[66] For the difference between trust in people and trust in government in the United States, see Joel D. Aberbach, "Alienation and Political Behavior," *American Political Science Review* (March, 1969), pp. 86–99.

TABLE III.7 *Trust of government, United States*

Question: How much of the time do you think you can trust the government in Washington? (SRC)

	1968	*1966*	*1964*	*1958*
Always	7%	17%	14%	15%
Most of the time	54	48	62	56
Some of the time	36	28	22	23
None of the time	0	3	0	0
DK, no answer	3	4	2	6

Source: ICPR.

and, certainly, not a complete trust in the government. Data gathered by Almond and Verba indicate that Americans trust but are at the same time somewhat wary. In response to a question asking if people are more inclined to help others or to look after themselves, 65 percent support the more pessimistic view.[67] In nature man is perceived as basically cooperative but tinged with some selfishness. Even with this qualification of man's nature, the American trusts his fellow citizen. Interestingly, this magnanimity does not seem to extend as much toward foreign peoples, and it extends to an even lesser degree to foreign governments.[68]

Lockean expectations for trust of authority are generally borne out in Table III.7, although perhaps there is somewhat more trust than many might think Locke would allow. Yet if the legislative power is constituted through the free consent of the community, Locke expects that authorities will be trusted as long as the rules of the consensus are followed, i.e., natural law is respected, property rights are allowed, and legislative power is not transferred.[69] Normally, then, trust is to be given to authorities. Indeed the whole idea of contract

[67] Speaking generally, would you say that most people are more inclined to help others, or more inclined to look out for themselves (Almond-Verba, ICPR)? Also see Madison's view, *The Federalist Papers* (New York: New American Library, 1961), No. 10, p. 79.

[68] Herbert H. Hyman and Paul B. Sheatsley, "The Current Status of American Public Opinion," in Daniel Katz *et al., Public Opinion and Propaganda* (New York: Holt, Rinehart and Winston, 1954), p. 47. However, the data come from a single poll of one large midwestern city.

[69] On trust, see *Second Treatise,* sec. 136, p. 80; sec. 142, p. 84.

rests upon it. On the other hand, of course, Locke emphasizes a suspicion of government power. In fact, if the authorities violate their trust, the regime is dissolved.[70]

In the United States, adult skepticism of government seems to be much greater than in childhood – suggesting that socialization may take place in the hypothesized direction. Hess and Torney report in a massive study of grade school children and their teachers that *complete* trust of the government decreases from 90 percent of the third graders, to 76 percent of eighth graders to a rather Lockean skepticism of 46 percent for their teachers.[71] The adult culture patterns would thus seem to be compatible with the Lockean hypothesis.[72] It should also be noted that, as Table III.7 indicates, there is some distrust of government – only about 15 percent trust the government all the time. On the other hand, there seems to be enough regime trust to allow government activity to be undertaken with a relatively high probability that the citizenry will comply with authority decisions.

The tradition would suggest, therefore, high amounts of community and regime trust together with limited amounts of authority trust. It is this distinction which is crucial. Certain competent authorities have not been totally aware of this difference in levels of trust and have, consequently, noted that trust is weak in the United States. However, the lack of trust mentioned by Dickens, and especially Bryce, relates to the political authorities and not to the regime and community levels.[73] At these levels, the data in Table III.6 suggests that trust is characteristic of Americans.

The foregoing considerations suggest that a historical de-

[70] *Ibid.*, sec. 221, p. 127.

[71] Robert D. Hess and Judith V. Torney, *The Development of Basic Attitudes and Values Toward Government and Citizenship During the Elementary School Years,* Part I, Cooperative Research Proposal No. 1078 (U.S. Office of Education, 1965), p. 132.

[72] That is, trust until the authorities provoke a state of war, i.e., a critical trust; see *Second Treatise,* secs. 149, 155, pp. 87, 91; sec. 221, pp. 127–129. It is relevant to note that in the United States, substantial minorities do not think that "the people running the government . . . know what they are doing" (SRC 1958 and 1964 surveys; in ICPR).

[73] See the discussion of these writers in Gabriel A. Almond, *The American People and Foreign Policy* (New York: Praeger, 1960; originally published 1950), pp. 33–34.

FIGURE III.2 *The development of an American sense of community identity*

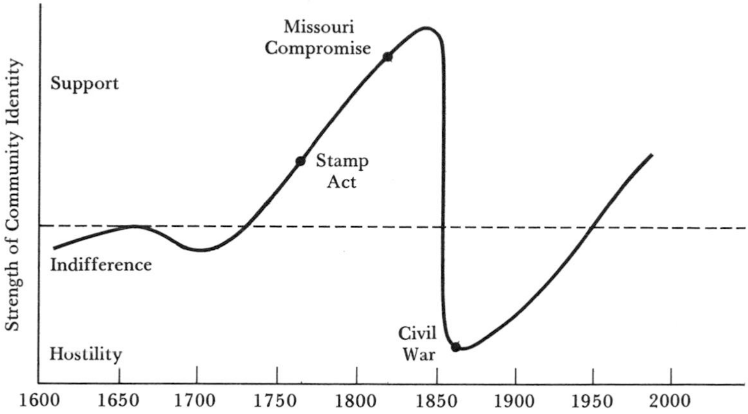

scription of the development of the sense of community identity in the United States may be made. Figure III.2 shows the assumed development of community identity over American history. The major assumptions are the same as made for Figure III.1, except that there are three dimensions of support – support, indifference and hostility. (The indifference category refers to indifference to a separate *American people.*) Figures III.1 and III.2 should be compared. It should be noted that while sense of national identity was probably recovered rather rapidly after the Civil War, sense of community was not. The intercommunity relations between North and South apparently took much longer to heal than the member-political system relationship of national identity.

POLITICAL CULTURE AND IDENTITY

Two dimensions of the identity system have been described in this part of my investigation of the political culture of the United States. Both sense of national identity and sense of community identity have been found to be widely diffused among the members of the American political system. Each is assumed to have an important, separate, culture-shaping influence. However, as they have both been related to the same system there should be a relationship between the two dimensions. They are related in Figure III.3. In addition to showing

FIGURE III.3 *Dimensions of the identity system*

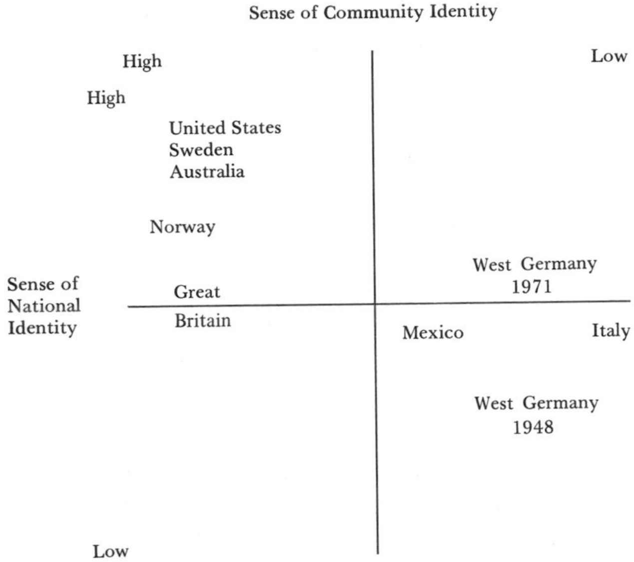

how the dimensions are interrelated, I have attempted to rank several countries in terms of identity system support (based upon the data in tables III.2, III.3, and III.6). Of course, these rankings are very crude (and for Sweden and Australia high trust is simply assumed for illustrative purposes), but the results are interesting. Most observations fall in the extreme high-high and low-low categories. This could explain why most analysts of identity only refer to one dimension. However, our analysis may suggest that other combinations are possible. Great Britain is the first of those countries studied which actually comes close to falling in one of the nonextreme categories. This country has high trust and a relatively low sense of national identity (though even it is higher than for most of the other countries for which we have data).[74] West Germany in 1971 is a second country which ap-

[74] Analysis of Great Britain's political culture is obviously much beyond the scope of the present work. However, some comment may be in order. It would appear that the low level of national identity found here has some substantiation in journalistic accounts of the so-called brain drain. When too much stress acts on the system, some members simply change political systems. On the other hand, other observations suggest that, for Britain, this one relatively weak aspect of the political culture is much

pears to fall in one of the nonextreme categories. Perhaps there are other examples.

In summary, it is most significant that for both sense of national identity and sense of community identity the United States ranks higher than any of the other countries for which data is available. If the identity system is an important preliminary political culture subsystem, these high levels of support are important for the political system. Identity support may provide some of the explanation for the persistence of the American political regime under stress. For a Lockean political culture the findings for the identity system are especially relevant. Data is provided which suggests that (1) a basic requisite for a consensual polity – realization of nationality – existed prior to the political system; (2) national identity was developmental, allowing for a connection to a tradition preexisting the creation of the political system; and (3) Americans have a relatively high degree of Lockean community trust. It appears, then, that the American political culture begins with a rather strongly supported value system. This identity value system is found not only to be widely accepted (and to thus provide a strong base for shaping) but the type of identity described was found to be compatible with the descriptive liberal tradition hypothesis.

THE AMERICAN SYMBOL SYSTEM

SYMBOL AND SOCIETY

The second subsystem of political culture is related to the first. Symbols often relate to the nation as a whole and help in the identification process.[75] But symbols are also important for the interpretation of rule and belief systems. Symbols serve to facilitate communication of complex reality. As a commu

---

more than offset by other elements in the culture. The strength of the horizontal dimension of identity and the strength of support for symbols (e.g., the monarchy), together with strong support for rules and beliefs (see Table IV.2) all lead to the conclusion that the total political culture in Great Britain is probably quite resistant to stress. See Richard Rose and Harve Mossawir, "Voting and Elections," *Political Studies* (June, 1967), pp. 173–201; and David Butler and Donald Stokes, *Political Change in Britain* (New York: St. Martin's Press, 1969), pp. 468, 484, 498.

[75] Gregor Sebba, "Symbols and Myth in Modern Rationalistic Societies," in Thomas J. J. Altiza *et al.*, eds., *Truth, Myth, and Symbol* (Englewood Cliffs, N.J.: Prentice-Hall, 1962), p. 147.

nication channel, the symbol system does not have intrinsic content. Rather it expresses the other three subsystems in terms of artifacts, and the hypothesis is tested through these symbols. The direct test of the hypothesis is the requirement that there be an attachment to liberal symbols which takes the form of a culture consensus.

Symbols are viewed as basic to any conception of man which involves comprehension and interaction. The more primitive symbols, such as language, are thought to be the building blocks which allow the individual to bring some order to social complexity.[76] Among the major symbols are those value-related ones which are the focus of the present study.[77] It is thought that without the symbol there could not be agreement or understanding. In this widest of conceptions of symbol, it is the basis for all human communication. It is *the* social mechanism which allows for basic human comprehension.

> Articulate speech means communication of ideas; communication means preservation — traditions — and preservation means accumulation and progress. The emerging of the faculty of symbolizing has resulted in the genesis of a new order of phenomena: an extra-somatic, cultural, order. All civilizations are born of, and are perpetuated by, the use of symbols.[78]

It is symbols, then, which provide the basis for human knowledge and comprehension. Language is one type of symbol, but the concept itself refers to any sign or indicator of something else which is more complex.[79] Language is certainly one of the most basic symbols,[80] but there are many other signs which make the world comprehensible.[81] All these simplifying signs provide *labels* for much more complex concep-

[76] S. I. Hayakawa, *Symbol, Status, and Personality* (New York: Harcourt, Brace and World, 1953), pp. 107, 143; also see Talcott Parsons, "Culture and the Social System: Introduction," in Talcott Parsons, *et al.*, eds., *Theories of Society* (New York: Free Press, 1961), 2:976.

[77] Hayakawa, pp. 106–108.

[78] Leslie A. White, *The Science of Culture* (New York: Farrar, Straus and Cudahy, 1949), p. 39.

[79] Frank Ford Nesbit, *Language, Meaning, and Reality* (New York: Exposition Press, 1955), p. 165.

[80] *Ibid.* The author indicates that signs may be more basic than language and are the means for learning language (pp. 166–167).

[81] W. Lloyd Warner, *The Living and the Dead* (New Haven: Yale University Press, 1959), p. 496.

tions which the individual may seek to communicate.[82] Since they simplify, symbols also allow the individual to understand his own ideas.[83] These labels, whatever form they may take — language, artifact, or myth — simplify and thus provide the basis for understanding outside stimuli. This is so for the individual and is conceived so for social systems.

For interaction to take place in a social system, it is assumed that there must be some communication. In this sense, the present theoretical view is that symbols are necessary.[84] Language is most basic in this sense. It provides the very basis for communication and concept formation.[85]

> We live in our language. It is, in a sense, our biography. If anyone wants to know what we think; what we have thought; what our beliefs, traditions, social perceptions, and public opinions are he can find his answers by studying our language. . . . Our language is the symbolic reservoir of the ideas, emotions, crises, disagreements, antipathies, compromises, loyalties, stereotypes, motives, etc., that are involved in the formation and expression of public opinion.[86]

And, yet, although symbols are necessary for interaction in focusing and identifying, they also necessarily limit understanding. By delineating complexity through symbols, symbols also exclude other aspects of reality. The very cognition aspects of a symbol system create "semantic blockages" which screen and shape certain perceptions.[87] Symbols allow certain stimuli and block or shape others. Social interaction depends upon the identifying and simplifying properties of symbols and tends to be limited by the boundaries set by symbols. In this manner the full set of symbols in a society tends to shape the broad references available for that society.

Symbols may also serve more detailed functions in a social

[82] William Albig, *Modern Public Opinion* (New York: McGraw-Hill, 1956), p. 77.

[83] Harold Lasswell, "World Politics and Personal Insecurity," in *A Study of Power* (Glencoe, Ill.: Free Press, 1950), p. 36. Also see Harold Lasswell, Nathan Leites, *et al., Language of Politics* (New York: Stewart, 1949).

[84] Clarence Schettler, *Public Opinion in American Society* (New York: Harper, 1960), pp. 29–30.

[85] Hayakawa, pp. 131–132 and ch. 1.

[86] Schettler, p. 29.

[87] *Ibid.*, pp. 38–49.

system. Support for symbols, since it is so central for communication, tends to provide certain agreed-upon elements even in the most diffuse social systems. These agreed-upon elements can then be viewed as providing a reason for social solidarity. It is these which provide the initial basis for coherence in societies.[88] Symbol support can preserve society "by adding emotion to instinct" to provide a more stable basis for support, and through the use of more stable symbols provide the foundation for rational support.[89] Symbols, thus, may not only aid in concept formation and perceptual screening but they may also aid in emotional and rational attachment which can act to maintain social systems. The vagueness of symbols provides an affective object which men can agree upon without the preciseness associated with specific affirmation of support.[90] Being basic, deep-rooted, and vague, and resulting from a very complex process, symbols are not easily changed.[91] As symbols become supported, they become major means of support for continued social interaction.

Symbolism is not only viewed as necessary for social interaction in the creation of societies, but it appears to remain important in modern society. It has been noted that modern man tends to fear symbolism because of its non-rational character and its relatively stable and deep-rooted nature. This fear leads modern rational man to attempt to deny and abolish it. Yet symbols tend to be maintained. Abolish the monarchical society and its attendant subordination and bowing and it reappears in the guise of the "republican Governor's reception" [92] (or in even less likely surroundings).[93] One monumental study of symbols traces the recurrence of certain common symbols through all recorded history.[94] It would appear that even though man seeks to rationalize his behavior, sym-

[88] Alfred North Whitehead, *Symbolism* (Cambridge: At the University Press, 1928), p. 81; also see Hayakawa, pp. 107–108.

[89] Whitehead, p. 82.

[90] Whitehead, p. 89.

[91] Murray Edelman, *The Symbolic Uses of Politics* (Urbana: University of Illinois Press, 1964), p. 20.

[92] Whitehead, p. 73; also see Marshal W. Fishwich, *American Heroes* (Washington, D.C.: Public Affairs Press, 1954), pp. 224–226.

[93] Milovan Djilas, *The New Class* (New York: Praeger, 1957), pp. 54–55.

[94] Elizabeth A. Goldsmith, *Life Symbols* (New York: Putnam 1928); also see Arthur S. Keller and Oliver J. Lissitzyn, *Creation of Rights of Sovereignty Through Symbolic Acts: 1400–1800* (New York: Columbia University Press, 1938).

bols have a strong tendency to persist. Apparently, persistence is not limited to only the most basic symbols. Even symbols which appear to be rather marginal have a remarkable longevity.[95] They seem to persist as representations of values under very different environmental conditions.[96]

In sum, symbols appear to have the characteristics of cultural values. That is, they influence over long periods of time, they shape, and they act as supports. All these are important characteristics of culture as it has been defined in this work. Most importantly, symbols appear to screen — they facilitate emotional attachment when other stimuli are identified with them, and symbols disguise when stimuli are not related to them. As such, symbols appear to be among the affective value attachments which should be included in the political culture.

SYMBOL AND POLITICAL SYSTEM

Symbol is not only important socially; its political role is basic. Three decades ago, one of the leaders of the modern behavioral movement in American political science noted its importance:

> No power could stand if it relied upon violence alone. . . . The eye, the ear, the aesthetic sense, must be attracted and enlisted also, if whole-souled admiration and loyalty are to be maintained.[97]

However, the importance of common symbols and myths has been recognized to some degree at least since Plato.[98] In America's formative period, its founders also considered their importance.[99]

The use of symbols does not appear to be historically related to a particular period or type of political system. Even in modern political systems,

> myth and symbol still hold their place, not as survivors of antediluvian or primitive states in man's development, but as es-

[95] Goldsmith, *passim*.

[96] Whitehead further argues that only the vagueness of symbols allows society to cohere without the demand for absolute conformity (pp. 77–78).

[97] Charles E. Merriam, "Political Power" (1934), in *A Study of Power* (Glencoe, Ill.: Free Press, 1950), p. 102.

[98] Harold Lasswell, Daniel Lerner, and Ithiel de Sola Pool, *The Comparative Study of Symbol* (Stanford, Calif.: Stanford University Press, 1952), p. 1.

[99] *The Federalist Papers* (New York: New American Library 1961), No. 49.

> sential autonomous functions of the body politic, be it healthy or diseased.[100]

Political symbols also can be viewed as having the other characteristics associated with societal symbolism above. They can be viewed as labeling, identifying, and screening. If political symbols are supported, they can focus attention upon the values and goals of the community and regime. If supported strongly enough, symbols may additionally provide commitment to these values and goals. Indeed, without this type of support, it is very unlikely that a regime could be maintained.[101] Most system members need symbols to reassure themselves of the value of the political system.[102]

The descriptive hypothesis requires that symbols act in the above manner for the American political system.[103] For it to be sustained, political symbols must be consensually supported. They must then act to maintain themselves as they shape stimuli which act upon them.

Symbols thus affect the persistence of the political community in a very fundamental way. Without the emotional support of community and regime symbols, the political system would be forced to rely solely upon rational commitment and coercion.[104] These are not enough to sustain it in periods of extreme stress. This type of support is especially insufficient for a liberal tradition regime which, in terms of its values, must deemphasize coercion. For this type of regime, attachment to symbols can be used to measure regime maintenance.

> Adults learn to treat many traditions as "just so" stories. It is not necessary to disabuse children of the idea that there is a Santa Clause, or that every founding father was a hero. Indeed, when members of a ruling group or of the community become exer-

[100] Sebba, p. 168.

[101] Whitehead, p. 89.

[102] Edelman, p. 2. For a totally different approach with a similar conclusion, see Thurmond W. Arnold, *The Symbols of Government* (New York: Harcourt, Brace and World, 1935), pp. 10–19.

[103] Ralph Henry Gabriel, *The Course of American Democratic Thought*, 2d ed. (New York: Ronald Press, 1956), pp. 91–92.

[104] William C. Mitchell, *The American Polity* (New York: Free Press, 1962), p. 124; Talcott Parsons, *The Social System* (New York: Free Press, 1951), p. 11.

> cised over the literal truth or falsity of the myth, it is evidence that the older bonds of civic cohesion have frayed, perhaps beyond repair.[105]

Without this symbol support, few regimes based upon consent are thought able to maintain their rational value-support.[106] In a broad sense symbols may be interpreted as representing all values. As a subsystem of political culture, however, their meaning is somewhat restricted. The symbol system emphasizes *artifacts* and *myths* which are politically relevant. These types of symbols are especially important in simplifying political objects for system members. They will be used to test the descriptive liberal tradition hypothesis as it relates to the symbol system. The hypothesis maintains that a single political tradition has been accepted as a cultural value system in the United States over the whole period of its history. This section of our study attempts to document current commitment to these values through symbols. For the symbol system, the hypothesis requires a high degree of continuity in symbol usage and a current culture-level affective orientation to these same political artifacts and myths.

EARLY AMERICAN SYMBOLISM

The events surrounding the creation of the American political system were not conducive to the creation of common symbols. The best study of early American symbolism, by Ralph Gabriel, notes that "the circumstances which surrounded the beginnings of the American nation were responsible for an unusual poverty in symbolism on the part of the early Republic." Gabriel lists six reasons why symbol attachment was especially difficult in the early United States: (1) the nation was the creation of a rebellion which presumably voided prior common political events; (2) no established or even mutually shared religion could provide the basis for common sacred symbols; (3) the Puritan attitude on singing and elaborate re-

[105] Lasswell, *et al.*, p. 1.

[106] Harold Lasswell and Abraham Kaplan, *Power and Society* (New Haven: Yale University Press, 1950), p. 121; and R. M. MacIver, *The Web of Government* (New York: Free Press, 1947; revised 1965), pp. 12–13, present more recent arguments for the necessity for symbols.

ligious symbolism was detrimental to conventional symbol development; (4) the rebellion was especially against monarchy – therefore, monarchy's characteristic forms of pomp and ceremony had to be ignored; (5) the founders were stopped from using military symbols because of a deep-seated liberal fear of standing armies; and (6) distance in time (rather than miles) effectively made the country larger than it is today – communication between isolated settlements was infrequent at best, and, thus, common symbol attachment was difficult. On the other hand, the colonists did not face the problems of many newly independent colonies today. The rebellion had not caused rejection of previous tradition. As Chambers points out, the acceptance of British legal institutions and liberal philosophy by the new American leaders allowed for a continuity which was favorable to symbol adoption. Second, many of the institutions, especially at the state level, were in continuous operation before and after independence. Finally, it should be noted that the atmosphere of interpersonal trust provided a firm basis for symbol acceptance at the community level.[107]

Yet stable institutions and community trust probably legitimized only the state governments and contributed to, at most, a stable local base for the future central government. The regime needed a symbol national in character. Gabriel maintains that the most important early symbol was Washington himself.[108] Washington has been linked above with the post-regime identification of the nation. Gabriel, seems to imply that Washington provided a *regime* symbol around which all could coalesce. With such a paucity of national symbols, the charismatic appeal of an individual was among the few alternatives available to the new American nation. With his unique position as the preeminent hero of the Revolution and a man highly regarded by all of the factions existing in the new nation, Washington was in an important position to serve as a unifying symbol.[109] In this sense, the selection of Washington

107 Gabriel, pp. 93–94; also see Kohn, p. 52 and ch. 2; and Lipset, p. 19; Chambers, *Political Parties in a New Nation,* pp. 8–9.

108 Gabriel, pp. 94–95. Also see Chambers, *Political Parties in a New Nation,* p. 36; and Easton, p. 305.

109 Lipset, pp. 21–26. See also p. 68 for Lipset's view of symbolism.

as presiding officer at the Constitutional Convention and later as first president undoubtedly represented (consciously or unconsciously) a linking of an accepted symbol to a new enterprise. The legitimacy of one symbol reinforced another – the regime itself was legitimized by the charisma of its first executive officer. Such syncretism seems to be a common process in symbol adoption for new nations.[110]

Once institutions were established, effort was expended to legitimize them. Symbols had to be found which would reinforce the ideals of the Constitution and at the same time be relevant to the American condition. One simple, effective and relatively direct means of symbol attachment is through the use of buildings.[111] In the United States

> Eighteenth century American republicans, establishing a new nation, just turned for their symbolism to the greatest republic they knew, Rome. They called the upper house of Congress a Senate. Led by the imaginative Jefferson, they used Roman architectural forms in the building of their first national Capitol and of the White House. These Classical Revival buildings were acceptable because they were not far removed from the Georgian structures with which Americans were familiar. The domed Capitol with its Roman columns expressed so well the dignity of the new nation that the type has become virtually indigenous to the United States.[112]

Buildings were not sufficient, however. The Capitol could symbolize the institutions of the regime. But what would symbolize the people, the private institutions, the values of the nation? "As a result of the poverty in symbolism, the flag became the chief representation of the Nation."[113] Together with patriotic songs, especially the national anthem and "America," the flag could become the emotional embodiment of the nation which a republic needs in the absence of the legitimacy of monarchy and its personal symbols. The urgency of having a flag is attested by the early attention given it by the Conti-

110 George M. Foster, *Traditional Cultures and the Impact of Technological Change* (New York: Harper, 1962), pp. 27–28.

111 Hayakawa, pp. 142–143.

112 Gabriel, pp. 95–96.

113 *Ibid.*, p. 96.

nental Congress. An official flag was adopted by 1777.[114] The flag now even has a special day set aside for its reverence. Popular observance of the day did not begin until the late nineteenth century, and official recognition did not take place until 1916.[115] Yet the flag has always been treated reverently as a special symbol. An elaborate courtesy and etiquette has traditionally been demanded — and usually accorded.[116] The flag was further reinforced through its daily use by local government. It has been noted that this type of locality reinforcement can be most effective in symbol attachment.[117] If only a limited central government used the symbol, reinforcement could not be strong. The federal structure allowed multiple loyalties which could be mutually reinforcing. Local government could combine relevance of government at a level of perceived activity with symbolic attachment to the nation by using the flag. Indeed, major American locality reinforcement rested in two conspicuous edifices of the community — the church and the courthouse. Each reinforced the idea of the other.[118] God and state, thus, both contributed to peace and order within the community, partially through this use of the most elaborate symbolism known to the average American.

To some extent all the symbols of nation were summarized in the patriotic holiday. For the people of the nation at an early period of development, the holiday was a welcome break in the monotony of agricultural existence. The epitome of the American holiday was the "Glorious Fourth." "The Fourth of July became inevitably the principal American ceremonial expression of homage to the nation.[119] The ceremony was important not only in the rural settlements, but also in the cities where the ward machines turned the occasion into ceremonies

114 Cleveland H. Smith and Gertrude R. Taylor, *Flags of All Nations* (New York: Crowell, 1946), p. 14.

115 J. Walker McSpadden, *The Book of Holidays* (New York: Crowell, 1958), p. 103; also see George William Douglas, *The American Book of Days,* rev. Helen Douglas Compton (New York: Wilson, 1948), pp. 342–347.

116 Preben Kannik, *The Flag Book* (New York: Barrows and Company, 1959), pp. 135–137; also see McSpadden, pp. 103–104.

117 Lasswell, "World Politics and Personal Insecurity," p. 33; Eckstein, pp. 175–176; and Morton Grodzins, *The Loyal and the Disloyal* (Chicago: University of Chicago Press, 1956), pp. 25–29.

118 Gabriel, pp. 101–102.

119 *Ibid.,* p. 98.

honoring not only nation but party.[120] The symbolic celebration of the adoption of the Declaration of Independence was fitting. Idea symbols are a higher development than material ones. They are not limited by geographical space.[121] The Declaration was thus superior to artifacts. It could even rival the somber Constitution since the Constitution's legalistic tone would have been incongruous in the boisterous celebrations of the period. The Declaration was also above the force of law and, therefore, did not carry the odium of political power. The Constitution was, after all, the law and it had to be interpreted.

> The Declaration, on the contrary, created no problems and roused no controversies. The Declaration was not law but rather rhetoric, brilliant rhetoric. For decades the famous phrases that all men are created equal and are endowed by nature with the right to life, liberty, and the pursuit of happiness provided the text for patriot homilies. When, however, it became possible in the 1850's to charge Americans who took their stand upon the Declaration with proposing equality and even intermarriage between the races, all national symbolism was falling and crisis was near.[122]

That is, when the Declaration turned from symbol to cause, it ceased to be a unifying symbol of political culture. The loss of this force of symbolism helped weaken the sense of national and community identity. A major support was, thus, not available for stress regulation. Decline in common symbol attachment in turn helped prepare the way for the Civil War. The War itself, however, helped create the Constitution as the common symbol of the regime. By nominally "settling" the nature of the regime the War raised the status of the document as a unity symbol.[123]

As the experience with the Declaration indicated, however, symbols can divide as well as unite. Unless there are commonly accepted symbols, dissolution always remains a threat. This was the consideration which led Washington to warn against

[120] William L. Riordan, *Plunkett of Tammany Hall* (New York: Knopf, 1948).

[121] Hayakawa, pp. 143–148.

[122] Gabriel, p. 99.

[123] *Ibid.*, pp. 443–444.

the "baneful effects of the spirit of party" which could lead to disorder and eventually to tyranny.[124] This view maintains that parties, as major symbols and structures of conflict, must be suppressed if a regime is to be maintained. It does not, however, consider the levels at which the political system can act. It may be that conflict can take place between authority symbols such as political parties, and that at the same time there can be consensus on overarching community and regime symbols which act to define the limits of the lower-level party conflict.

Based upon the preliminary analysis done above, this would appear to be the case for the United States.

> The consensus of Americans, in short, unites all its symbolic system, Democrats and Republicans alike responding to the same symbols in essentially the same ways. *There are few sharply differentiated symbolic systems in American politics.* Even the few that do exist, as between the two major parties, are well within the more inclusive set uniting the people.[125]

At the regime level, the unifying symbols seem to greatly outweigh the divisive ones. The study of verbal symbols used in presidential inaugural addresses (see Table II.7) indicates that with all the changes, verbal symbols remained substantially the same for the first fourteen decades of American history.[126] Our analysis further suggests that the early symbols of the American regime were not only widely held but were held as culture values which helped sustain the early years of the system. Symbols served to aid identity and helped to provide emotional attachment to the infant liberal regime. However, it still remains to be seen if rule, identity, and belief symbols are actually supported in twentieth-century America.

THE CONSTITUTION: BASIC RULE SYMBOL

What kind of rule symbols would one expect to be most important based upon the liberal tradition hypothesis? Certainly the most important artifact in the development of a

124 J. D. Richardson, ed., *Messages and Papers of the Presidents, 1789–1897* (Washington, D.C.: Government Printing Office, 1896), 1:218–219.

125 Mitchell, p. 127.

126 John McDiarmid, "Presidential Inaugural Addresses — A Study in Verbal Symbols," *Public Opinion Quarterly* (July, 1936), pp. 80–81.

Lockean civil society would be the original contract itself.[127] In some societies this document is antecedent to history. In the United States, however, the contract exists in the form of the written Constitution. The hypothesis would suggest that this document should be especially revered. As mentioned above, the Declaration of Independence rivaled the Constitution as the original contract, but after the Civil War the Constitution took a special place among American symbols. The importance of this symbol has been apparent to both domestic and foreign observers. Apparently, Europeans are especially aware of the importance of this American symbol.[128] Until the Russian Revolution in 1917, the older symbols, including the Declaration, remained relatively important regime symbols. But the age of revolutionary *coup d'état* – in Russia, Germany, Italy, etc. – brought the Constitution into a position of unrivaled preeminence. The Constitution became the symbol of stability. American reaction to the revolutionary era took the form of a great emphasis upon legality, order, and the rules of the Constitution.

> In the years between the two wars the Constitution as symbol grew in power as the European social religions of communism and of fascism waxed in importance. This role of the Constitution continued after World War II. . . . The Founding Fathers who met at Philadelphia sought to make the individual secure against extension to the United States of . . . hereditary aristocracy; their twentieth-century descendants hoped to defend the individual against . . . totalitarian mass movements. In the twentieth century the Constitution had become a symbol equal in emotional power to those about which the peoples of Europe have rallied: the British Crown, the fasces, the swastika, the hammer and sickle.[129]

Unfortunately, direct attitudinal data on emotional commitment to this basic symbol is not available except for the atypical 1930's and early 1940's. Table III.8 presents an attempt to measure adherence to the constitutional regime symbol. It

[127] Lasswell *et al.* report (p. 14) that constitutions are important symbols for most societies that have used them.

[128] Edward W. Chester, *Europe Views America* (Washington, D.C.: Public Affairs Press, 1962), pp. 152, 48.

[129] Gabriel, p. 445.

TABLE III.8 *The Constitution as symbol*

Question: Which of the following most nearly represents your opinion of the American form of government? (FOR)

	1940	*1943*[a]
1. Our form of government, based upon the Constitution, is as near perfect as it can be and no important changes should be made in it.	64.2%	68.2%
2. The Constitution has served its purpose well but it has not kept up with the times and it should be thoroughly revised to make it fit present-day needs.	19.2	23.7
3. The system of private capitalism and democracy are breaking down and we might as well accept the fact that sooner or later we shall have to have a new form of government.	5.2	3.5
4. Don't know.	11.4	4.6

Source: *Public Opinion Quarterly* (June, 1940), p. 349; and Cantril, p. 984.
[a]Factory workers only. For the exact wording, consult the source.

shows that about two-thirds of the people are highly supportive of the Constitution. If one places the commitment shown in Table III.8 within the environment which existed at the time, the depth of commitment becomes clearer. The two polls were taken during America's most severe depression and at a time when the system was being seriously challenged from abroad. Yet two-thirds believed that no important changes should be made in the Constitution. Attachment may even be stronger in a more stable period. Some evidence from the Almond and Verba study suggests that attachment is still very high in the 1960's. A sample was asked: "What are the things about this country that you are most proud of as an American?" Almost nine out of ten of the respondents were most proud of their political and legal system.[130]

In 1937 over two-thirds of the people disagreed that the Constitution should be made easier to amend.[131] Even ques-

130 Almond and Verba, p. 67.
131 AIPO, 1937; in Cantril, p. 939.

tions phrased in extreme form do not greatly affect the support of the Constitution as a symbol. In 1944, a national sample was asked if the "Constitution of the United States should ever be changed in any way." Presumably this would include amendments made to the Constitution in accordance with Article V. Yet 54 percent said that the Constitution should *never* be changed in *any* way.[132] With this amount of support under such strong limitations, the Constitution must be accepted as a very basic Rule support symbol. When danger threatens the regime, the tremendous symbolic reserves of attachment to the Constitution can and are tapped. Acceptance of the main Rule symbol seems to be sufficiently high to provide a major support beyond the day-to-day policy conflict. The Constitution, in other words, acts as an important regime component of the political culture.

ARTIFACTS OF THE NATION

If the shaping effect of political culture acts as hypothesized, one could predict that acceptance of symbols of the nation would be widespread. In a culture with a strong national identity, the hypothesis suggests that the United States should have symbolic attachment to representations of the nation such as flags, anthems, buildings representing basic institutions, and other national artifacts.

The flag appears to be the primary artifact symbol of nations.[133] Commitment to the flag has been so widely assumed, however, that attitudes on the subject have never been sampled among adult Americans. The Hess and Torney study of grammar school children does offer some assistance. As is shown in Table III.9, children identify highly with the flag.[134] It is interesting to note the teacher's symbol acceptance. A greater percentage of the teachers relate to the flag symbol than do the children! As a relatively sophisticated group, teachers may very likely have less attraction to the flag symbol than the population as a whole. At least, there is no reason to suppose

[132] NORC, 1944; in *Public Opinion Quarterly* (Summer, 1944), p. 283.

[133] Regarding the universality of the phenomenon, see Kannik, p. 134 and *supra;* and Smith and Taylor, "Preface."

[134] This is not the situation in every political system. For an interesting example of children who do not identify highly with their nation's flag, see Table VII.2 and the related discussion.

TABLE III.9 *The flag as symbol*

Question: The American flag is the best flag in the world (1960).

School grade	*N*	*Yes*	*Don't know*	*No*
Grade 2	1,631	90	7	3
Grade 3	1,670	89	8	3
Grade 4	1,731	86	9	5
Grade 5	1,796	88	7	5
Grade 6	1,743	86	9	5
Grade 7	1,716	87	8	5
Grade 8	1,688	88	6	6
Teachers	382	92	4	4

Source: Hess and Torney, p. 80.

TABLE III.10 *Knowledge of the national anthem*

Knew all words to "The Star-Spangled Banner"	13%	65%
Knew part of "The Star-Spangled Banner"	52	
Knew name but no words	3	
Said anthem was "America"	11	
Did not know	21	

Source: Derived from Cantril, p. 508 (AIPO, 1939).

that the full adult population would be any less attached than teachers to the flag as an important national symbol.[135]

A closely related identity symbol is the national anthem. Fortunately, polls of adults have been taken on this subject, though both known surveys were taken in the same year. Knowledge of the anthem is shown in Table III.10. As can be seen, popular acquaintance with this symbol seems relatively high. Over two-thirds of the population knew the name of the anthem, and almost as many knew at least some of the words (and 11 percent thought it was the closely related "America").

[135] For two empirical examples of community symbol support in the United States, see Marian Roth and G. R. Boynton, "Communal Ideology and Political Support," *Journal of Politics* (February, 1969), pp. 167–185; and G. Cleveland Wilhoit, "Political Symbol Shifts in Crisis News," *Midwest Journal of Political Science* (May, 1969), p. 318.

TABLE III.11 *The national anthem as symbol*

Question: Do you think people should stand when "The Star-Spangled Banner" is played in public? (AIPO, 1939)

Yes	90%
No	6
DK	4

Source: Derived from Cantril, p. 508.

The rate of knowledge of the concrete symbol should be contrasted with the knowledge of the Bill of Rights shown in Table II.1. Not only is knowledge relatively high, but Table III.11 shows that nine out of ten Americans feel that public respect should be paid to the anthem. Symbol support, thus, seems to be relatively high for this basic regime symbol.

In their study of childhood socialization, Hess and Torney found that children in all grammar school grades chose the flag and the Statue of Liberty as the best representations of America.[136] Of course, those objects are conspicuously honored in the classrooms of most schools in the United States. In few cases do school-system efforts at symbol acceptance seem more directly related to political socialization. The flag, anthem, and other basic symbols are also reinforced through adult life at mass-attended athletic events such as football and baseball games. Many civic associations and fraternal organizations begin their meetings with the pledge to the flag or with the national anthem. An indication of the emotional attachment to the flag can be gathered from the reaction to the Vietnam War protestors who publicly burned American flags. Although the emotion was apparently spent by the time the Senate could deal with the bill, the House of Representatives passed by a 385–16 vote a strict bill ($1,000 fine and a year in jail) to punish those who defiled the flag.[137]

High recognition also extends to the major place symbols. When asked, "What is the capital of the United States?" 94

[136] Hess and Torney, p. 86.

[137] *Congressional Quarterly Weekly Report* (June 23, 1967), p. 1059. Also see John D. Lees, *The Political System of the United States* (New York: Harper and Row, 1969), p. 98.

TABLE III.12 *The White House as symbol*

The White House should be kept standing and everything done to strengthen it on the inside and restore the whole building to good condition.	55%
The White House should be torn down and an entirely new building put up that would look just like it and which the best materials would be used.	29
Qualified (mostly suggestions to build a new one and keep the old as a monument).	4
No opinion	12

Source: *Washington Post* (May 29, 1949), sec. III, p. 5B (AIPO).

percent could give the correct answer.[138] Again, the comparison with the Bill of Rights question is instructive. When compared with other school-learned social science knowledge, it is quite high.[139] Mitchell and Merriam [140] both emphasize the importance of buildings for political symbolism. This type of symbol support is most clearly shown by a poll which asked if the White House should be restored or rebuilt. Table III.12 shows the opinion distribution for this poll question. Over half of the population wished to save the White House even after it was widely reported that it was so dilapidated that it would not be worth restoring. Less than one-third would even agree to build a new White House which would look just like the original. However, certain buildings undoubtedly have more association with myth than others. The White House and Capitol are undoubtedly very important. Yet Sebba indicates that neither of them – nor even the Jefferson or the Washington monuments – reach the warmth of personality needed for intense symbolism. Only Lincoln provides the focus needed for a complete myth symbol.[141]

Interestingly, a completely different symbolism surrounds other government buildings. The poorly maintained govern-

138 AIPO, 1945; in Cantril, p. 107.

139 Hazel Gaudet Erskine, "The Polls," *Public Opinion Quarterly* (Spring, 1963), pp. 136–141.

140 Mitchell, p. 125; Merriam, p. 105.

141 Sebba, p. 163.

ment offices throughout the country have a separate symbolic significance. They represent the American feeling that government is not very central or important. More directly, the impression is given that government does not waste money on frills. "This combination of Greek temples and shabby buildings symbolizes important political values, namely, our ambivalence toward power." [142] The Lockean tradition wants a limited government but also wants an efficient one when it fulfills its "proper" functions.[143]

But when Americans want to symbolize the regime, they will spend resources. One need only consider the District of Columbia and its wealth of historic monuments.[144] Federal government interest in the preservation of national symbols did not begin until 1906 but it became substantial after 1935.[145] Table III.13 gives an indication of the amount of federal land used as historical symbol today. Over 10 million acres of land are reserved for symbol expression by the federal government alone. Most of the land is in undeveloped so-called National Monuments, but over a million acres are devoted to essentially historic symbols – and these holdings have been substantially increased in recent years.

To some analysts the major symbols are related to events. One notes that "the political group has appropriated more days of the calendar than any other except the ecclesiastical." [146] Yet, to some extent, present holidays do not seem to hold the attraction that historical evidence suggests was common just a few decades ago. No longer is the holiday the main inexpensive amusement event. The advent of mass communication, especially television, and the mobility provided by the automobile have provided competition for the political event. On the other hand, proclamations of new holidays proceed unabated. While direct participation in the events of the

[142] Mitchell, p. 125.

[143] See the rule system in Chapter IV.

[144] Mitchell, p. 125. The architect of the Capitol alone spent over $22 million in 1967; see *The Budget of the United States Government: Fiscal Year 1969* (1968), p. 204.

[145] *Congress and the Nation* (Washington, D.C.: Congressional Quarterly Service, 1964), p. 1075.

[146] Merriam, p. 105. To determine that he is not exaggerating, see the complete daily calendar in Douglas, pp. xi–xxii.

TABLE III.13 *Federal investment in historic symbols: total acres of land within the boundaries of historic properties*

	1948	*1954*	*1960*	*1966*
National Monuments	9,691,592	9,227,272[a]	9,129,537	9,062,987[b]
National Historical Parks	16,251	17,220	37,202	39,619
National Military Parks	27,289	26,907	28,708	31,938
National Battle-field Parks	4,578	8,749	7,488	13,477
National Memorial Parks	70,369	71,191	70,374	70,436
National Battle-field Sites	736	736	736	785
National Historic Sites	5,094	1,317	1,494	3,371
National Memorials	2,096	4,484	4,600	5,526

Source: National Park Service, *National Parks and Landmarks* 1948, 1954, 1960, 1966), p. 1; also called *Areas Administered by the National Park Service.*

[a]Almost the entire loss was accounted for by two congressional restrictions of what was claimed to be original overextension of park land (262,624 acres from Joshua Tree Monument and 222,929 acres from Jackson Hole Monument). See *National Parks and Landmarks* (1950, 1951). Also see *Congress and the Nation* (Washington, D.C.: Congressional Quarterly Service, 1964), pp. 1082–1083.

[b]In 1967 National Monuments consisted of 9,596,212 acres. See *National Parks and Landmarks* (1967), p. 4.

holiday have undoubtedly lessened, the ceremonies are still a significant symbol for many and, perhaps, an indirect symbol for many others.[147] Voting is another event with symbolic importance. To most citizens it is the most direct way to indicate regime support. By voting, the citizen indicates that he thinks the process of selection is meaningful.[148] This type of support has been substantial in the United States.[149]

147 Warner maintains that Memorial Day, for example, is still an important symbolic event; see W. Lloyd Warner, *American Life* (Chicago: University of Chicago Press, 1953), ch. 1. Also see Robert Meyer, Jr., *Festivals U.S.A. and Canada* (New York: Inez Washburn, 1967), pp. 132–159. Perhaps the holidays are still important symbolically — but perhaps, with the change of structural differentiation in the society, they are just observed differently.

148 Sebba, pp. 146–147.

149 Robert E. Lane, *Political Life* (New York: Free Press, 1959), ch. 2. Also see William G. Andrews, "American Voting Participation," *Western Political Quarterly* (December, 1966), p. 650.

ICONS AND HEROES: BELIEF SYMBOLS

Symbols not only represent the nation and its rules but also its important beliefs. Although beliefs will not be considered in detail until the following chapter, certain symbols are important to the liberal tradition because they relate to these values. Values such as achievement and liberty are especially important to the tradition. Icons and heroes often serve to represent these important values.

Icons have always had a limited usefulness in the United States. A Freudian has made much of a distinction between "Uncle Sam" and the "Goddess of Liberty," maintaining that a male Uncle Sam was used to demand sacrifice while female Liberty was seen as the motherly expression of the rewarding aspects of life.[150] However, in my opinion, neither icon has been very important. In the United States the closest approximations to icons seem to be mythical figures of the rugged frontiersmen. Daniel Boone, Leatherstocking, Davy Crockett, Buffalo Bill, Kit Carson, etc., seem important as they represent values such as liberty, strength, virtue, and hard work.[151]

In the United States heroes are drawn overwhelmingly from the political or governmental arenas. Table III.14 indicates that governmental leaders are the hero-symbol choice of two and a half times as many people as is its nearest competitor — the three fields of entertainment, culture, and sports. Although the military was represented by only 2 percent of the total responses in 1960, another poll indicates that they may be more important under certain circumstances than this percentage would suggest. Of the five most admired people in 1947, four had military backgrounds and the fifth was the president of the United States.[152] Undoubtedly, this amount of support can be attributed to the interest generated by the then recently completed war. In more normal times, the military does not seem very popular. Of the ten most admired men in 1965 and

150 Geoffrey Gorer, *The Americans* (London: Crescent Press and Bank Society, 1948), pp. 34–36.

151 Henry Nash Smith, pp. 57–114; and Dixon Wechter, *The Hero in America* (Ann Arbor: University of Michigan Press, 1963).

152 AIPO, 1947; in *Public Opinion Quarterly* (Summer, 1947), p. 309.

TABLE III.14 *Occupation of admired heroes*

Question: Of all the people you hear or read about, could you name one or more individuals you admire very much? (A V, 1960)

	Percent of total responses
Political, governmental, military	45%
Entertainment, cultural, sports	18
Admire no one	16
Religious	13
Other, DK	8
N	1,236

Source: Derived from ICPR.

1966, only one had a military background and he was also a former president.[153]

Presidents, in fact, provide the major symbol system emphasis. Locke maintained that this was the normal situation in a republic.[154] This is especially true for Abraham Lincoln. As already noted, Sebba maintains that Lincoln provided the most believable credentials for true mythical hero. Gabriel maintains that "twentieth-century generations, who never saw Lincoln in the flesh, who thought they were scientific, and who were sure they were sophisticated, elevated the dead Lincoln to the status of a demigod." [155] Even scientific and sophisticated people apparently value symbols, and the symbol of one who had been flesh and blood appears to be very appealing. Lincoln, the common man from the log cabin, is easy for the common American to identify with.[156] He is, in many ways, the ultimate symbol of liberal tradition beliefs in the United States. The myth of Lincoln provides the identification and simplification needed for effective symbol identification. The symbol of Lincoln, by providing something understandable for

153 AIPO, 1965–1966, in *Gallup Opinion Index* (January, 1967), p. 12.

154 *Second Treatise,* sec. 151, p. 89.

155 Gabriel, p. 446.

156 An AIPO poll in 1945 showed that more than half of those highly favorable to Lincoln associated his greatness with his commonness; Cantril, p. 562. Europeans also view Lincoln as the epitome of the American; see Chester, p. 51.

TABLE III.15 *Lincoln and Washington as symbols*

Question: Who do you think was a greater man, George Washington or Abraham Lincoln? (AIPO, 1945)

Washington	22%
Lincoln	42
Equally great	28
Uncertain	8

Source: *Public Opinion Quarterly* (Spring, 1945), p. 87.

the average citizen to identify with, makes government meaningful to the ordinary system member.

Note the reinforcement given to Lincoln identification in newspapers each year – even to the extent that many carry special supplements recreating his life and deeds for new generations. Lincoln's special place has, indeed, been compared with a religious cult:

> A cult requires a sanctuary. That of Lincoln has three – the birth-place in Kentucky, the grave in Illinois, and the great memorial in Washington. Of these the last is most important. Analysis reveals significant characteristics. It is a Greek temple. Within it is a graven image. Daniel Chester French's figure is a romanticized Lincoln. [Certain] devices enhance the religious atmosphere; on the walls in bronze are the words of the hero; and above the brooding figure is an inscription. . . . In such temples and with similar inscriptions the citizens of ancient Greece placed statues of Apollo. By so little is the twentieth century after Christ separated from the fifth before His coming.[157]

The popularity of Lincoln can be seen by reference to some pertinent attitudinal data. Table III.15 indicates that Lincoln is more important than Washington as a political symbol. In fact, Lincoln is chosen by almost twice as many individuals as choose Washington.[158] It is also interesting to note that more than one-quarter refused to choose between these two high-order symbols, both of whom relate closely to the belief

[157] Gabriel, pp. 449–450.

[158] For further substantiation of the greater popularity of Lincoln, see Cantril, p. 565, q. 91; and p. 590, q. 12.

TABLE III.16 *Hero symbols in American history*

Question: In all the history of the United States, whom do you regard as two or three of the greatest men who have ever lived?

	Percent mentioning
Abraham Lincoln	57%
George Washington	46
Thomas A. Edison	11
Woodrow Wilson	8
General Eisenhower	7
Thomas Jefferson	6
General MacArthur	5
Theodore Roosevelt	5
Benjamin Franklin	4
Other mentions	92
No one	3

Source: *Public Opinion Quarterly* (Winter, 1945–1946), p. 520. Franklin Roosevelt also figured prominently among those mentioned. He was removed because his death only eight months before (while in office during a war) certainly overestimated his long-run cultural importance. A poll taken in 1948 shows this rather dramatically. In a question relating to men in public affairs only in the *past fifty years*, Roosevelt dropped eighteen percentage points in admiration in just three years. See Elmo Roper, *You and Your Leaders* (New York: Morrow, 1957), p. 22.

tradition. Another example of the greater support given to Lincoln and to the special place given these two presidents is given in Table III.16. As can be seen, Lincoln leads Washington by a substantial amount, and the two together are much more highly regarded than are any of the other "great men." [159] Washington received more than four times the number of mentions that are recorded by the next highest rated man. Clearly, as the frequent reference to them (especially Lincoln) in political speeches adequately attests, these two figures are important in the American political culture.

AUTHORITIES AND SYMBOLS

If the regime is to be maintained from generation to generation, the political culture must provide a connection between

[159] Table III.16 also suggests that in a supposedly "business civilization" businessmen receive little adulation.

regime and authority levels. Symbols provide one connection. Regime symbols provide for a screening effect which allows transfer of the cultural values to decision-makers who come to power through the regime's rules. The regime is legitimized before any government takes office, but each authority must legitimize his own position through symbol referents legitimizing his particular leadership position. One author refers to the set of symbols which legitimize this process as the "rites of passage." [160] Oaths (including swearing on a Bible), the appearance of the former office-holder to show continuity, and the other elaborate ceremonies of inauguration are the symbols invoked to allow regime legitimacy to be passed to the authorities. Since the symbols of partisanship are inferior to those of the regime, even the opposition defers to the new authorities.

Authorities find their power cloaked under various mantles. Rule may be classified as traditional, charismatic, or rational.[161] While each of these categories is an ideal type and, therefore, real structures may only approximate them, they can serve rather nicely to distinguish the different types of symbols attached to the main institutions of American government. Mitchell finds that distinctive patterns exist for the different regime structures.[162] The judiciary has many of the attributes of *traditional* power. The emphasis of the robes, the icons of classical justice, the impressive traditional buildings, the seating of the judges, the flags, and the gavels is upon finding justice from past wisdom. The practice (or myth) of *stare decisis* is a good example. Contrast this reverence for precedent with the stark symbolism in the output-oriented setting for bureaucratic decision-making. The emphasis is upon efficiency and *rationality*. Consider the American president: Constitutional provisions demand that the executive be singular, and practice has dictated that he take the position of policy advocate. This situation has forced the president to rely upon his personal powers of persuasion. Mass media communications

[160] Mitchell, pp. 132–136.

[161] Max Weber, *The Theory of Social and Economic Organizations,* trans. A. M. Henderson and Talcott Parsons (Glencoe, Ill.: Free Press, 1947), pp. 324–363; also see Mitchell, p. 24.

[162] Most of the following is taken from Mitchell, pp. 128–129.

has even intensified the development. This has led one authority to conclude that, although rationality is not absent, the president relies on charisma more than do other authorities. Legislative power is more difficult to classify. Its power seems to rest on all three elements in more of a balance than does the power of the other main institutions. It is more rational and charismatic than the judiciary, more traditional and rational than the executive, and more traditional and charismatic than the bureaucracy.

Symbols seem important to the polity because they serve social goals. From the individual's perspective they provide understanding and shaping protection. From the perspective of the political system, symbols serve the same functions but they emphasize community, regime, and authority screening. For example, it appears that in the United States symbolism becomes more formal as one proceeds from the limited power of the local government to the greater power of the state government, to the center of power in the federal government.[163] This is especially true of the only institutions which can legally take away one's life or freedom — the courts. Symbolism is at its highest in a court of law, and the Supreme Court of the United States presents an ultimate in symbolism. As symbolism becomes more formal it tends to hide better the unpleasant aspects of power that society and the individual want to ignore but that seem necessary for maintenance of regimes — especially those based upon consent.[164]

The American political culture has been shown to use many of the generally accepted symbolic devices.[165] It has used architectural style, formalized language, and romanticized distortions of concepts and persons to achieve attachment to the community and its regime. Temples, constitutions, flags, heroes, etc., evoke a commitment to the more basic divisions of the political system. In a day when the president of the United States has the power to literally destroy the world, the Lincolnesque symbol of the president as common man is most employed. Comparative research on the United States and France suggests that chief executives with great power

[163] Mitchell, p. 127.
[164] Merriam, p. 103.
[165] Edelman, pp. 11, 20.

are generally perceived in this personal, symbolic sense.[166] Human contact with complexity appears to create a tendency in social interaction to rely upon the identifying, simplifying, shaping, affective, and reassuring properties of symbols. Support for basic political symbols in the United States does not appear to be an exception to this tendency.

PRELIMINARY SCREENING: IDENTITY AND SYMBOL FUNCTION

The approach of this chapter has been descriptive. It was hypothesized that identity was formed early in the history of a nation. Symbols were likewise assumed to have been adopted at an early time. Immediately upon formation both would begin being slowly incorporated into the political culture. Both of the subsystems were, thus, analyzed in terms of the *development* of support. For this, a brief historical review seemed appropriate. Analytically, the identity subsystem focused upon the vertical and horizontal dimensions of the basic object of the community; the symbol system focused upon attachment to the valued artifacts of the community and regime. Both identity and symbol were measured in terms of the amount of support for the political culture they evoked from members of the political system.

The attitude data which exists for the United States suggests that support for these two basic subsystems is presently quite strong. Indeed, the operational criterion set for cultural support is more than met by the identity and symbol measures. Substantial support was presented for both identity dimensions. For the sense of national identity, about 90 percent of the American population (whether measured in a purely attitudinal manner or from a behaviorally-related attitudinal perspective) support nationality as a cultural value. Eighty percent saw fellow citizens as cooperative members of the community, and the United States (of the countries for which data was available) was found to have the highest support for trust or sense of community identity.

166 Philip E. Converse and George Dupeux, "DeGaulle and Eisenhower: The Public Image of the Victorious General," in Angus Campbell *et al.*, eds., *Elections and the Political Order* (New York: Wiley, 1966), pp. 298–301.

Symbol system support was also found to be high. More than six out of ten Americans fully supported the Constitution as a regime symbol in a period of crisis. In periods when stress is not high, support may even be more widespread. Although much of the data on symbols of the nation are not directly applicable to adult attitudes, there does appear to be substantial support for such symbols as the flag and building symbols such as the White House. Attitude data on respect for the national anthem does show that nine out of ten support this national symbol. Affective orientation to hero symbols of belief is also high. Almost six Americans in ten spontaneously mention Lincoln as an important reference symbol.

In summary, the support found here appears to meet the criteria set for the acceptance of a descriptive hypothesis. The evidence of culture-level support in the United States for basic national identity, community, and symbol objects is substantial. There is also evidence that these objects are related to the liberal tradition. Yet, since these two subsystems are more universal in content than the rule and belief systems, they are not as related to a particular belief tradition as are the latter. Certain of the findings, however, suggest the liberal nature of the identity and symbol systems. A developmental tradition which allows for the possibility of a consensual political culture was found through the Merritt analysis. More importantly, the Lockean value of trust is widely supported in the United States. It was suggested above that a liberal tradition regime, deemphasizing coercion, would have an especially high acceptance of symbols. It was also suggested that the liberal tradition would place an emphasis upon attachment to the original contract, to symbols of the nation, and to symbols of liberal beliefs. Each of these expectations was fulfilled. For both the identity and symbol systems, consensus support for a political culture consistent with the Lockean liberal tradition appears to exist.

The identity system and the symbol system are, thus, conceived as important subsystems of the American political culture. The identity system is concerned with the affective orientation to two aspects of the political community. As a cultural system it serves a screening function. Stimuli are screened to see if they relate to the nation or community.

FIGURE III.4 *Division of political culture: basic and developed subsystem inputs*

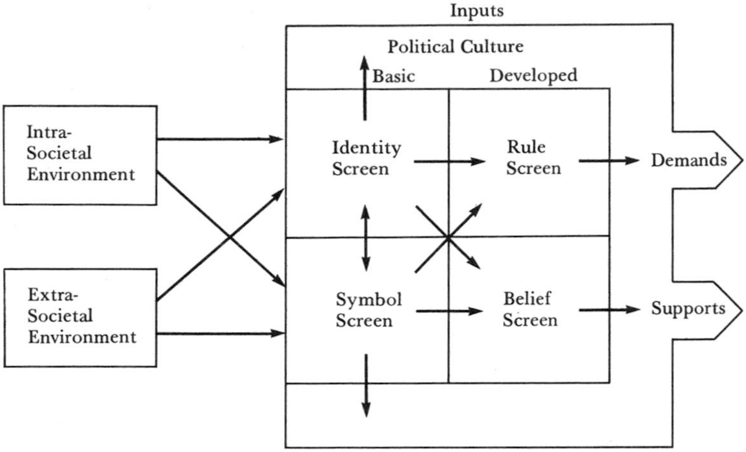

Those that do should receive a more favorable reception, and the screening process should shape accordingly. The symbol system acts to facilitate emotional attachment to the nation, the community, the rules, and the beliefs. At the same time they act to screen out unfamiliar symbols. Both subsystems are fundamental to the creation of attachment to the political system.

The identity and symbol subsystems are, furthermore, prerequisite to detailed commitment to the regime implied by the rule and belief systems. They allow basic commitment and attachment to be formed prior to more demanding adherence to rules and beliefs. In this sense the first two systems are basic requisites for the full development of a political culture. Since these two systems shall be viewed as the more universal and elementary subsystems, political culture may be classified into basic and developed components. The belief and rule systems are the more developed subsystems which depend upon the basic formation of the identity and symbol systems. Figure III.4 shows this division. It will be noted that environmental stimuli must first strike either the identity or symbol systems before they can be referred to the more developed subsystems (some impulses could undoubtedly miss the preliminary screen and directly affect the developed screens, but this is probably a very small percentage of the total stimuli aimed at the system). Normally, for any stimuli to be processed in any way they must be recognized by the cognizance screens of identity or symbol. Once recognized, the impulse is then subject to the screening operation within the identity and symbol subsystems and forwarded with this supplementary information to the rule and belief subsystems. After being processed by the subsystems a product emerges. Then, after interacting with other input structures, the product of the stimuli–political culture interaction is relayed to the political system as demands and supports. At any single point in time the composition of the subsystems describes the political culture.

In the United States the identity and symbol subsystems appear liberal and strongly supported. However, the crucial test of the liberal tradition hypothesis must be made through empirical description of the developed rule and belief subsystems.

CHAPTER IV

Developed Political Culture

THE BELIEF AND RULE SYSTEMS are conceived to be more directly related to the values of a specific regime than are the basic subsystems of identity and symbol. As such they must be used as the crucial descriptive test of the first of our four major hypotheses. The hypothesis holds that the political culture of the United States is based on the liberal tradition. It is necessary therefore to relate the values of Locke and Madison systematically and directly to the available opinion data. It is only in this way that a convincing empirical test can be made. The present work will not attempt, however, to provide an authoritative interpretation either of the entire works of either author or of his "real" psychological disposition. Since the hypothesis is concerned with the shaping influence of these writers as they represent the tradition, it will consider these authors only as they are represented by their most influential political works. In the case of Locke the obvious choice is the *Second Treatise on Civil Government: An Essay Concerning the True Original Extent and End of Civil Government.*[1] The choice is no more difficult for Madison. *The Federalist Papers* is, unquestionably, the most influential work on politics written in the United States. Although Madison did not write all

[1] John Locke, *Second Treatise on Civil Government,* in Sir Ernest Barker, ed., *Social Contract* (New York: Oxford University Press, 1962), pp. 3–143; hereafter referred to as *Second Treatise.*

the papers himself, the work will be treated as unitary — with a concentration upon those papers attributed to Madison.[2]

Since the developed subsystems must be more directly related to the descriptive hypothesis, there must be a systematic attempt to determine if the liberal tradition is culturally accepted in the United States. In addition to using the political, cultural, and value criteria mentioned in Chapter II, a content analysis of the *Second Treatise* and *The Federalist* was undertaken to find the values expressed in these works. Each expression of values was then combined with like values and these combinations were used as major tests of the hypothesis. At least one quotation for each author is used to state the expectation of the liberal tradition. Attitudinal data is then used to test whether the expectation is met for that major value. If the attitudinal data show a cultural consensus support for the values of the rule and belief systems, the descriptive hypothesis is accepted.

THE RULE SYSTEM IN THE UNITED STATES

POLITICAL CULTURE AND REGIME RULES

Easton holds that

> Unless there is a minimum convergence of attitudes in support of these fundamental rules — the constitutional principles, as we call them in Western society — there would be insufficient

[2] *The Federalist Papers* (New York: New American Library, 1961). Actually the references will be confined to the Madisonian papers with the exception of the first and last papers, which Madison must certainly have approved; and the papers concerning the Executive and Judiciary, since Madison did not write about them. These latter, however, were necessarily inhibited by Madison's earlier papers — especially those on the legislative and Nos. 10, 39, and 51. Even a creative interpreter of Madison like Riemer notes that Madison's theory rests mainly in *The Federalist Papers;* see Neal Riemer, "Two Conceptions of the Genius of American Politics," *Journal of Politics* (November, 1958), p. 699. There is also the question of whether Madison is the relevant focus for our present purpose. As Kendall and Carey note, Hamilton, Madison, and Jay did not write the *Federalist;* rather they assumed a common *role* — that of Publius, the real author. Each writer submerged his personal views to express the consensus between them; Willmoore Kendall and George W. Carey, "Introduction," *The Federalist Papers* (New Rochelle, N.Y.: Arlington House, Heirloom Edition), p. xiii. For ease of identification and to emphasize his preeminent contribution, however, Madison will be used to identify Publius.

> harmony in the actions of the members of a system to meet the problems generated by their support of a political community. The fact of trying to settle demands in common means that there must be known principles governing the way in which resolutions of differences of claims are to take place.[3]

Anthropologists have found that, though rules are often unspoken and "informal" they appear to exist even in the most primitive political systems.[4] These rules may only be informal understandings — even only among elites. Yet it would appear that to be most effective, they must also be widely accepted. Indeed, anthropologists indicate that the consent of the entire group is needed.[5] However, the present study views rule support in terms of degree. It looks at the *extent* of support for rules by system members.[6]

James Buchanan and Gordon Tullock present an excellent logical basis for rule acceptance. Why should an individual freely consent to inclusion in the community under rules which may affect him adversely? Essentially he will only do so when the benefits exceed the costs.[7] But at the earliest stage — the constitutional stage — "the individual is uncertain as to what his own precise role will be in any one of the whole chain of later collective choices that will actually have to be made." [8] All men, in other words, have a common interest in uncertainty. It is, therefore, in everyone's interest to agree on rules — which, after being agreed upon unanimously, shall determine the conditions under which authoritative decisions are to be made and how they are to be made.

3 David Easton, "An Approach to the Analysis of Political Systems," *World Politics* (April, 1957), p. 392.

4 Marc J. Swartz, Victor W. Turner, and Arthur Tuden, eds., *Political Anthropology* (Chicago: Aldine, 1966), pp. 4–7 and *passim*.

5 *Ibid.*, p. 4.

6 David Easton, *A Systems Analysis of Political Life* (New York: Wiley, 1965), pp. 200–204 (hereafter referred to as *A Systems Analysis*); Norton E. Long, "The Local Community as an Ecology of Games," *American Journal of Sociology* (November, 1958), pp. 251–261; and, Paul A. Smith, "The Games of Community Politics," *Midwest Journal of Political Science* (February, 1965), pp. 37–60.

7 James M. Buchanan and Gordon Tullock, *The Calculus of Consent: Logical Foundations of Constitutional Democracy* (Ann Arbor: University of Michigan Press, 1965), ch. 5.

8 *Ibid.*, p. 78.

The situation is similar to that which exists prior to a game.

> Since no player can anticipate which specific rules may benefit him during a particular play of the game, he can, along with all the other players, attempt to devise a set of rules that will constitute the most interesting game for the average or representative player. It is to the self-interest of each player to do this. Hence, the discussion can proceed without the intense conflicts of interests that are expected to arise in the later playing of the game itself.[9]

The benefits of community membership, which were high anyway for survival reasons before the agreement on rules, come to outweigh heavily the costs after the agreement. When all have had a voice in the making of the basic rules, it becomes rational (because of the high uncertainty costs of not joining) for man to join a political system which may do him some harm but will, in total, be expected to be just for all. Member acceptance of rules would seem most relevant for democratic regimes. Indeed they may be *defined* in terms of individual support for rules.

> It is perfectly reasonable of me to say that I do approve of the process by which certain kinds of decisions are made, even if I do not always like the specific results. Thus the consent of the governed may be interpreted to mean their approval of the processes by which decisions are arrived at and their willingness to abide by these decisions even when these seem wrong.[10]

Yet there are special problems for democratic regimes. They seem to be severely restricted to situations where a certain equality of power prevails. For if one group with great power acts from self-interest it would not agree to rules favorable to all, i.e., effectively voting to take its own power away. Even if it acts from altruistic motives, the other members of the community are not likely to believe it did.[11] Democratic support of

[9] *Ibid.*, p. 80.

[10] Robert A. Dahl, *Pluralist Democracy in the United States: Conflict and Consent* (Chicago: Rand McNally, 1967), p. 17. Also see Ernest Barker, *Reflections on Government* (London: Oxford University Press, 1942), pp. 63–72.

[11] Buchanan and Tullock, p. 80. Also see Ernest Barker, *The Politics of Aristotle* (New York: Oxford University Press, 1962), 5:xii.

rules also assumes the development of an identity and symbol system, at least in rudimentary form. When these conditions exist, viz., power equality and at least a somewhat cohesive basic political culture, the making of a constitution is rational *and* possible. Rules may then be created which are relevant to the particular cultural, social, and physical environment.

It seems reasonable that the rules must also act to shape subsequent activity of the political system and, thus, meet the main behavioral criterion for political culture. For although no one knows *how* the rules will affect him in the future, it seems obvious that they must have an influence on how decisions are made, and therefore on what decisions are made.

THE TRADITION AND RULES

Support for rules, then, appears to be important for different regimes. For a liberal tradition regime, however, they are of especial importance. A liberal regime eschews coercion and bases its legitimacy upon popular consent.[12] Support for rules must be higher for a liberal regime than for others because without popular support the regime is illegitimate. In fact, it cannot be liberal. This type of regime must have support from its very creation, and this was evident to the founders of the American Constitutional regime. Madison realized that the people were liberal (republican) and that "it is evident that no other form would be reconcilable with the genius of the people of America. . . ." [13] Indeed, the first of *The Federalist Papers* gave as one of the primary purposes of publishing the collection the opportunity to show "the conformity of the proposed Constitution to the true principles of republican government." [14] Without showing the conformity of the Constitution to Lockean values, consent could not have been obtained and a liberal regime could not have been instituted.

Rules are, then, central for a liberal tradition regime. Without rules man has nothing to consent to. Indeed, as Locke

[12] *Second Treatise,* secs. 22, 149, pp. 15, 87.

[13] *The Federalist Papers* (citations are from the New American Library edition), No. 39, p. 241.

[14] *Ibid.,* No. 1, p. 36; also Nos. 37–40.

noted, there cannot be a republican civil society at all without rules.

> For wherever any two men are who have no standing rule and common judge to appeal to on earth, for the determination of controversies of right betwixt them, there they are still in the state of nature.[15]

In fact, rules prescribe the very nature of man under the civil society of a liberal tradition regime. According to the *Second Treatise:*

> freedom of men under government is to have a standing rule to live by, common to every one of that society, and made by the legislative power erected in it. A liberty to follow my own will in all things where the rule prescribes not, not to be subject to the inconsistent, uncertain, unknown, arbitrary will of another man, as freedom of nature is to be under no other restraint but the law of nature.[16]

Rules are of such basic importance to the liberal tradition that if the authorities break them, the government is dissolved.[17] A liberal regime, thus, must have an especially high commitment to the rules of its polity.

Americans show substantial support for rules. In 1968, only 15 percent thought that *unjust* laws should not be obeyed—even if the person was willing to go to jail. One state poll found that over 90 percent of those questioned believed that the law should be upheld to the letter no matter what the reasons of the lawbreaker were.[18] Support for the rule enforcement officer may also be used as an indication of acceptance of the necessity of rules. Hess and Torney reveal that the policeman is seen by school children of all ages as being concerned

[15] *Second Treatise,* sec. 91, p. 53; also see F. A. Hayek, *Constitution of Liberty* (Chicago: University of Chicago Press, 1960), chs. 11, 13; and Samuel P. Huntington, *Political Order in Changing Societies* (New Haven: Yale University Press, 1968), p. 196.

[16] *Second Treatise,* sec. 22, p. 15; also see sec. 57.

[17] *Ibid.,* secs. 197–242.

[18] SRC, 1968; in ICPR. *Polls* (Summer, 1966), p. 84. Another indication of this attitude is shown by a Gallup Poll in 1969: 82 percent believed that college students who broke laws in demonstrations should be expelled. *Gallup Opinion Index* (April, 1969), p. 9. Also see the Appendix.

TABLE IV.1 *Public support for rule-enforcers*

	Police treatment (A V, 1960)[a]	*Respect police (AIPO, 1967)*[b]	*Police holiday (AIPO, 1967)*[c]
Favorable	85%	77%	73%
Somewhat favorable	–	17	–
Unfavorable	8	4	19
No opinion, etc.	7	2	8

Source: Gabriel A. Almond and Sidney Verba, *The Civic Culture* (Boston: Little, Brown, 1965; originally published 1963), p. 70; *Gallup Opinion Index* (September, 1967), p. 19; *ibid.* (July, 1967), p. 19.

[a]Actual question: "If you had some trouble with the police – a traffic violation, maybe, or were accused of a minor offense – do you think you would be given equal treatment? That is, would you be treated as well as anyone else?"

[b]Actual question: "How much respect do you have for the police in your area?"

[c]Actual question: "Would you favor or oppose setting aside one day each year throughout the nation to honor police and law enforcement officials?"

about the child almost as much as is his own father.[19] Adult attitudes are shown in Table IV.1. As can be seen substantial majorities support these most visible of representatives of the rule system. Over three-quarters of the population seem to have a high degree of respect for the police.

THE CONSTITUTION

A tradition which places so much emphasis upon rules and in which the necessity for them is thought to be such a basic element of its regime might bc thought to devise special procedures to express and preserve its basic rules. The use of a written constitution would seem to be one expression of this aspect of the tradition. Certainly until the time of the American federal Constitution few nations were enamored of written documents. Interestingly, the only older written constitution still in use is that of the Commonwealth of Massachusetts.[20]

[19] Robert D. Hess and Judith V. Torney, *The Development of Basic Attitudes and Values Toward Government and Citizenship During the Elementary School Years,* Part I, Cooperative Research Project No. 1078 (U.S. Office of Education, 1965), p. 120.

[20] Louis Hartz, *The Liberal Tradition in America* (New York: Harcourt, Brace and World, 1955), p. 85.

What better way to emphasize the basic rules than to have them written and thus explicitly preserved for posterity?

Chapter III has already considered the Constitution as a symbol. It is also an important object of rule belief. Europeans are apparently amazed at the influence of the Constitution as the basis of rule values.[21] It is not the less appreciated by the American people as *the* prime expression of the rule system. Apparently strong support for the Constitution began at an early stage in American history.

> Within an unbelievable short time, hardly more than a decade, Americans no longer debated seriously whether their constitutional system was good or bad; it appeared to be a postulate almost universally subscribed to that the American Constitution was a good one, if not in fact the best in the world. Americans ceased to be divided over questions of constitutional structure.[22]

This acceptance of constitutional rules is still apparent in current public opinion polls. As previously indicated in Table III.9, two-thirds of the population – and about seven out of ten people who have an opinion – feel that "Our form of government, based upon the Constitution, is as near perfect as it can be and no important changes should be made in it." In a time of depression and on the eve of war only about five out of one hundred were in favor of a new form of government. As was noted above, large majorities want to keep the rules the same by not allowing the Constitution to be made easier to ammend. This repository of the central rules of the American regime appears to be very widely accepted.

The Constitution and the rules it enumerates represent the core of the rule system of the United States. This rule system is viewed as compatible with the liberal tradition hypothesis. "All of the great ideas and institutions of our English heritage were embodied in the Constitution as adopted in 1788 and amended by the First Congress and the states." [23] Its summary nature as a liberal document is well attested by the liberal nature of its defense in *The Federalist Papers.* Even

[21] Edward W. Chester, *Europe Views America* (Washington, D.C.: Public Affairs Press, 1962), p. 152; also see Chapter V.

[22] Dahl, p. 72.

[23] William H. Young, *Ogg and Ray's Introduction to American Government,* 12th ed. (New York: Appleton-Century-Crofts, 1962), p. 30.

the Lockean emphasis upon the central importance of the original contract influenced its written form. Support for the Constitution as surrogate for the whole Rule System seems very high. This value acceptance is certainly important evidence in support of the hypothesis. The liberal tradition, as represented by the Constitution, is supported. Yet the data offered up to this point are too general to be used as a complete test of the hypothesis. Individual rule-values of Locke and Madison must be systematically investigated in great detail before one can have a firm basis for hypothesizing an American liberal rule system.

POPULAR RULE

Of all the rules within the liberal tradition, the one that the people should ultimately control is the most central. It is not that conceptions of popular rule within the liberal tradition do not differ — they do. Nonetheless, popular consent as an ideal is universally accepted under the tradition. Although Locke is not primarily interested in ultimate sovereignty, popular rule is of central importance to his value-system. As noted, civil society itself rests upon the consent of the people:

> Yet the legislative being only a fiduciary power to act for certain ends, there remains still in the people a supreme power to remove or alter the legislative, when they find the legislative act contrary to the trust reposed in them. . . . And thus the community perpetually retains a supreme power of saving themselves from the attempts and designs of any body, even of their legislators, whenever they shall be so foolish or so wicked as to lay and carry on designs against the liberties and properties of the subject.[24]

Popular rule should also be central for Madison. Indeed Madison does base republican rule on the support of the people. To him, this relationship is what distinguishes the true liberal government.

> We may define a republic to be, or at least may bestow that name on, a government which derives all its powers directly or indirectly from the great body of the people. . . .[25]

[24] *Second Treatise,* sec. 149, p. 87; also see sec. 168.
[25] *The Federalist Papers,* No. 39, p. 241; also see Nos. 37, 46, 57, and 58.

In cases where a legitimate civil society has been compacted based upon popular consent, Locke writes, the individual is then bound to follow majority rule:

> And thus every man by consenting with others to make one body politic under one government, puts himself under an obligation to everyone of that society to submit to the determination of the majority. . . .[26]

Yet the rule of the majority is not unlimited. It should act only in cases when the government "must move one way." [27] Majority rule is restricted by the law of nature,[28] and especially by restrictions against the expropriation of property.[29] Furthermore, majorities may not use their power to coerce minorities unjustly.[30] Madison also emphasizes restrictions on popular rule. The liberal tradition regime stresses the importance of limiting the sphere of government.[31] Thus, the tradition emphasizes restrictions both for minorities and even for popular majorities. For the tradition maintains that either can threaten liberal values through factional tyranny.

> If a faction consists of less than a majority, relief is supplied by the republican principle, which enables the majority to defeat its sinister views by regular vote. It may clog the administration, it may convulse the society; but it will be unable to execute and mask its violence under the forms of the Constitution. When a majority is included in a faction, the form of popular government, on the other hand, enables it to sacrifice to its ruling passion or interest both the public good and the rights of other citizens. To secure the public good and private rights against the danger of such a faction, and at the same time to preserve the spirit and the form of popular government, is then the great object to which our inquiries are directed.[32]

Madison and the other framers of the American regime of 1789 resolved this conflict between the fear of majority tyranny

[26] *Second Treatise,* sec. 97, p. 57. Also see *The Federalist Papers,* esp. No. 58, p. 361; but also Nos. 37, 46, and 57.

[27] *Second Treatise,* sec. 96, p. 57.

[28] *Ibid.,* sec. 135.

[29] *Ibid.,* sec. 138.

[30] *Ibid.,* sec. 76, p. 103; sec. 81, p. 107.

[31] *The Federalist Papers,* No. 10, p. 78; *Second Treatise,* sec. 22, p. 15.

[32] *The Federalist Papers,* No. 10, p. 80; also see No. 51.

(against republican liberty)[33] and the desire for popular government by *institutionalizing popular rule* through representation.[34] The people could rule by electing representatives who would actually make the decisions. Popular rule was provided for by elections and liberty was provided for by institutions which necessitated a process of deliberation, compromise, and minority representation. This balance between the values of liberty and popular rule was institutionalized through the structures of the new regime. Locke had made reference to the value of separation of powers;[35] but it is the Constitution which institutionalizes the value.[36]

To Madison, liberal regimes are based upon popular rule. The liberal tradition view of human nature, however, leads to the use of additional regime supports.

> But what is government itself but the greatest of all reflections on human nature? If men were angels, no government would be necessary. . . . In framing a government which is to be administered by men over men, the great difficulty lies in this: you must first enable the government to control the governed; and in the next place oblige it to control itself. A dependence upon the people is, no doubt, the primary control on the government, but experience has taught mankind the necessity of auxiliary precautions.[37]

Separation of powers, checks and balances, federalism, and elections represent the major regime structures which represent these auxiliary precautions. But would the members of a liberal regime accept such a limit to their power? Could they support a norm which stated that policy should only be indirectly determined by the people through elections and that direct control should be given to representative institutions? This is the fundamental rule hypothesis that must be tested.

The People Rule. Unfortunately, it is quite difficult to find direct attitudinal poll data on this basic liberal tradition rule norm. Apparently American opinion analysts feel that this

[33] See pages 190–192.
[34] *The Federalist Papers,* No. 10, p. 81.
[35] *Second Treatise,* sec. 144, p. 85; sec. 143, p. 85.
[36] *The Federalist Papers,* No. 47.
[37] *Ibid.,* No. 51, p. 322. Also see Nos. 47, 48, 51, 14, 45, 46.

TABLE IV.2 *Comparative popular rule*

Question: Which people do you think have more to say about the way their government is run, the British people or the American people? (NORC, 1943)

American	66%
The same	11
British	9
Don't know	14

Source: Hadley Cantril, *Public Opinion, 1935–1946* (Princeton: Princeton University Press, 1951), p. 985.

question has such an obvious answer that it would be senseless to ask the question. There is evidence, however, that Americans think that they rule. Table VI.14 shows that 88 percent of the American people believe that their country is a democracy. This is higher than for most countries. In chapter II it was noted that members believe that they are able to affect government through voting (see Table II.6).

Table IV.2 shows further evidence that Americans believe that they rule. Two-thirds believe that Americans have more control over their government than do British citizens and another 11 percent believe that they have at least as much control.

There is also direct evidence that Americans support the value of popular rule. A study of documents spanning the whole of American history found that Americans have been committed to democratic rule through every time period.[38] In an in-depth psychological study of working class men, Robert E. Lane also found a commitment to democracy.[39] Data are also available from opinion polls. Table IV.3 shows that Americans almost unanimously reject an alternative rule. More than nine out of ten would not support a dictatorship

38 Lee Coleman, "What is American? A Study of Alleged American Traits," *Social Forces* (May, 1941), p. 497. Also see Alexis de Tocqueville, *Democracy in America,* trans. Henry Reeve (New Rochelle, N.Y.: Arlington House, Heirloom Edition), 1:387; James Bryce, *The American Commonwealth* (New York: Putnam, Capricorn Books, 1959), vol. 2, part 4, ch. 4; Robin M. Williams, Jr., *American Society,* 2d ed. rev. (New York: Knopf, 1960), pp. 461–462.

39 Robert E. Lane, *Political Ideology* (New York: Free Press, 1962), pp. 84–96. It is interesting to note that Lane finds a commitment to the rules even among the "undemocrats" (p. 174).

TABLE IV.3 *Would you like to see a dictatorship established in this country? (AIPO, 1937)*

Yes	3%
No	91
No opinion	6

Source: Cantril, p. 869.

even when they were asked at the time of the Great Depression and at the beginning of a period of world instability which would lead very shortly to World War II. Table IV.4 presents a recent and direct expression of American faith in popular rule. Over three-fourths of the population express a great or good deal of confidence in the people's capacity to make general political judgments. Americans, thus, seem to believe in popular rule. Not only do the mass of people believe in popular rule, they also think that the government responds to them. Table IV.5 indicates that seven out of ten Americans feel that the government pays attention to their desires.

Since it was found that Americans believe that the means by which they affect politics is through voting in elections (see Table II.6), one would expect that support for popular rule should be very apparent for that measure. The data confirm that suspicion. Almost nine out of ten Americans

TABLE IV.4 *Popular rule*

Question: In general, how much trust and confidence do you have in the wisdom of the American people when it comes to making political decisions? (1964)[a]

Great deal	14%	76%
Good deal	62	
Not very much	19	
None at all	1	
Don't know	4	

Source: Lloyd A. Free and Hadley Cantril, *The Political Beliefs of Americans: A Study of Public Opinion* (New Brunswick, N.J.: Rutgers University Press, 1968), p. 193.

[a]$N = 1611$.

TABLE IV.5 *Government responsiveness to popular rule*

Question: How much attention do you feel the government pays to what the people think when it decides what to do? (SRC)

	1964	*1968*
Some, or good deal	70%	66%
Not much	24	29
DK, no answer	6	5

Source: ICPR.

TABLE IV.6 *Support for elections*

Question: How much do you feel that having elections makes the government pay attention to what the people think? (SRC)

	1964		*1966*		*1968*	
Good deal	65%	90%	62%	87%	60%	88%
Some	25		25		28	
Not much	6		9		8	
Don't know	4		4		4	

Source: ICPR.

believe they should vote even when their party does not have *any* chance to win.[40] Support for campaigns is also high. Seventy-two percent feel that even with the bitterness and unreliable information that campaigns engender, they are nonetheless needed.[41] Table IV.6 shows direct support for elections. As is reported, approximately nine in ten Americans believe that elections make the authorities at least somewhat responsible to popular wishes. This is significant support for this most central instrument of popular rule. Other data suggest that elections are also highly supported as effective, al-

[40] SRC, 1952: 86 percent, 1956: 89 percent, 1960: 90 percent; in ICPR. Interestingly, almost 60 percent of Americans view voting as a duty rather than a right (NORC, 1944; in ICPR).

[41] AV, 1960, in ICPR.

TABLE IV.7 *Limited popular rule*

Question: Which group do you think is more often right in its views on broad national issues — Congress or the people? (AIPO, 1939)

Congress	42%
The people	40
Don't know	18

Source: Cantril, pp. 929–930.

though at a somewhat lower level, for local government.[42] Indeed, no matter how operationalized, poll data show substantial support for popular rule in the United States.

Qualified Popular Rule. Americans therefore seem to believe in popular rule. However, the tradition also indicates that this support should be modified. Some attitudinal data exist that suggest this may be so. When asked if the "majority of the people in this country are usually correct in their ideas on important questions," Americans are almost evenly divided in their views. Thirty-eight percent said the people were usually correct, while 42 percent believed the people were not usually correct.[43] From other attitudinal data, it can be noted that substantial minorities would restrict the franchise to those who were literate.[44] On the other hand, they would not deny it to those on welfare — at least during the Depression.[45] Also note that in Table IV.4 only 14 percent have a *very* high confidence in the people.

It seems to me, however, that a question asked in 1939 epitomizes some of the restrictions Americans see on the popular will. Table IV.7 indicates that a plurality of Americans in 1939 actually thought Congress was right more often

[42] SRC, 1964; SRC, 1952, 1956, 1960; in ICPR. For additional evidence of American support for elections, see Jack Dennis, "Support for the Institution of Elections by the Mass Public," *American Political Science Review* (September, 1970), pp. 819–835. Also note the lower level of support for elections in Britain; David Butler and Donald Stokes, *Political Change in Britain* (New York: St. Martin's Press, 1969), pp. 477, 485, 500.

[43] AIPO, 1939; in Hadley Cantril, *Public Opinion, 1935–1946* (Princeton: Princeton University Press, 1951), p. 692.

[44] AIPO, 1965; in *Polls* (Summer, 1965), p. 76.

[45] AIPO, 1938; in Cantril, p. 831.

on *broad* national issues than they were themselves. This division of opinion emphasizes the strong commitment to institutions inherent in a Lockean value system.

The special place of political institutions in a liberal tradition regime is dramatically illustrated with comparative data. Table IV.8 shows support for various characteristics of nations. Of the five nations for which data are presented, the United States most fulfills the tradition expectation. The respondents were asked to mention aspects of their country of which they were proud. Americans tended to stress their political institutions. From the perspective of the *total number of responses* given, over 50 percent of those in the United States referred to the rule system (over 70 percent if the economic system is considered part of rule-making), as the aspect of the nation of which they were proud. This contrasts with only slightly more than 30 percent in the country with the next most highly rated rule system, Great Britain, down to only about 3 percent for Italy.

From the perspective of *support,* 85 percent of the American members mentioned a political institution as that part of their country of which they were most proud. This appears to be a very strong support for the regime—indeed it approaches unanimity. Only 15 percent, then, either gave other

TABLE IV.8 *Commitment to institutional rule system, 1960*

Percentage who say they are proud of:	*U.S.*	*U.K.*	*Mexico*	*Germany*	*Italy*
Governmental, political institutions	85%	46%	30%	7%	3%
Economic system	23	10	24	33	3
Social legislation	13	18	2	6	1
Characteristics of people	7	18	15	36	11
Position in international affairs	5	11	3	5	2
Physical attributes of country	5	10	22	17	25
Other aspects	20	35	48	44	73
Total percent responses[a]	158	148	144	148	118

Source: Almond and Verba, p. 64.
[a]Percentages exceed 100 percent because of multiple responses.

aspects of the nation of which they were proud to the exclusion of the political or were not proud of anything. In no other country was political institution support nearly as widespread. Even the United Kingdom, which has similar cultural values, has 54 percent who do not mention political institutions. In West Germany and Italy almost no one mentions these structures. In regimes which rely upon institutions, widespread support of these rule structures would appear an important aspect of the political culture.[46]

The American people, thus, appear to support widely the liberal tradition value of popular rule. Only 1 percent (see Table IV.4) have no faith at all in popular rule. Three-quarters of the population have substantial trust in it. Americans also appear to believe that authorities are responsive to popular rule. On the other hand, there does not appear to be support for unlimited popular rule. Few Americans have a very high trust in the people's wisdom. Institutions are seen as important limits to direct popular rule. Political institutions are more highly supported in the United States than in the other countries for which data are available. Americans appear to value both popular rule and representative institutions. The people are expected to rule indirectly through control of these representative institutions.

RULE INSTITUTIONS

From the above, it appears that institutions have a special place in the American value system. The first and most basic institutional rule of the liberal tradition is that the legislative predominates. For the political culture of the United States to be described correctly as the liberal tradition, in the sense used here, Americans must believe in legislative predominance. They must also support the other central institutions of the regime, however. Legislative support cannot be of such a nature as to threaten the existence of the other regime institutions.

[46] On the special place of institutions in the United States, also see Daniel J. Boorstin, *The Genius of American Politics* (Chicago: University of Chicago Press, 1958), p. 9; and Nelson W. Polsby, *Congress and the Presidency* (Englewood Cliffs, N.J.: Prentice-Hall, 1964), pp. 6–8.

The Legislative. Locke is emphatic on the position of the legislative,

> There can be but one supreme power, which is the legislative, to which all the rest are and must be subordinate. . . .[47]

The legislative is subordinate to the people, but of the institutions the legislative is supreme. It is the institution "which has a right to direct how the force of the commonwealth shall be employed for preserving the community and the members of it." [48] This legislative power is best put into "the hands of many who assemble to make laws but who do not administer them." [49] Indeed, civil society itself did not exist until power passed to legislative bodies.[50] And once the legislative power is placed in one institution, Locke maintained that it could not be shifted to another.[51] Therefore, for American political culture to be called the liberal tradition the people must still believe, with the Constitution, that "All legislative powers herein granted shall be vested in a Congress. . . ." [52] Madison follows the tradition when he maintains that "in republican government, the legislative authority necessarily predominates." [53]

The legislative, then, must be highly supported if the liberal tradition hypothesis is to be sustained. Table IV.5 reported how much attention people felt government pays to what they think. Table IV.9 shows the data for the same question except that Congress was substituted for "the government." It should be noted that 9 percent more thought that Congress was responsive who had not thought the government was. The difference is not very great, but the questions were asked of the same people, so we are sure that for at least a tenth of the population Congress is deemed more representative than the government as a whole. It should also be noted that the original high percentage who thought government responsive did

[47] *Second Treatise,* sec. 149, p. 87.

[48] *Ibid.,* sec. 143, p. 85.

[49] *Ibid.,* and sec. 218; also see *The Federalist Papers,* No. 37, p. 227.

[50] *Second Treatise,* sec. 94, p. 56; also see secs. 132 and 89.

[51] *Ibid.,* sec. 134, p. 77.

[52] The Constitution of the United States of America, article 1, sec. 1.

[53] *The Federalist Papers,* No. 51, p. 322. Of course, the legislative is limited in the same way the whole regime is; see *Second Treatise,* secs. 132, 135, 149, 201; and *The Federalist Papers,* No. 48.

TABLE IV.9 *Congressional responsiveness*

Question: How much attention do you feel that Congress pays to what people think when it decides what to do? (SRC)

	1964	*1968*
Some or good deal	79%	75%
Not much	16	19
DK, no answer	5	6

Source: ICPR.

not leave much room for higher support, so the 9 percent may be qualitatively important. In any event, almost eight Americans out of ten report that Congress does pay at least some attention to what they think.

Boynton, Patterson, and Hedlund have done a study which is relevant for present purposes. Although the data is from one state only, there is little reason to believe that studies in other states would be markedly different. In their study of Iowa, these investigators found that support for the legislative was high. Two questions focus upon long-run cultural commitment to the legislative institution per se. The replies are shown in Table IV.10. About seven out of ten of those interviewed supported the legislative as a basic political institution. Of course, short-term perceptions of an institution's *authority level* performance oscillates substantially over time in the

TABLE IV.10 *Legislative support in Iowa, 1966*

	Pro-Legislative	*Anti*	*DK*
One should be willing to do everything that he could to make sure that any proposal to abolish the state legislature was defeated.	72%	14%	14%
It would not make any difference if the constitution of Iowa was rewritten to reduce the powers of the state legislature.	66	12	22

Source: G. R. Boynton, Samuel C. Patterson, and Ronald D. Hedlund, "The Structure of Public Support for Legislative Institutions," *Midwest Journal of Political Science* (May, 1968), p. 166.

United States.[54] It appears, however, from the data presented (and more will be considered below) that over time the legislative, as a regime level rule institution, has been supported by percentages which can be considered cultural.

The Executive. Both Locke and Madison are suspicious of executive power, but they also state that executive power is necessary.[55] For them, of course, executive power is naturally subject to the legislative as is all power in a civil society. The executive is additionally limited by the wishes of the people.[56] Foreign relations is mentioned by Locke as an area in which the executive will predominate – still, of course, ultimately subject to the legislative and the people.[57]

Although data are limited, the two available surveys which deal with the presidency (rather than with a particular president) show general satisfaction with that value. However, as Table IV.11 shows there is also a substantial minority which feels that the president should have less power. Most of the other data available show support from children, and this support is high.[58] Indications of support for the executive value per se can also be inferred from support given to particular presidents. Most of the time this support is substantial – though, of course, it fluctuates with policy changes.[59] Support is high but probably limited by liberal fear of executive power. In certain crisis situations, however, substantial numbers of Americans are willing to allow great power. For example, Table IV.12 shows that almost one-third of the population in 1938 was willing to give the president dictatorial power if a war came. Even then, however, under substantial stress 61 percent did not believe that such power should be given. Since 1943 Americans have also believed in a limited number of

[54] Roger H. Davidson, David M. Kovenock, and Michael K. O'Leary, *Congress in Crisis* (Belmont, Calif.: Wadsworth, 1966), p. 53.

[55] *Second Treatise,* sec. 144; *The Federalist Papers,* No. 37.

[56] *Ibid.,* No. 70.

[57] *Second Treatise,* sec. 147, 153.

[58] See page 156, n. 65.

[59] Fred I. Greenstein, "Popular Images of the President," *American Journal of Psychiatry* (November, 1965), pp. 523–529; John E. Mueller, "Presidential Popularity from Truman to Johnson," *American Political Science Review* (March, 1970), pp. 18–34; and *Gallup Opinion Index* (February, 1970), pp. 8–16.

TABLE IV.11 *Presidential support*

Question: Do you think the president should have more or less power than he now has? (AIPO)

	1937	*1938*
More	18%	17%
Same	47	41
Less	35	42

Source: *Public Opinion Quarterly* (July, 1938), p. 376.

TABLE IV.12 *Limited executive*

Question: Do you believe the president of the United States should have the power of a dictator over this country in time of war? (AIPO, 1938)

Yes	31%
No	61
No opinion	8

Source: Cantril, p. 590.

terms of office for presidents – again in keeping with the tradition.[60]

The tradition suggests that executive power may be suspect, and the data generally support this inference. But it also expects that the executive will be supported, and this also appears to be the case.

Legislative Predominance. In the American political system the regime rules require not only separation of powers but also checks and balances. That is, the very nature of the regime requires that there shall be interaction. By the comment of most current political scientists, the executive has by far the more commanding position in the exchange with the legislative. Clinton Rossiter presents the typical position when he refers to the president as "chief legislator."

> Congress has a wealth of strong and talented men, but the complexity of the problems they are asked to solve by a people who assume that all problems are solvable has made *external* leader-

[60] Also see Table VI.5 and the related discussion.

> ship a requisite of effective operation. The President alone is in a political, constitutional, and practical position to provide such leadership and he is therefore expected, within the limits of constitutional and political propriety, to guide Congress in much of its lawmaking activity.[61]

Many would say that the president dominates Congress.[62] Foreign observers, especially, would.[63] Probably most intellectuals believe that the president is more important in decision-making although a minority maintains that on the most important issues, Congress works its will.[64]

Whatever the reality, it is clear that presidential dominance is inconsistent with the liberal tradition. As previously mentioned, the legislative must have the ultimate power for a government to be within the tradition. It does not mean, of course, that the presidency should be powerless but merely that the Congress must "predominate" in cases of direct conflict. After all, the tradition believes that the executive has a legitimate role. It should again be kept in mind that the interest at the present is not whether Congress actually predominates but whether the people think that it should. In other words, for the liberal tradition hypothesis to be sustained, Americans must believe that Congress has the right to predominate in areas of overlapping jurisdiction.

Socialization studies have found that children focus upon the president as the preeminent authority in American government.[65] As the child matures, however, there seems to be a

61 Clinton Rossiter, *The American Presidency,* 2d ed. (New York: Harcourt, Brace and World, 1960), p. 28; also see ch. 2 and p. 249, and Bertram M. Gross, *The Legislative Struggle* (New York: McGraw-Hill, 1953), p. 391.

62 "Congress declines as the presidency participates jointly with Congress in the initiation of legislation"; see James A. Robinson, *Congress and Foreign Policy-Making,* rev. ed. (Homewood, Ill.: Dorsey Press, 1967), p. 9.

63 Chester, pp. 49–50.

64 Willmoore Kendall, "The Two Majorities," *Midwest Journal of Political Science* (November, 1960), p. 329. Also see C. A. Amlund, "Executive-Legislative Imbalance," *Western Political Quarterly* (September, 1965), pp. 640–654.

65 Fred I. Greenstein, "The Benevolent Leader: Children't Images of Political Authority," *American Political Science Review* (December, 1960), pp. 934–944; Fred I. Greenstein, "More on Children's Images of the President," *Public Opinion Quarterly* (Winter, 1961), pp. 648–654; Fred I. Greenstein, *Children and Politics* (New Haven: Yale University Press,

TABLE IV.13 *Children's perception of "who runs the country"*

	Congress	*President*	*Supreme Court*	*Don't know*	*N*
Grade 2	3.9%	86.3%	3.3%	6.5%	1,627
Grade 3	6.7	85.4	3.1	4.7	1,662
Grade 4	13.2	77.0	3.4	6.5	1,725
Grade 5	20.0	71.8	3.8	4.3	1,796
Grade 6	24.9	66.2	4.5	4.4	1,744
Grade 7	27.8	64.0	5.3	2.9	1,711
Grade 8	35.1	58.4	3.6	2.9	1,683
Teacher	61.4	35.8	3.0	0.0	383

Source: Hess and Torney, p. 94.

very substantial increase in the perception of Congress as important. Table IV.13 indicates that the percentage of students who think Congress runs the country increases almost nine times from the second to the eighth grade. At the eighth grade over half still perceive that the president rules, but 61 percent of the teachers surveyed maintained that Congress ran the country. The Hess and Torney study found an even more dramatic increase in who elementary students thought made the *laws* in the United States. At grade 2 only 4.8 percent thought Congress made the laws. By the eighth grade 85.3 percent believed that the national legislature did.[66] Easton and Dennis show very similar results. Positive association with images of government dramatically increase for Congress and decrease for the president. In their study a picture of President Kennedy was shown to increase the association with the executive. Even under their circumstances Congress became the favorite image of schoolchildren by the fifth grade. The ratio between favorable images of president and Congress goes from nine-to-one in favor of the president in the second grade to two-to-one in favor of Congress in the eighth.[67] From this so-

1965); Roberta S. Siegel, "Image of a President: Some Insights into the Political Views of School Children," *American Political Science Review* (March, 1968), pp. 216–226. Also see footnote 47 in Chapter I and David Easton and Jack Dennis, *Children in the Political System* (New York: McGraw-Hill, 1969).

[66] Hess and Torney, p. 93.

[67] David Easton and Jack Dennis, "The Child's Image of Government," *The Annals* (September, 1965), p. 45; and *Children in the Political System*, pp. 113–121.

TABLE IV.14 *Belief in legislative predominance*

Question: Some people say the president is in the best position to see what the country needs. Other people think the president may have good ideas about what the country needs but it is up to the Congress to decide what ought to be done. How do you feel about it? (SRC, 1958)

President is in the best position	17%
It is up to Congress	61
Depends, DK, no answer	22

Source: ICPR.

cialization data one could conclude that as the child becomes more socialized into the liberal tradition, he perceives Congress as a more important institution vis à vis the president.

Additional evidence suggests that adults also accept the legislative predominance suggested by the hypothesis. In 1958 the Survey Research Center asked a series of questions relating to choices between the president and Congress. As Table IV.14 shows, three and one-half times as many American believe that Congress is in a better position to evaluate and act upon the country's needs than believe the president is. The Lane study previously referred to finds a similar popular disposition. In his study of working-class men, Lane determined that "more than two-thirds of the men said of the three institutions, the Congress, the Supreme Court, and the presidency, it is the Congress that is most important." [68] He reports that they choose Congress because it is representative of the people while the president is autonomous and awesome. The president does what he thinks is right while Congress listens more to the desires of the people.[69] The American people believe not only that Congress should be supreme, but also that Congress actually has more power in deciding what should be done (see Table IV.15). In both cases the people support Congress more highly than the presidency.[70]

[68] Lane, p. 147.

[69] *Ibid.*, p. 149.

[70] Although many political scientists have expressed the opinion that the people support the president more, surprisingly little data has been presented. The only case where data is offered seems to support the position presented here rather than the interpretation offered by the authors.

TABLE IV.15 *Perception of congressional predominance*

Question: In general, which do you think has the most say in the way our government is run – the Congress, the president, or are they about equal? (SRC, 1958)

Congress	52%
Equal	23
President	11
Depends, DK, no answer	14

Source: ICPR.

It is possible, however, that people give the most support to Congress in a very general manner and do not transfer this support to important subpolicy areas. Data are available for two important domestic problems and various foreign policy areas where the president might be expected to play a dominant role. The Survey Research Center supplies data showing that more people believe that Congress and the president are equally important in setting economic policy than believe they are equal in setting general policy. However, a plurality still thinks Congress is most important (Table IV.16). Three times as many think Congress is most important in economic policy as think the president is most important. This pattern remains the same with a radical time-shift and an emphasis upon balancing the budget. The Gallup organization took a poll two decades earlier and asked people which branch of government was more responsible for balancing the budget. As Table IV.17 shows, Congress was again chosen as the more powerful – by more than 60 percent of those who had an opinion.

Does foreign policy reverse the picture of popular belief in congressional predominance? It was here that even Locke expected more executive power.[71] In most literature on the

Davidson, Koven, and O'Leary offer data on popular conceptions of Congress but only one figure compares Congress to the president. In that figure, with what appears to be one exception, more people believe that the president does a poor job than believe the Congress does. With the addition of the present data their conclusions should be substantially modified. My data suggest that the "Wig" theory is most popular, followed by the "literary" theory. The "executive force" theory has little support; see Davidson *et al.*, p. 60.

[71] *Second Treatise*, sec. 147, p. 86.

TABLE IV.16 *Congressional predominance: economic policy*

Question: When it comes to doing something about problems like unemployment and recessions, which would you say is more important – the Congress or the president – or are they about equally important? (SRC, 1958)

Congress	39%
Equal	34
President	13
Depends, DK, no answer	14

Source: ICPR.

United States the president is presumed to have total dominance in this area. A student of congressional policy-making maintains that "Congress possesses relatively little influence in foreign policy. . . ." [72] The attitudinal data indicate, however, that legislative predominance is also supported for foreign policy. Table IV.18 shows that just prior to World War II a majority would have more confidence in Congress declaring war than they would have if the president did it. A similar attitude is apparent for acts which could lead to war during the Cold War era. In 1951 almost two-thirds of the population felt that Congress had to approve the sending of troops overseas, and a majority supported Congress even during World War II (Table IV.19). Neutrality policy prior to the war again indicates higher faith in congressional action (Table IV.20). Although the question used could have been more directly related to legislative-executive relations (as it was in 1937), the

TABLE IV.17 *Congressional predominance: budget policy*

Question: In your opinion, whose responsibility for balancing the budget is greater, the president's or Congress'? (AIPO, 1936)

President	31%
Congress	50
No opinion	19

Source: Cantril, p. 58.

[72] Robinson, p. 15.

TABLE IV.18 *Congressional predominance: war*

Question: If it comes to a question of the United States declaring war, in whose judgment would you have the greatest confidence? (FOR, 1940)

The president	26%	Congress	51%
Depends on president	6	Neither	1
Both	7	Don't know	9

Source: *Public Opinion Quarterly* (March, 1941), p. 159.

TABLE IV.19 *Congressional predominance: troop placement*

Question: Some people say the president should not be allowed to send troops overseas unless Congress first approves it. Other people say the president should have the power to send soldiers overseas when he feels the situation calls to it. (AIPO)

	1951	*1945*[a]	*1945*[a]
Congress approve	64%	54%	51%
President have power	28	41	38
Undecided	8	5	11

Source: *Public Opinion Quarterly* (Summer, 1951), p. 383; 1945 data from Cantril, p. 587.

[a]In 1945, the wording was slightly altered. For the exact wording, consult the source.

TABLE IV.20 *Congressional predominance: neutrality laws*

Question: Which plan for keeping out of war do you have more faith in — having Congress pass stricter neutrality laws, or leaving the job up to the president? (AIPO)

	Jan. 1937[a]	*March 1938*	*Sept. 1938*	*Oct. 1938*
Congress	51%	56%	65%	66%
President	23	27	24	29
No opinion	26	17	11	5

Source: Cantril, p. 966.

[a]In January, 1937, the wording was slightly altered. For the exact wording, consult the source.

TABLE IV.21 *Congressional predominance: peace treaties*

Question: Which one of these three ways would you personally favor as the best way to have peace treaties approved after the war? (AIPO)

	Oct. 1943	*May 1944*	*Jan. 1945*	*June 1945*
President only	7%	7%	8%	7%
President and majority of Congress	54	60	58	58
President and two-thirds of Senate	25	19	22	21
Don't know	14	14	12	14

Source: *Public Opinion Quarterly* (Spring, 1945), p. 101; *ibid.* (Summer, 1945), p. 248.

cumulative evidence indicates great support for the hypothesis.

Fortunately, substantial data is even available on popular perceptions as to which rule institution is more appropriate for formally ending a war. Table IV.21 indicates that on four separate occasions less than 10 percent of the population would allow the executive to make the decision himself.[73] Three-quarters of the population believe that the national legislative body should participate in treaty ratification.[74] This attitude existed during a time when the president possessed the extraordinary powers of war.

Other data on the legislative-executive relationship is inconclusive. Free and Cantril's study used a visual rating device, on which Congress and the president were separately rated on a scale. The median trust scores for each institution were essentially the same.[75] However, the two institutions were

[73] A *Fortune* poll indicated that if President Roosevelt's name was included in the question, a majority would allow the president to keep Germany and Japan from rearming without congressional approval. This question is an excellent indication of the problems involved when using personalities rather than institutions. Fifty-four percent of the Democratic voters backed Roosevelt while 69 percent of the Dewey voters said Congress should decide; see *Public Opinion Quarterly* (Spring, 1945), p. 94.

[74] Note that a majority would prefer participation by both houses of Congress.

[75] Lloyd A. Free and Hadley Cantril, *The Political Beliefs of Americans* (New Brunswick, N.J.: Rutgers University Press, 1967), p. 193.

TABLE IV.22 *Popular belief in congressional predominance, 1936–1958*

Kind of support	*Percent preferring Congress*	*Percent preferring president*
Belief	78	22
Perception	83	17
Economic	75	25
Budget	62	38
War	61	39
Troops	61	39
Neutrality	70	30

Source: This table is an attempt to summarize Tables IV.14 through IV.20. It considers only the percentage of people who directly support *either* Congress *or* the president. Those who chose both institutions or those who did not express an opinion were ignored. The scores were averaged for Tables IV.19 and IV.20.

not directly compared. In a study of Detroit, Sigel found that a majority actually preferred presidential leadership.[76] However, there is little reason to believe that Detroit, being a highly urban and industrial central city, is typical of the whole nation.[77] The overwhelming evidence indicates that the mass of the adult population supports the requirement of the liberal tradition that the legislative predominate. Congress is preferred over the president in all seven situations outlined above. These are summarized in Table IV.22. For each area of conflict, Congress has substantial popular predominance.

The Judiciary. Locke's only stricture to the judiciary is that the judges be impartial; [78] and, of course, be subject to the legislative power as are all parts of the government. *The Federalist Papers* view the judiciary as a subordinate power. It has neither "force nor will." In fact, "the judiciary is beyond comparison the weakest of the three departments of power." [79] However, the Court is deemed an important part of the system. It is the duty of the Supreme Court "to declare all acts contrary to the manifest tenor of the Constitution

[76] Roberta S. Sigel, "Image of the American Presidency," *Midwest Journal of Political Science* (February, 1966), p. 127.

[77] Indeed, Table V.7 suggests that the Detroit findings are not even representative of highly urban places.

[78] *Second Treatise,* sec. 136, p. 80.

[79] *The Federalist Papers,* No. 78, pp. 465–466.

TABLE IV.23 *Judicial support*

Question: As a general principle would you favor limiting the power of the Supreme Court to declare acts of Congress unconstitutional? (AIPO)

	Sept. 1935	*Nov. 1936*
Yes	31%	33%
No	53	48
No opinion	16	19

Source: Cantril, p. 148.

void." [80] The courts are to act as bulwarks against legislative encroachment. Yet it was believed that the Court could not act against the will of the people for long. The republican principle could only be *delayed* by the Court.[81] The liberal tradition would lead one to hypothesize a respected court which in the short run could even frustrate popular views and remain respected. *In the long run,* however, the liberal tradition would lead one to expect that the legislative would predominate.

Two surveys taken before the Roosevelt "court packing" controversy indicate the general acceptance the public gives to the Supreme Court as an institution (Table IV.23). It should be recalled that in 1935 thirteen acts of Congress were declared invalid.[82] Yet support for the judiciary remained high. A few years later 70 percent of the population maintained that federal judges were honest – 6 percent more than thought state judges were honest and 11 percent more than thought municipal judges were honest.[83] Table IV.24 reports more recent opinion on the Court. A substantial plurality believe that the Court is impartial even after the many controversial decisions of the 1950's and 1960's (though "impartiality" does not technically meet the consensus criterion – 30 percent do not believe). Murphy and Tanenhaus report that a study done

80 *Ibid.,* p. 466.

81 *Ibid.,* p. 469.

82 Harvey M. Karlen, *The Pattern of American Government* (Beverly Hills, Calif.: Glencoe Press, 1968), p. 264. Note that Table IV.23 includes the values of both the legislative and the judiciary.

83 AIPO, 1939; in Cantril, p. 390. Also see Table IV.30.

TABLE IV.24 *Supreme Court impartiality*

Question: Do you think the Supreme Court has been impartial in its decisions or do you think it has tended to favor one group more than another? (AIPO, 1967)

Impartial	47%
Favors one group	30
No opinion	23

Source: *Gallup Opinion Index* (August, 1967), p. 16.

for them in 1966 indicates that twice as many people think the Court is doing its job very well as think that it is not performing very well.[84]

But how does the Court rate when compared to the other rule institutions? Socialization studies show that children tend to perceive the Court as "most important" substantially less than the legislative or executive.[85] Table IV.13 indicated that the Court and Congress were perceived as "running the country" by about the same number of second graders. By the eighth grade, however, almost ten times as many students thought Congress ran the country as compared with the Court. More than twenty times as many teachers perceived Congress as more powerful and ten times as many perceived the president as more important than the Supreme Court.

Other opinion data also suggest that although the Court is supported, it is not as widely supported as the legislative or executive. After the Court had declared the National Recovery Act unconstitutional, on two occasions a majority of those with an opinion thought that Congress and the president should enact another NRA.[86] In a national poll taken in 1964,

[84] Walter F. Murphy and Joseph Tanenhaus, *Constitutional Courts, Public Opinion, and Political Representation,* University of Iowa Report No. 12 (Iowa City: Laboratory for Political Research, 1967), p. 26. Another poll in 1957 also indicates that most people are satisfied with the Court. See Walter F. Murphy, *Congress and the Court* (Chicago: University of Chicago Press, 1962), p. 265. Also see John H. Kessel, "Public Perceptions of the Supreme Court," *Midwest Journal of Political Science* (May, 1966), pp. 167–191.

[85] Easton and Dennis, "The Child's Image of Government," p. 45; and Hess and Torney, p. 93.

[86] AIPO, 1937; in *Public Opinion Quarterly* (July, 1938), p. 381; also see Cantril, p. 344.

TABLE IV.25 *Legislative and judicial support*

	1966[a]	*1957*[b]
Supreme Court	26%	30%
Congress	44	29
Pro and con, equal	8	23
DK, no answer, no opinion	33	18

Source: Murphy and Tanenhaus, p. 28; and *Washington Post and Times Herald* (July 28, 1957), p. E5.

[a]Actual question: "Some people tell us they trust the Supreme Court more than Congress. Others disagree with that. Which do you trust more?" (SRC, 1966)

[b]Actual question: "Which do you have greater respect for – Congress or the Supreme Court?" (AIPO, 1957)

the Court was given a significantly lower rating than either the president or Congress.[87] A study done in Wisconsin found that 50 percent had confidence in actions taken by Congress, 35 percent in actions taken by the president, and only 28 percent in decisions taken by the Court. The Court was the only institution which received more unfavorable than favorable opinions.[88] Table IV.25 shows two national samples comparing support for Congress and the Court. In 1957, support is about equally divided between the two institutions. In 1966, a number almost three-quarters again as large supports Congress. Note, however, that where the two rule values are equally supported, the question refers more to respect. It appears that Americans respect the two institutions equally. In viewing these values as rule institutions, though, the 1966 data appear more relevant. Americans trust Congress's decisions more. This interpretation is reinforced by the other studies cited.[89] The legislative does appear to predominate over the judicial when both are viewed as rule institutions to be

87 Free and Cantril, p. 193.

88 Kenneth M. Dolbeare, "The Public Views the Supreme Court," in Herbert Jacob, ed., *Law, Politics and the Court* (Boston: Little, Brown, 1967), p. 197; also see p. 196.

89 Also see Edward N. Muller, "Correlates and Consequences of Beliefs in the Legitimacy of Regime Structures," *Midwest Journal of Political Science* (August, 1970), pp. 406–407; and the same author's "The Representation of Citizens by Political Authorities," *American Political Science Review* (December, 1970), esp. p. 1164.

trusted. Yet the judicial value does appear to be supported as a culture value. Most appear to support it as one of the necessary tradition values.

Federalism. Federalism presents the uniquely American contribution to the liberal tradition. The system which was outlined in *The Federalist Papers* envisioned a government where

> the general government is not to be charged with the whole power of making and administering laws. Its jurisdiction is limited to certain enumerated objects which concern all the members of the republic, but which are not to be attained by the separate provisions of any. The subordinant governments, which can extend their care to all those other objects which can be separately provided for, will retain their due authority and activity.[90]

There was to be a division of powers where "the State governments would clearly retain all the rights of sovereignty which they before had, and which were not, by that act, *exclusively* delegated to the United States." [91] Both governments were expected to act in their respective areas of competence but there were also areas of overlap. It was expected that the state governments would be more closely tied to the people since they were closer to them.[92] The only way the central government legally could take power from the states would be by performing services more effectively, and even then it could do this only in certain enumerated areas where powers were shared.[93]

The liberal tradition hypothesis would therefore lead one to hypothesize that the people would have respect for the state and local rule institutions. In areas where people believe that the central government should act, it will be seen as the rightful rule institution.[94] In areas not specifically assigned to the federal government, people will tend to choose the state and local governments. In general, federalism should be seen positively — as an aid to liberal or republican government.[95]

90 *The Federalist Papers,* No. 14, p. 102.
91 *Ibid.,* No. 32, p. 198.
92 *Ibid.,* No. 46, p. 295.
93 *Ibid.*
94 *Ibid.,* No. 44.
95 *Ibid.,* Nos. 10, 51.

TABLE IV.26 *Support for state legislature*

Question: What kind of job do you think our state legislature does? (AIPO, 1967)

Excellent	12%	73% (Excellent + Fair)
Fair	61	
Poor	9	
No opinion	18	

Source: *Gallup Opinion Index* (February, 1967), p. 19.

TABLE IV.27 *Local and national political behavior*

Question: Have you ever done anything to try to influence a local (congressional) decision? (AV, 1960)[a]

	Local	*Congressional*
At least once	28%	17%
Never	72	83
Other, DK, no answer	b	b

Source: ICPR.
[a] $N = 970$.
[b] Less than 1 percent.

Coleman maintains that his documentary study indicates a long history of popular support for local government.[96] Bryce also remarks upon this popular belief.[97] More recent evidence is available in Table IV.26. Although only a small percentage think the legislatures are doing an excellent job, almost three-quarters of the population believe they are providing at least fair governmental performance. Less than 10 percent think state legislatures perform poorly. Not surprisingly, Americans seem to understand local issues much better than national and international ones. Table II.4 showed that this difference is rather substantial. Almost one-fifth of the population understands local issues and does not comprehend national and international ones. More interestingly, Table IV.27 indicates that citizens also *act* more on the local level. More than one and one-half times as many Americans apparently try to influ-

96 Coleman, p. 497.
97 Bryce, vol. 2, part 4, ch. 4.

ence local decisions as try to influence national congressional decisions.

Support for the state and local rule institutions, then, appears high enough to substantiate the hypothesis expectation of cultural-level support. The hypothesis can only be sustained, however, if there is support for state and local institutions when they come in conflict with the national government. Table IV.28 presents data for this very complex area of opinion interpretation. Four questions on polls taken within two years of each other seem to suggest opposite conclusions. In the same year (1937) people seemed to support the federal government one time and the state government the other time. Of course, it is possible that changes take place over time. However, columns two and three in the table represent views taken from the same people in the same sample! I would suggest that the apparent confusion can be resolved by reference to the tradition. The "Locus of Power" question was apparently interpreted to mean a belief in concentration of power in the federal government as that concentration already existed at the time the question was asked. In other words "federal government" seems to be interpreted as a support for the *status quo*. Certainly the second question in the 1937 survey indicates that a majority did not believe the states should transfer

TABLE IV.28 *Federal-state relations*

	Locus of power[a]			
	1936	*1937*	*Increase federal power? 1937*[b]	*State or federal more power? 1938*[c]
Federal	42%	43%	26% (yes)	27%
State	34	33	54 (no)	32
Same as now	—		—	20
No opinion	23	23	20	21

Source: Cantril, pp. 815–816.

[a]Actual question: "Which theory of government do you favor, concentration of power in the federal government or concentration of power in the state government?" (AIPO, 1936, 1937)

[b]Actual question: "Should state governments transfer more of their powers to the federal government?" (AIPO, 1937)

[c]Actual question: "In the division of government power between the federal and state governments, do you think the federal should have more power and the state less, or the state more and the federal less?" (FOR, 1938)

TABLE IV.29 *Federal-state relations*

Question: Which do you think spends the taxpayer's dollar more wisely — the state government or the federal government? (AIPO, 1966)

State	49%
Federal	18
Neither	17
No opinion	16

Source: *Gallup Opinion Index* (January, 1967), p. 6.

more power to the federal government. In fact, the 1938 survey indicated that those who wanted more state power and less federal power had a slight plurality over those who wanted more federal power at the expense of the states. Perhaps one further general observation would help to explain this complex relationship between state and federal values. It should be noted that the four polls mentioned above were all taken during the depression of the 1930's. Although substantial support was shown for the states at that time, one would suspect from the shaping influence of culture that as this strong environmental stimulus was reduced support for the states would increase. Table IV.29 indicates that in more "normal" periods the state is very highly supported vis à vis the federal government. Almost three times as many people in 1966 believed that the states allocated more wisely than did the central authorities.

In some specific policy areas, the people seem to show substantial support for the federal government. In areas where both governments have jurisdiction Madison advised that the central government could predominate if its service was better than the states'.[98] Table IV.30 shows an area where federal activity was deemed superior to state action. It should be noted that all levels were supported rather well but that the federal judges were thought more honest by more people. In such areas of obvious concurrent jurisdiction, the people would seem to make judgments upon the Madisonian criteria of performance. It may well be, however, that in periods of fed-

[98] *The Federalist Papers,* No. 46, p. 295.

TABLE IV.30 *Perceived national judicial superiority*

Question: Do you think judges in the federal (state, local) courts of this country are honest? (AIPO, 1939)

	Federal	*State*	*Local*
Yes	70%	64%	59%
No	11	20	23
Refused, DK	19	16	18

Source: Cantril, p. 390.

TABLE IV.31 *State vs. federal: relief*

Question: Which government should care for people on relief?

	1936[a]	*1937*[b]	*1939*[c]
Federal	45%	35%	36%
State-local	55	46	64
Combination	—	6	—
None	—	3	—
Don't know	—	10	—

Source: Cantril, pp. 893–896.
[a]Actual question: "Should the responsibility of caring for all persons be returned now to the state and local governments?" (AIPO, 1936)
[b]Actual question: "Which kind of government – federal, state, or local – do you feel should take care of relief?" (FOR, 1937)
[c]Actual question: "Do you think unemployment relief should be handled by the federal government or by state and local governments?" (AIPO, 1939)

eral judicial activism the relative support levels may change.[99]

In areas of more questionable federal jurisdiction, however, the people seem to give more support to the state and local governments. Data is available in only two policy areas. The first concerns relief measures during the late 1930's. As Table IV.31 makes clear, the majority of those who chose between the two kinds of government chose the state and local levels as the ones which should handle a program which the Constitution does not assign specifically to the federal government. Another area where the Constitution does not directly assign a role to the federal government is education. Polls from 1938 to the present have shown that when asked if they favored

99 See Chapter VI.

TABLE IV.32 *State vs. federal: education*

	1968 (SRC)[a]	*1964 (SRC)*[a]	*1943 (NORC)*[b]
Federal	28%	31%	21%
State and local	48	47	68
No interest, other, DK	24	22	11

Source: ICPR; NORC, 1943; in Cantril, p. 187.

[a]Actual question for 1968 and 1964: "Some people think the government in Washington should help towns and cities provide education for grade and high school children; others think this should be handled by the states and local communities. Have you been interested enough to favor one side over the other?"

[b]Actual question: Do you think that public school systems should be mostly controlled by the federal government or mostly controlled by each state government?

federal aid to education, substantial majorities of the people have answered that they did.[100] However, three polls were taken which gave a choice between state and local aid only and supplementary federal aid. As is shown in Table IV.32 a substantial plurality favors state and local aid only.

In general, the value of federalism appears to be rather widely supported by the American mass public. In areas of concurrent jurisdiction either the federal or state could be perceived as the primary institution depending upon who performs the duties best. In areas where the national government has not been given specific authority, however, the states seem to be supported as preeminent.

Political Parties. The idea of political party is usually thought to postdate *The Federalist Papers* and, of course, to a great degree this is so.[101] But the question deserves to be investigated in somewhat more detail. The idea of ideological, mass-based parties certainly was not in the the minds of Madison and the other Constitutional fathers. However, what Duverger calls

[100] For an excellent review of this opinion data, see Frank J. Munger and Richard Fenno, Jr., *National Politics and Federal Aid to Education* (Syracuse, N.Y.: Syracuse University Press, 1962), pp. 91–96.

[101] Robert A. Dahl, *Political Oppositions in Western Democracies* (New Haven: Yale University Press, 1966), p. xi; also see Seymour Martin Lipset and Stein Rokkan, *Party Systems and Voter Alignments* (New York: Free Press, 1967), pp. 26–30.

cadre parties[102] were familiar in the late eighteenth century. Not only were they emerging in Britain at the time but, more importantly, they were evident in the American states.[103] The most important aspect of the cadre party is that it is decentralized. Most scholarly opinion in the United States maintains that this is the characteristic description of American political parties.[104]

If we turn to *The Federalist Papers,* it will be noted that Madison anticipated a decentralized party system. He held that American legislators tended to be locally oriented and that they would remain so under the new Constitution.

> For the same reason that the members of the State legislatures will be unlikely to attach themselves sufficiently to national objects, the members of the federal legislature will be likely to attach themselves too much to local objects.[105]

The Constitution itself also left the prescription of the times, places, and manner of holding elections to the states.[106] The combination of local control and local orientation of legislators was considered major regime protection of the "rights of the individual States" and "the prerogatives of their governments."[107] A decentralized legislative party was one of Madison's essential balances which would protect the liberal nature of the republic.

Support for the American type of political parties seems rather strong. As indicated in Table IV.33, over three-fourths of the American population are generally satisfied with the nature of the political parties. As Table IV.34 shows, five polls taken between August, 1937, and April, 1947, showed substantial opposition to a restructuring of the party system to have

102 Maurice Duverger, *Political Parties,* trans. Barbara and Robert North (New York: Wiley, 1954), ch. 1.

103 Wilfred Binkley, *American Political Parties,* 4th ed. (New York: Knopf, 1964), ch. 1.

104 V. O. Key, Jr., *Politics, Parties and Pressure Groups,* 4th ed. (New York: Crowell, 1958), p. 361, and ch. 12; Frank J. Sorauf, *Political Parties in the American System.* (Boston: Little, Brown, 1964), pp. 39–43; Austin Ranney and Willmoore Kendall, *Democracy and the American Party System* (New York: Harcourt, Brace and World, 1956), ch. 21; Lipset and Rokkan, p. 31.

105 *The Federalist Papers,* No. 46, p. 296.

106 Article 1, sec. 4.

107 *The Federalist Papers,* p. 297.

TABLE IV.33 *Support for political parties*

	1944 (FOR)[a]	*1964 (Minn.)*[b]
Favor present parties	78%	89%
Would like to see change	14	7
Other, no opinion, DK	8	4

Source: *Public Opinion Quarterly* (Winter, 1944–1945), p. 571; *Polls* (Summer, 1965), p. 80.

[a]Actual question: "On the whole how do you feel about the present setup of the political parties in the United States? Do you find that you are usually satisfied with the stands taken by one or the other of the present big parties, or would you like to see a strong new party entirely different from either of the present parties?"

[b]Actual question: "Would you say it is important that the United States have a strong two party system, or not too important?" Although this question comes only from one state (Minnesota), there is little reason to believe that there is a geographical basis for this opinion distribution. Poll by Minneapolis *Tribune*'s Minnesota Poll.

two "rational" parties – one Liberal and one Conservative. Each time the question was asked a consensus distribution of opinion was opposed to such a party realignment.[108] In 1950 opinion on a realignment between Republican and Southern Democrats was similarly rejected.[109] Again in 1968, over two-thirds of those questioned said they were satisfied with the present parties as compared to only 27 percent who wanted a new party more in line with the way they themselves thought.[110] In the 1960's, two-thirds said that political parties helped the government pay at least some attention to what people wanted.[111]

There seems, therefore, to be general support for the present parties. But is there direct support for the American-type, decentralized cadre party? The available data, shown in Table IV.35, indicate that Americans do not expect their congressman to follow his party leadership. Only about one-quarter of

108 George H. Smith and Richard P. Davis, "Do the Voters Want the Parties Changed?" *Public Opinion Quarterly* (Summer, 1947), pp. 236–243.

109 AIPO, 1950; in *Public Opinion Quarterly* (Spring, 1950), pp. 183–184.

110 AIPO, 1968; in *Gallup Opinion Index* (October, 1968), p. 27. At the same time only 8 percent thought labor should have its own political party; 80 percent favored the present system; *ibid.*, p. 28.

111 SRC, 1964 and 1968; in ICPR. A majority even supports the convention system (SRC, 1952).

TABLE IV.34 *Support for "rational" party realignment*

Question: It has been suggested that we give up the present Republican and Democratic parties and have two new parties — one for the Liberals and one for the Conservatives. Would you favor this idea? (AIPO)

	Aug. 1937	*April 1938*	*Sept. 1938*	*Jan. 1946*	*April 1947*
Support new Liberal and Conservative parties	22%	24%	15%	12%	13%
Oppose realignment of parties	52	55	65	67	72
No opinion	26	21	20	21	15

Source: Austin Ranney and Willmoore Kendall, *Democracy and the American Party System* (New York: Harcourt Brace Jovanovich, 1956), p. 503, derived from the Smith and Davis article.

the population believes in congressional accountability to some kind of "responsible party" leadership. Very substantial majorities believe that the congressman should be allowed to make up his own mind. Indeed, although the evidence is somewhat ambiguous and dependent upon question wording, the Survey Research Center's 1958 poll indicates that although some voters want a congressman to vote on a bill against the way his own conscience would bid him to vote, a slight plurality of those interviewed believe that the congressman should

TABLE IV.35 *Support for decentralized parties*

	1958 (SRC)[a]	*1946 (AIPO)*[b]
Congressman should support party	25%	21%
It depends	7	–
Congressman should make up own mind	57	69
Don't know, no opinion, etc.	11	10

Source: ICPR; and Cantril, p. 134.

[a]Actual question: "On most issues should a congressman support the position his party takes when something comes up for a vote, or should he make up his own mind regardless of how his party wants him to vote?"

[b]Actual question: "If a congressman is elected on the Democratic ticket and he does not vote with his party on all major issues, should he be prevented from running for office again as a Democrat?" (Response, Democrats only: yes, 25 percent; no, 64 percent; no opinion, 11 percent.)

TABLE IV.36 *Roosevelt's "purge" failure*

Question: Do you think it was a good or bad thing that certain senators, who were recently opposed by President Roosevelt in his so-called purge attempt, were renominated?

Purge failure was good thing	47%
Purge failure was bad thing	11
No opinion	42

Source: Elmo Roper, *You and Your Leaders: Their Actions and Your Reactions, 1936–1956* (New York: Morrow, 1957), p. 33.

vote the way he thinks best rather than the way the voters want him to vote.[112]

A further indication of popular support for the liberal tradition idea of decentralized parties, can be gained from popular attitudes on the Roosevelt "purge primaries." In 1938 President Franklin Roosevelt attempted to influence Democratic primaries in the hope of preventing the renomination of anti-New Deal Democrats. That is, as leader of the party he attempted to impose some ideological constraints upon the fragmented party system. As Table IV.36 indicates, more than four times as many people believed that the purge should have failed as believed that Roosevelt should have been able to discipline his fellow party members.[113] Apparently the idea of a locally controlled political party is rather well accepted within the American rule system.[114]

112 SRC, 1958; in ICPR. One poll (Cantril, p. 133) does not confirm this finding. However, belief that legislators should exercise independent judgment (while also considering public wants) is often implicit rather than explicit. See Roger H. Davidson, "Public Perceptions of the Representative's Job," paper delivered to the American Political Science Association (September, 1969), pp. 6–7.

113 When asked if they approved of "President Roosevelt's campaign to defeat Democrats who oppose him," 23 percent approved, 58 percent disapproved, and 19 percent had no opinion; see Charles M. Price and Joseph Boskin, "The Roosevelt 'Purge': A Reappraisal," *Journal of Politics* (August, 1966), p. 663. However, Democrats were more evenly divided — 34 percent approved and 43 percent disapproved.

114 I.e., support is high for the cadre party viewed here but not for the organized "responsible" party; see Jack Dennis, "Support for the Party System by the Mass Public," *American Political Science Review* (September, 1966), esp. pp. 605–606.

TABLE IV.37 *Length of congressional terms*

Question: What do you think is the ideal length of time for which they [congressmen] should be elected? (AIPO, 1934)

Two years or less	31%
Two to four	44
Four to six	6
More than six	5
Don't know	15

Source: Cantril, p. 134.

The tradition also indicates that the electoral process should be based on short terms. For republican government to have liberty, Madison held that incumbents in office must be within the periodic control of the people.[115] Terms should be long enough, however, to "provide the requisite stability and energy in government" and short enough to be true to "liberty and to the republican form." [116] Table IV.37 indicates that three-quarters of the people chose a term of four years or less. Of course, the Constitution set the terms for representatives at two years but *The Federalist Papers* argued for four-year terms for president and six years for senator. Presumably, then, six years would be the permissible limit for terms of office under the liberal tradition.[117] According to Table IV.37, only 5 percent of the population would favor terms of more than six years.

THE LIBERAL TRADITION AND RULES

The liberal tradition hypothesis seems to suggest that a rule system which is congruent with its descriptive thesis must meet nine criteria. The people must believe that: (1) a rule system is needed; (2) there should be formal consent; (3) the basic rule should be popular majority rule; (4) majorities should be restrained, as should all government; (5) the legislative predominates; (6) the other two formal national institutions be respected as long as they do not violate their trust; (7) there

115 *The Federalist Papers,* No. 37, p. 227.
116 *Ibid.,* p. 226.
117 *Ibid.,* No. 39, p. 242.

be substantial belief in the values of federalism; (8) political parties should remain decentralized; and (9) terms of office should be limited. Each of these basic rule norms was investigated using the available public opinion data. In no case did the national data contradict the hypothesis. On the contrary, the basic rules propounded by Locke and Madison seem remarkably well accepted in the United States.

Of those with an opinion, over 70 percent support Constitution rules. Support for regime institutions is diffused among 85 percent of the system members. Popular rule is supported by 76 percent (while 91 percent reject dictatorship). However, only 14 percent accept an unqualified view of popular wisdom. Popular rule seems most evident, though, in support for rule institutions. Almost eight out of ten think Congress is responsive to their wishes. On the average, in a direct conflict situation, over three-quarters value legislative predominance. All the major institutions seem to be highly supported. The president and the Court appear to be less supported than Congress but appear to be accepted as cultural. Federalism is also valued. In terms of efficiency, 73 percent choose the state over the central government. The political parties also appear to be supported — by almost eight out of ten. In addition, 75 percent believe that legislative terms should be less than four years. Based upon this and the other data reported in this section of the present study, it appears that liberal rule values are widely supported in the United States. That is, the rule subsystem of the American political culture appears to sustain the liberal tradition hypothesis.

It should be mentioned that it was not at all obvious before the investigation that all these rule beliefs were valued by the public. Consider the following analysis by an eminent student of the affective relationships people develop with authorities.

> The qualities of age, wisdom, impartiality, inaccessibility, and irrevocability associated with the upholders of the highest law can smoothly blend the Supreme Court into the image of the "Lord of Law," as the *ruler* is called in some communities. On the other hand, a legislative body has weaknesses in this respect — its numbers, its publicized internal disagreements, and sec-

> tional interests. The greatest potentiality among governmental branches rests in the office of the chief executive, where lie the punitive (police and army) and the administering (policy-making and executing) powers.[118]

Certainly the author of this analysis makes an excellent case. The weaknesses of the legislative referred to are especially evident in the United States where Congress is often unfavorably characterized by popular journalists and other intellectuals. The data, however, seem to indicate that the legislative is most revered of the American rule institutions. Perhaps the fact that Congress is most revered when speculative analysis indicates it should not be is simply an indication of the strength of the screening influence of the rule subsystem of a liberal tradition political culture.

THE BELIEF SYSTEM OF THE UNITED STATES

BELIEF SYSTEMS AND POLITICAL CULTURE

The final subsystem of the American political culture is called the belief system. Of course, to a great extent, all of the subsystems are belief systems. Indeed there is a residual flair to the final subsystem. To some extent, what is not identity, symbol, or rule is belief. Yet the belief system is distinctive in that it refers to regime moral values, i.e., values which are more directly related to the moral principles deeply held by the members. For any particular regime beliefs represent the central affective goals toward which political actions should be directed. They are the moral guides for the regime.[119]

The belief system is, of course, a subsystem of the political culture. As such, it is presumed to act as a screening organism. However, the screen is not necessarily ordered on a single hierarchical dimension. It can be empirically held that beliefs do seek consistency.[120] However, "no individual is ever completely

118 Sebastian De Grazia, *The Political Community* (Chicago: University of Chicago Press, 1948), pp. 96–97.

119 Easton, *A Systems Analysis,* pp. 194–195.

120 Talcott Parsons, *The Social System* (New York: Free Press, 1951), p. 379; also see Cora DuBois, "The Dominant Value Profile of American Culture," in Michael McGiffert, ed., *The Character of Americans* (Homewood, Ill.: Dorsey Press, 1964), pp. 224–232.

consistent in his evaluation of an object through time." [121] The system model developed earlier suggested that outputs are the result of interactions between inputs and the system. It has been further suggested that demands and supports are the result of interaction between environmental stimuli and the interaction between the subsystems of political culture (see Figure III.4 above and the related discussion). Each of the value subsystems also has a complex of single value components which can conflict. For example, we have already seen that legislative and executive values can conflict. Component conflict, however, is especially important for the belief system. There is less agreement on the order of value preference in this subsystem than in the more rationalistic, systematized, behavior-directed, written rule system. Belief values are more vague and, thus, demand less internal hierarchical ordering.

In general, values have "a wide range of obligatoriness, explicitness, and specificity." [122] Value systems are not simple but can be considered "value constellations" — values that are grouped together.[123] The constellation is composed of interrelated values, which undoubtedly have some internal rank ordering.[124] However, the rank-ordering can change:

> values, and systems of beliefs *do not operate as single and separate units* but *are in continually shifting and recombining configurations* marked by very complex inter-penetration, conflict and reformation.[125]

Certain value systems, however, are more stable than others. The basic subsystems of the political culture (identity and symbol) are viewed as having relatively non-conflicting value constellations. The developed subsystems (rule and belief), however, tend to display value conflict.[126] The more diffuse

[121] William C. Mitchell, "The Ambivalent Status of the American Politician," *Western Political Quarterly* (September, 1959), p. 694.

[122] Williams, p. 373.

[123] Bruce L. Smith, "Communications Research on Non-Industrial Countries," in William Schramm, ed., *The Process and Effects of Mass Communications* (Urbana: University of Illinois Press, 1965), p. 173.

[124] Gabriel A. Almond, *The American People and Foreign Policy* (New York: Praeger, 1960; originally published 1950), p. 48.

[125] Williams, p. 468.

[126] Max Lerner, *America as a Civilization* (New York: Simon and Schuster, 1957), pp. 72–73; Almond and Verba, p. 366; and DuBois, *passim.*

the society, apparently the more complex the value system ordering.[127] As noted above, though, since the rule system is so directly related to system operation, this developed system tends to be ordered. For example, the legislative always tends to predominate. It is the belief system, then, which is the least integrated (and in the United States, at least, this inconsistency in individual ordering is recognized, accepted, and even valued [128]).

The totality called the belief system is, thus, viewed as a complex interaction between value components in which value support will tend to be unstable in ordering but still, like any culture subsystem, will tend to provide commitment to a stable *group* of value components. The choice of belief system values will, thus, present a more complex problem than for the previous subsystems, and so a special approach will be taken.

The total political culture can now be outlined. Figure IV.1 shows how the mass political culture process is viewed. The political culture is an input social structure related to the political system. Environmental stress acts upon input structure – predominately upon political culture. In general, stress stimuli act first upon the basic systems of identity and symbol. There is some stress between symbols but very little, and only a little more between the components of the identity system. More conflict exists within the rule system, though the legislative is supported more than the other rule values. The belief system values are more equally supported resulting in more internal stress. Besides internal subsystem stress major interaction takes place between subsystems, and between subsystems and environmental stimuli. The product of this interaction in turn acts upon the other input structures. Finally this product is processed to the political system as demands for action and as specific support. The residue of the political culture which acts at any point in time upon the political system is a major determinant of diffuse support. Positive diffuse support can act to maintain regimes.

127 Williams, p. 375.

128 *Ibid.*, pp. 375–395. Williams also makes the interesting point that values are not expected to be followed completely. They are supposed to be moderated. Saints, i.e., those who follow the norms completely, are often greatly admired but socially ostracized *(ibid.*, p. 380).

FIGURE IV.1 *The member political culture system*

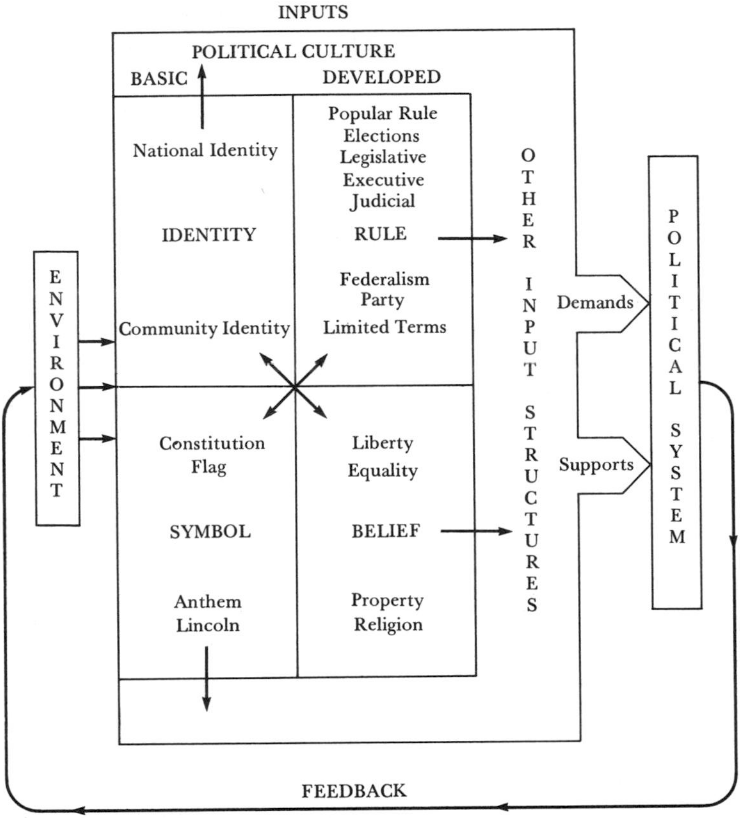

THE TRADITION AND AMERICAN BELIEFS

Since belief values are such complex phenomena, we have noted that they will present special problems of identification. Even in a hypothesis such as the present one, which reduces a tradition to two works, beliefs would be relatively difficult to abstract. This is especially true when one realizes what a large number of beliefs have been attributed to Americans through a broad literature. In a symposium reported in the pages of the journal of the American Political Science Association, three of the most eminent political scientists of the time tried to define the values necessary for a successful democracy. Ernest Griffith said that religion was the necessary basis, while John Plamenatz said that a desire for freedom was the essential value. J. Roland Pennock maintained that it was based upon both of these values in addition to what I have called rule values, plus a concern for justice, tolerance, and compromise, and agreement on fundamentals.[129]

The present hypothesis holds that beliefs must be found in the values of the liberal tradition. Locke and Madison must again provide the ground upon which a belief system can be identified — and then tested. Because of the complexity of beliefs, however, an intermediate process will be performed. In order to abstract the central principles of the American political culture a most direct connection between the values and the political system must be found. This connection can be provided by the Declaration of Independence. It is widely recognized that Locke is most directly related to the American community through this system-creating and justifying document.[130] A document that expresses the very *raison d'être* of a system would seem a propitious point to begin. As one authority maintains, "The basic ideas of the American political 'mind' received their most explicit statement in the Declaration of Independence." [131] These values are summarized in the second paragraph of that document.

> We hold these truths to be self evident, that all men are created equal, that they are endowed by their Creator with certain unalienable Rights, that among these are Life, Liberty and the pursuit of Happiness.

The beliefs the hypothesis will maintain that Americans support are that (1): men are created equal; (2) rights are bestowed by a Creator; (3) liberty is a prime goal; and (4) preservation

[129] Ernest S. Griffith, John Plamenatz, and J. Roland Pennock, "Cultural Prerequisites to a Successfully Functioning Democracy: A Symposium," *American Political Science Review* (March, 1956), pp. 101–137.

[130] Dan Nimmo and Thomas D. Ungs, *American Political Patterns,* 2d ed. (Boston: Little, Brown, 1969), pp. 65–68; Marian D. Irish and James W. Prothro, *The Politics of American Democracy,* 4th ed. (Englewood Cliffs, N.J.: Prentice-Hall, 1968), pp. 73, 97, 546; Karlen, pp. 53, 85; Mitchell, p. 107; among many others.

[131] Mitchell, p. 107.

of individual property provides the means for happiness.[132] In simplified form these resolve into belief in liberty, equality, property, and religion.

Even though beliefs are normally conceived as relatively unstable, most analysts of American social values have identified these same four values as prominent American beliefs through all of its history. De Tocqueville was an early observer who noted these values and how widely they were supported.

> The observer who examines what is passing in the United States . . . will readily discover that their inhabitants, though divided into twenty-four distinct sovereignties, still constitute a single people; and he may perhaps be led to think that the Anglo-American Union is more truly a united society than some nations of Europe which live under the same legislation and the same prince.
>
> Although the Anglo-Americans have several religious sects, they all regard religion in the same manner. They are not always agreed upon the measures that are most conductive to good government, and they vary upon some of the forms of government which it is expedient to adopt; but they are unanimous upon the general principles that ought to rule human society. From Maine to the Floridas, and from the Missouri to the Atlantic Ocean, the people are held to be the source of all legitimate power. The same notions are entertained respecting liberty and equality, the liberty of the press, the right of association, the jury, and the responsibility of the agents of government.[133]

Turner noticed the pervasiveness of the beliefs at the beginning of the twentieth century.[134] Indeed, most of those who have empirically viewed the beliefs of the United States have found that these four have predominated from the very beginning. Figure IV.2 presents an outline of the views of twenty-eight of the most prominent writers on values in the United States and their findings on American beliefs. All writers tend

[132] There is disagreement as to whether "happiness" in the Declaration directly refers to Locke's idea of property. In my opinion the linkage is direct. However, since we are interested more in Locke than we are in the Declaration, this dispute does not have to be resolved here; see the works cited in note 130.

[133] Tocqueville, pp. 386–387.

[134] Frederick Jackson Turner, *The Frontier in American History* (New York: Holt, 1921), p. 37.

FIGURE IV.2 *Views of American belief*

	Liberty[a]	*Equality*[b]	*Religion*[c]	*Property*[d]
Gabriel A. Almond	x		x	x
Denis W. Brogan		x		x
Lord Bryce	x	x	x	x
Lee Coleman	x	x	x	x
Henry Steele Commager	x	x		x
Robert Dahl		x		x
K. Davis et al.	x	x		x
Cora DuBois		x		x
Daniel Elazar	x		x	x
Ralph Gabriel	x		x	
John Gillin	x	x	x	x
Saxon Graham	x	x		
Louis Hartz	x	x		x
V. O. Key, Jr.		x		x
Hans Kohn	x	x		
Robert Lane	x	x	x	x
Max Lerner	x	x	x	x
Seymour Martin Lipset		x		x
John McDiarmid	x	x	x	x
Curtis MacDougall	x	x	x	x
Margaret Mead			x	x
Charles Morris		x		x
Talcott Parsons	x			x
David Potter	x	x		
David Reisman		x		
Charles Schettler	x			x
Goodwin Watson	x	x	x	x
Robin Williams	x	x	x	x

[a]Freedom, individualism, free from government, self-reliance, absence of external force.

[b]Conformity, low class-consciousness, equality of opportunity, fear of inferiority, egalitarianism, tolerance of minorities.

[c]Belief in God, morality, subject to higher law, moral people, conscience.

[d]Achievement, free enterprise, progress, competition, materialistic, commercial, business oriented, pursuit of money, Protestant ethic, bargaining, capitalist success ethic, work as a value, effort.

to emphasize at least one of the belief attitudes mentioned above, and most mention two or more of the attributes as descriptive. One very extensive survey of American values finds a strong agreement between all writers studied.[135] As the

[135] Almond, p. 39.

TABLE IV.38 *Primary value: the home*

Question: If you could be sure of only one of the following, which would you choose? (AIPO, 1940)

Wealth	13%
A happy home	68
An interesting job	19

Source: Cantril, p. 281.

survey indicates, this does not guarantee that these are the relevant values but it is at least quite suggestive that they are.[136]

AMERICAN VALUES

In order to try to describe American beliefs, the emphasis will again be on empirical data – especially concentrating upon public opinion polls. It is a commonplace of empirical attitude research that Americans, or most people for that matter, are not primarily interested in politics. Some of this data was cited in Chapter II. In 1940 Gallup asked a question which seems to suggest where primary American interests lie. Table IV.38 indicates that more than two-thirds of Americans value a happy home over wealth or an interesting job – at a time when 15 percent of the population was unemployed.[137] Immediately after the war, Gallup asked the people of four countries what happiness meant. Table IV.39 again suggests that Americans are more interested in the world of personal and family contentment than the world of material well-being or political affairs. The Stouffer surveys of almost 6,000 people found that "most Americans answered in terms of personal, family or business problems. Only a small minority expressed anxiety about problems outside the orbit of their immediate personal life – even the threat of war." [138] More recently in

[136] *Ibid.*, p. 47.

[137] It could be argued that the value of the family is part of the political culture rather than just part of the general culture. If so, it is certainly consistent with the liberal tradition. See *Second Treatise,* ch. 6, esp. secs. 52, 56, 58, and 61.

[138] Samuel A. Stouffer, *Communism, Conformity, and Civil Liberties* (New York: Wiley, 1966), p. 69. Also see the Almond and Verba data for question 9A, in ICPR.

TABLE IV.39 *Meaning of happiness*

Question: Will you tell me in your own words what the word "happiness" means to you? (AIPO, FIPO, CIPO, BIPO, 1944)[a]

	U.S.	*France*	*Canada*	*Great Britain*
Contentment, peace, etc.	44%	12%	32%	14%
Marriage, home, etc.	17	9	19	33
Money, wealth, etc.	12	52	38	12
Health, etc.	16	22	18	15
Success, etc.	15	–	6	14
Miscellaneous or vague	22	28	25	28
No answer, etc.	5	30	6	7
Total	131%	153%	144%	123%

Source: Cantril, p. 281.
[a]Percentages add to more than 100 because some respondents gave more than one answer.

1964, through the auspices of the Gallup organization, Free and Cantril found that the great majority of Americans were primarily oriented to concerns of family, economic status, and personal and world peace.[139]

In a most systematic study of individual values, Charles W. Morris advises that American college students rank higher on being flexible and open to other ways of life, on participating and preserving through moderate actions, on mastering the environment, and on taking charge than do similar students in India, Japan, Nationalist China, and Norway.[140] These latter values appear to relate to the tradition belief values but they are not fundamentally political. All of the cited studies on American values appear to have one common theme: The American is not primarily oriented to the political. His main interests seem centered in the home and its economic and social preservation; and on personal advancement and manipulation of the environment.

[139] Free and Cantril, p. 97.

[140] Charles W. Morris, *Varieties of Human Values* (Chicago: University of Chicago Press, 1956), pp. 45–61 and *supra*. Other evidence gives further indications that Americans feel that they can achieve mastery over the environment. Optimism is perennially high in the United States – even during the Great Depression (Cantril, p. 894, qq. 8, 9, 11, 12, and p. 897, q. 49) and during World War II, though less so (*ibid.*, p. 898).

TABLE IV.40 *Citizen values to be transmitted to the next generation*

Question: If you think you as a citizen have an obligation to pass on anything to the next generation, what is it? (FOR, 1940)

Good citizenship and political environment	30%
Moral and ethical qualities	30
Economics and social environment	21
Culture and education	20
Material wealth, property	10
Health and other	15
Don't know, etc.	5
Total	131%[a]

Source: *Public Opinion Quarterly* (June, 1940), p. 355.
[a]Multiple responses were allowed.

Of course, the present work is not primarily interested in general goals and principles but in specifically political values. When Americans are forced to orient themselves to national rather than personal problems and hopes, political concerns become more manifest. In 1940 *Fortune* magazine's survey asked, "Do you feel that you, as a citizen, have an obligation to pass on anything to the next generation?" The 70 percent who said they did feel an obligation (an interesting finding on socialization dynamics, in itself) were then asked what they would transfer, *as citizens*. Table IV.40 reports that when the individual's citizen role is emphasized there are more political and cultural responses. Indeed all of the responses except health could, in a broad sense, be considered part of the political culture. When Free and Cantril asked about *national* problems, they too find different patterns of response. When asked what fears they have for their country, people mention such things as war, communism, depression, racial matters, disunity, socialism, etc.[141]

When questions are focused even more toward the political, those values which have been associated with the belief system

141 Free and Cantril, p. 105. A survey of high school students taken in 1945 shows similar emphasis when educational problems are mentioned; see *Public Opinion Quarterly* (Spring, 1945), pp. 85–86.

TABLE IV.41 *The belief system and American values*

	Percent of responses favorable to value[a]				
	1941[b]	*1942*[c]	*1946*[d]	*1948*[e]	*1949*[f]
Liberty: freedom; specific freedom of press, speech etc.; do as I please	49%	46%	42%	27%	63%
Equality: in voting, jury trial, equal rights, etc.	0	9	20	37	8
Religion: God, freedom of religion, etc.	9	37	4	–	7
Property: including freedom to earn a living, to take jobs, to have equal opportunity; free enterprise, etc.	22	5	9	28	14
Other	11	–	5	2	5
Don't know, no answer	9	3	20	6	3

Source: OPOR, 1941; Cantril, p. 870; FOR, 1942, *Public Opinion Quarterly* (Spring, 1943), p. 165; AIPO, 1946, *ibid.* (Winter, 1946–1947), p. 608; *Time*-Roper, 1948, *ibid.* (Fall, 1948), p. 533; AIPO, 1949, *ibid.* (Winter, 1949–1950), p. 711.

[a]All responses converted into 100 percent of total responses.

[b]Actual question: "If you were living under a dictatorship instead of under our present form of government, in what way do you think you, personally, would be affected?" (OPOR)

[c]Actual question: "Which of the following [list given, all four belief systems used] would you be the least willing to give up?" (FOR, youth study)

[d]Actual question: "What would you say is the greatest advantage of our type of government?" (AIPO)

[e]Actual question: "Which of the following rights [list given, religion not used] is the most important?" (*Time*-Roper)

[f]Actual question: "If a person who is not a citizen of the United States were to ask you what is the one best thing about being an American citizen, what would you say is the greatest advantage of our type of government? (AIPO)

begin to emerge as important,[142] Table IV.41 shows an attempt to organize responses to five relatively open-ended questions regarding political values (unfortunately they were all asked in the 1940's but they were the only ones available for this type of analysis) under the four belief categories. Of course, this type of categorization is subjective and the use of different

[142] This may suggest that these beliefs are relatively deep within the individual's reality world.

response situations further confounds the procedure's usefulness. However, the table presents a defensible classification of the political values of the American populace. The interesting aspect of Table IV.41 is that such a great percentage of the replies could be placed under one of the four belief categories noted above. In the least descriptive situation only a maximum of eleven percent had to be placed under the residual "other" category. I do not mean to imply that the four beliefs mentioned represent 89 percent of the beliefs of Americans. Rather the table should be interpreted as simply suggesting that a great percentage of American political beliefs can be subsumed under these four categories.

LIBERTY

Liberty is preeminent among the beliefs of the tradition. Indeed Locke's conception of man and government are based upon this value.

> To understand political power aright, and derive it from its original, we must consider, what state all men are naturally in, and that is, a state of perfect freedom to order their actions, and dispose of their possessions and persons, as they think fit, within the bounds of the law of nature, without asking leave, or depending upon the will of any other man.[143]

Locke does not say, however, that man has complete liberty. As the quotation makes clear man is not totally free even in the state of nature. In the state of nature he still must act "within the bounds of the law of nature." When writing about the state of nature, he notes, "But though this be a state of liberty, yet it is not a state of license . . ." [144] (e.g., even in nature man is not free to enslave himself by contract [145]).

When he enters civil society man's liberty is even more limited:

> He is to part also with as much of his natural liberty, in providing for himself, as the good, prosperity, and safety of the

143 *Second Treatise,* sec. 4, p. 4; also see sec. 95.
144 *Ibid.,* sec. 6, p. 5.
145 *Ibid.,* sec. 22, p. 15.

> society shall require, which is not only necessary but just, since the other members of the society do the like.[146]

On the other hand, man only gives up so much of his liberty to the authorities as is necessary to preserve his life and his property.[147] He agrees to transfer a limited number of powers to the authorities only to escape the arbitrary force existing in the state of nature.[148] In civil society, liberty means the subjection to just rules, freedom from the arbitrary extinction of the rules and provisions of the compact by the authorities, and autonomy for the individual from the authorities within the rules.[149]

The Federalist Papers also focus on liberty. The very first one gives liberty as the primary reason why a person should vote to adopt the Constitution.[150] Although it is true that the framers of that document were primarily interested in instituting a strong government of rules to "break and control the violence of faction," [151] it is also true that Madison supported a large area of individual autonomy from the state as a central value. Government has a legitimate sphere within which it can operate but this is conceived to be limited. The necessity for order cannot justify governmental elimination of man's freedom. Order-disturbing factions are caused by individual liberty, but allowing government to eliminate liberty creates order at too high a price.

> It could never be more truly said than of the . . . remedy that it was worse than the disease. Liberty is to faction what air is to fire, an aliment without which it instantly expires. But it could not be a less folly to abolish liberty, which is essential to the political life, because it nourishes faction than it would be to

[146] *Ibid.*, sec. 130, p. 75. Indeed, the whole *Second Treatise* can be interpreted as an attack upon Hobbes' unrestricted freedom, as the *First Treatise* was against Fulmer's contention that man is not free; see Richard Ashcraft, "Locke's State of Nature," *American Political Science Review* (September, 1968), p. 903.

[147] *Second Treatise,* sec. 131, p. 76.

[148] *Ibid.*, sec. 57., p. 33.

[149] *Ibid.*, sec. 22, p. 15; sec. 202, p. 117.

[150] *The Federalist Papers,* No. 1, p. 36.

[151] *Ibid.*, No. 10, p. 77.

wish the annihilation of air, which is essential to animal life, because it imparts to fire its destructive agency.[152]

Liberty is as essential to a liberal regime as air is to life in Madison's value system. To use government to suppress faction would be to eliminate the essential end of the polity – liberty. This emphasis is made through *The Federalist Papers*.[153] In a liberal society faction must be regulated through rules and limited government – not through abolition of individual prerogatives within the rules.

Liberty is, in sum, for both Locke and Madison the *sine qua non* of the tradition. If the liberal tradition is the political culture of the United States, the value of liberty should be strongly supported. If the hypothesis is to be sustained, Americans should believe that man should be relatively free to order his actions independent of the state – within the rules and agreements of the compact.[154]

Most of the writers on American characteristics (Figure IV.2) mention liberty (or freedom, or individualism in the sense of autonomy from government) as a belief of the people. Table IV.41 showed that when people are asked to identify the benefits of the American type of government without being given structured response categories, most choose liberty as the most important value. Hyman and Sheatsley note that about four out of five Americans define democracy either by the idea of freedom or the concept of popular rule.[155] Specific attitudinal data are provided by two polls over a decade apart. In 1948 a sample of Americans was asked whether they would prefer a government whose main purpose was to provide freedom or one whose purpose was to give economic security. As Table IV.42 shows, more than eight out of ten chose a government which provided liberty. In the same year in Germany 60 percent desired economic security rather than political freedom.[156] The 1960 Almond and Verba study ap-

152 *Ibid.*, p. 78.

153 *Ibid.*, Nos. 37–57; esp. 37, 39, 47, 51, 57.

154 Support will be studied assuming commitment to rules. See the data on the rule system above.

155 Hyman and Sheatsley, p. 41.

156 *Public Opinion Quarterly* (Fall, 1948), p. 535. Of course, this survey was taken immediately after World War II.

TABLE IV.42 *Belief in freedom vs. economic security*

Question: If you had to choose between these two types of government, which one would you choose: a government whose main purpose is to provide the people with economic security – that is the possibility of a steady income; a government whose main purpose is to insure free elections, freedom of speech, press and religion? (Time, 1948)

Economic security	12%
Freedom	83
Qualified answers	2
No opinion	3

Source: *Public Opinion Quarterly* (Fall, 1948), p. 535.

proached the question of liberty from another perspective. Respondents were asked whether they believed that individuals owed their first duty to the state or whether they owed their first duty to themselves. Over two-thirds of those Americans questioned chose the more Lockean response (see Table IV.43). Sixty-eight percent believe with Locke that the individual is more important than the state. Of course, both are accepted values, but the individual value is more widely supported. This should be compared with the data of the other four countries studied. Commitment to individual liberty seems relatively widely supported by Americans – by eight out of ten when compared to economic security and by almost seven out of ten when compared to duty to the state. The hypothesis can be said to be met for the value of liberty.

TABLE IV.43 *Individual freedom vs. duty to the state*

Question: The individual owes his first duty to the state and only secondarily to his personal welfare. (A V, 1960)

	U.S.	*U.K.*	*Germany*	*Italy*	*Mexico*
Agree	25%	38%	41%	48%	92%
Disagree	68	55	45	32	5
Don't know, etc.	7	7	14	20	3

Source: ICPR.

TABLE IV.44 *Government press censorship*

	U.S. (AIPO, 1937)[a]	*U.S. (FOR, 1939)*[b]	*Great Britain (BIPO, 1938)*[c]	*France (FIPO, 1945)*[d]
Allow censorship	11%	16%	16%	22%
Not allow censorship	82	76	70	66
No opinion	7	9	14	12

Source: Cantril, pp. 416, 418.
[a]Actual question: "Do you think the government should tell the newspapers what news to print and what not to print?"
[b]Actual question: "Do you think the government should or should not establish a bureau to supervise what should be printed in newspapers and magazines?"
[c]Actual question: "Do you think that the state should impose a political censorship of the British press?"
[d]Actual question: "Do you favor the censoring of political news?"

Freedom of the Press. Americans also appear to support specific liberty beliefs. Most seem to believe in the principle of freedom of the press at least when defined as not permitting government control of the news. Table IV.44 shows popular attitudes toward government *censorship* in three major western democracies. All are rather strongly opposed to government censorship of political news.[157] This seems especially true for the United States. Government *ownership* of mass media is also rather strongly opposed (Table IV.45). Only about 5 percent of the American population apparently favored government ownership when the question was asked in 1939. Table IV.45 also shows that almost three-fourths of the people feel that newspapers, and two-thirds that radio, should be "left alone" by the government.[158] Ownership and censorship, however, may be seen as rather drastic examples of limits on press freedom. Do the people actually support freedom of the press to criticize the authorities? Table IV.46

157 The figures for France are, no doubt, somewhat lower because the question was asked just after the war. Perhaps the decline shown for the United States also results from similar influences.

158 Radio seems generally less subject to popular support pressures for liberty. Whereas over three-fourths of one *Fortune* sample believed that a censorship bureau should not be established for newspapers, only 51 percent of the same sample were opposed to the establishment of a similar bureau for radio; see FOR, 1939; in Cantril, p. 715. Also see AIPO, 1938; FOR, 1938; in Cantril, p. 714.

TABLE IV.45 *Government media ownership and control*

Question: Do you believe the government should own or more closely control the newspapers (radio)? (FOR)

	Radio 1938	*Newspapers 1939*	*Radio 1939*
Own	–	4%	5%
More closely control	17	10	14
Leave alone	66	74	69
Don't know	17	13	12

Source: Cantril, pp. 417, 714.

TABLE IV.46 *Press allowed to criticize government?*

Question: In peacetime do you think newspapers should be allowed to criticize our form of government?

	1943	*1946*	*1948*
Yes	66%	64%	70%
No	30	31	22
Undecided	4	5	3

Source: *Public Opinion Quarterly* (Summer, 1946), p. 248; *ibid.* (Winter, 1948–1949), p. 765.

indicates that about two-thirds of the people would actually allow criticism of the form of government. (However, "form of government" is probably more applicable to the authority than the regime level.)

The tradition also suggests, however, that Americans should not believe in absolute freedom of the press. As Chapter VI will indicate, Americans have a hierarchy of limits on this freedom. As freedom of the press is made more absolute, larger and larger minorities refuse to support it. As Table IV.46 indicated, a little less than one-third of the people opposed press freedom to criticize. This minority increases to 39 percent when the question asks if newspapers should be allowed to "print anything they choose except libelous matter," [159] and increases to 42 percent when asked if the press can say "anything it pleases" about public officials.[160] Finally it should be

159 FOR, 1937; in Cantril, p. 416.
160 AIPO, 1936; in *ibid.*

TABLE IV.47 *Do you believe in free speech?*

	1938 (OPOR)	*1938 (AIPO)*	*1940 (AIPO)*
Believe	96%	95%	97%
Do not believe	3	3	1
No opinion	1	2	2

Source: Cantril, pp. 244–245.

noted that Americans do not believe that censorship actually takes place. Seventy-four percent, in one survey, indicated that the press has been able to print anything or almost anything they wished. Even among those who thought that some limitations existed, the majority believed that controls were exercised by nongovernmental means. Only 10 percent believed that politicians or the government prevented the free flow of printed opinion.[161]

Freedom of Speech. This more personal of liberties also appears to be highly supported in the United States. Table IV.47 shows that support for the value of free speech is greater than for any value reported in this study so far. More than ninety-five persons in a hundred believe in free speech. Yet, as the tradition would lead one to expect, even this strong commitment is limited. Table IV.48 shows that, although in most cases reported pluralities still favor free speech, the number who support it in an unlimited sense is drastically reduced. This is especially evident if one considers that the wording is more absolute in the 1940, 1941, and 1949 surveys. In these samples, almost as many do not believe in unlimited free speech as believe in it. Certainly the belief is not at a level here which could be considered political culture. (In Chapter VI this will be explored further.) However, general support for the value is unquestionably very widespread. Yet it is supported as a limited type of liberty. Both the support and its limitations are consistent with the tradition. It values liberty but not "license." [162]

161 FOR, 1939; in *ibid.*

162 In 1969, 75 percent in one survey thought that there should be a law against trying to disrupt a speech; see *Gallup Opinion Index* (May, 1969), p. 16.

TABLE IV.48 *Unlimited free speech?*

	1940 (FOR)[a]	*1941 (OPOR)*	*1943 (NORC)*[b]	*1946 (NORC)*	*1949 (AIPO)*[c]	*1953 (NORC)*
Yes	49%	46%	63%	64%	50%	53%
No	44	52	34	32	45	45
No opinion	7	2	3	4	5	2

Source: *Public Opinion Quarterly* (June, 1940), p. 350; data for January, 1941, from Cantril, p. 245; data for 1943 and 1946 from *Opinion News* (August, 1946); *Public Opinion Quarterly* (Winter, 1949-1950), p. 726; Herbert H. Hyman and Paul B. Sheatsley, "Trends in Public Opinion on Civil Liberties," *Journal of Social Issues* (1953), p. 15.

[a]Actual question for 1940 and 1941: "Do you think that in America anybody should be allowed to speak on any subject he wants to, or do you think there are times when free speech should be prohibited or certain subjects or speakers prohibited?"

[b]Actual question for 1943, 1946, 1953: "In peacetime, do you think people in this country should be allowed to say anything they want to in a public speech?"

[c]Actual question: "Do you believe in freedom of speech for everyone – that is, for example, permitting anyone to say anything at any time about our government or our country?"

Other Liberty Values. Americans seem to believe in liberty either when considered as a universal value or when looking at the two specific freedoms discussed above. However, freedom of the press and speech are only two aspects of the idea of liberty. In his in-depth psychological study of the common man Lane suggests that the liberties discussed here are mainly elite values.[163] Even speech which is at least within the reach of most Americans as a liberty is not used politically to any great degree by the average citizen. In summarizing findings on political speech, Lester Milbrath finds that only about 25 to 30 percent talk to others to convince them how to vote.[164] Lane found that freedom of religion, movement, consumption, and job choice were the preeminent concerns of the common man.[165] Religion, property use, and job choice will be considered below. Unfortunately, there does not seem to be much attitudinal data on Lane's other important values and this is especially true of support for the value of migration. Perhaps

163 Lane, pp. 20–25.

164 Lester Milbrath, *Political Participation* (Chicago: Rand McNally, 1965), p. 19.

165 Lane, pp. 22–24.

it is expressed in the generalized response to questions on values (such as reported in Table IV.41) in responses like "I can go and do what I please." If so, the 1949 poll reported above would indicate that about 10 percent would feel that it was a very important value. Certainly, it would be of some importance for many others.[166]

One further belief should be mentioned under liberty. This is the commitment of Americans to education. Education is a central value to the tradition because of the importance attached to reason. Locke mentions that man cannot be free if he is not able to use reason.[167] Children, the insane, and others incapable of reason must always be subject to others and, therefore, are incapable of giving their consent to government and thus cannot be members of civil society.[168] Madison was also a product of the Enlightenment, though less sanguine than others about education's worth. Yet Madison believed that it is essentially the knowledge and spirit of the people which, in the long run, will determine whether liberty survives.[169]

In order to make the connection between liberty and education clearer, it is necessary to depart somewhat from the exclusive use of our two sources to represent the tradition. Madison, in a letter written in 1822, best expresses what I believe represents the central position the tradition gives to education as a value and its importance in a liberal regime.

> A popular government, without popular information, or the means of acquiring it, is but a prologue to a farce or a tragedy; or, perhaps, both. Knowledge will forever govern ignorance; and a people who mean to be their own governors must arm themselves with the power which knowledge gives.[170]

But why is education assigned to the liberty subsystem of belief? An argument could easily be made that education relates to equality or even the achievement value of property. In my opinion, however, the tradition is clear on the primary

166 Migration is another value mentioned in the tradition; see Locke, *Second Treatise,* secs. 121, 173, 115.

167 *Second Treatise,* sec. 57, p. 33.

168 *Ibid.,* secs. 170–174.

169 *The Federalist Papers,* No. 57, p. 353.

170 *Letters and Other Writings of James Madison* (New York: published by order of Congress, R. Worthington, 1884), 3: 276.

relationship with liberty. Madison, in the same letter, further notes:

> Learned institutions ought to be favorite objects with every free people. They throw that light over the public mind which is the best security against crafty and dangerous encroachments on the public liberty.[171]

And again: "What spectacle can be more edifying or more seasonable than that of liberty and learning, each leaning on the other for their mutual and surest support?" [172]

The attachment of Americans to education is easily documented. In 1967, 71 percent of a national survey thought that schools do a good job.[173] Table IV.49 suggests that about eight out of ten people in the United States have a belief that more education is needed. Most believe that they have not had enough education (though they think it is too late to get it),[174] and most want their children to have more education than they themselves had. Even when the questions are phrased so that education is pictured as less than essential, people still support it. One question noted that Henry Ford had had little formal education and was still able to become a success. People were then asked if "education is absolutely necessary to be a success in life." In another people were asked if college was really worth the cost for those not entering the professions. In a third, a sample was told that some successful people believed that too much emphasis was placed upon high school and college education, and they were asked if they agreed.

171 *Ibid.*, p. 277.

172 *Ibid.*, p. 279. Further substantiation of this point is given by Madison's life-long collaborator, Thomas Jefferson; see Adrienne Koch, *Jefferson and Madison: The Great Collaboration* (New York: Oxford University Press, 1964). In a letter to Washington he stated: "It is but an axiom in my mind that our liberty can never be safe but in the hands of the people themselves, and that too of the people with a certain degree of instruction. This it is the business of the state to effect, and on a general plan;" see *The Papers of Thomas Jefferson,* Julian P. Bond, ed. (Princeton: Princeton University Press, 1954), 9:150. Also see Jefferson's *Notes on the State of Virginia.*

173 AIPO, 1967; in *Gallup Opinion Index* (August, 1967), p. 23. Of course, this is a poor operationalization. People could value education and not think the schools were doing a good job.

174 AIPO, 1944; in Cantril, p. 189.

TABLE IV.49 *Need more education?*

	Self need more (AIPO, 1938)[a]	*Self need more (AIPO, 1939)*[a]	*Child need more (AIPO, 1939)*[b]	*Son need more (FOR, 1945)*[c]
Yes	76%	81%	87%	81%
No	24	19	7	15[d]
No opinion	–	–	6	3

Source: Cantril, pp. 178, 186.
[a]Actual question: "Do you feel you have had enough education?"
[b]Actual question: "Do you want your children to have more education than you did?"
[c]Actual question: "After the war, if you had a son graduating from high school, would you prefer that he go on to college, or would you rather have him do something else, or wouldn't you care one way or the other?"
[d]Do something other than college = 5 percent; wouldn't care = 10 percent.

In all three cases almost six out of ten still chose education as an important value.[175]

In summary, the preeminent value of the liberal tradition seems highly accepted by the American people. The hypothesis expected rather strong commitment to liberty and each measure of it that has been used tends to support the expectation. Individual freedom, freedom of the press, freedom of speech, and education are all found to be culture beliefs are consistent with the views of Locke and Madison. Even the limitations on these freedoms suggested by the tradition are somewhat evident in the data. The expectations of the tradition hypothesis appear to be met for the value of liberty. From a comparative perspective, one may see certain themes relating to value conflicts within nations. One of these important themes is the value conflict between liberty and coercion.[176] In terms of the mass beliefs considered here, Americans seem to be much closer to the liberty end of the continuum than to the coercion end. From the perspective of the hypothesis, liberty seems supported as one of the values of the Lockean political culture.

175 AIPO, 1943; in *ibid.*, p. 179; AIPO, 1936; in *ibid.*, p. 85; AIPO, 1940; in *ibid.*

176 Lucian W. Pye, "Introduction," in Lucian W. Pye and Sidney Verba, eds., *Political Culture and Political Development* (Princeton: Princeton University Press, 1965), p. 22.

EQUALITY

The second value to be considered part of the belief system is equality. According to Locke, in the state of nature there exists "a state also of equality, wherein all the power and jurisdiction is reciprocal, no one having more than another. . . ." [177] Yet the English philosopher emphasizes that this equality is not absolute even in the state of nature.

> Though I have said above . . . *that all men by nature are equal,* I cannot be supposed to understand all sorts of *equality:* Age or virtue may give men a just precedency. Excellency of parts and merit may place others above the common level. Birth may subject some, and alliance or benefits others, to pay an observance to those to whom Nature, gratitude, or other respects, may have made it due; and yet all this consists with the equality . . . I there spoke of as proper to the business in hand, being that equal right that every man hath to his natural freedom. . . .[178]

In civil society, equality is of a different but still restricted type. The distinctive act upon entering civil society is the act of consent to general rules. All are equal in this act. All, thus, are equally subject to the same laws and all have the same rights under the rules.[179] In the liberal tradition everyone has the right to political equality, i.e., to be treated equally by government.

To Madison, equality of men before the law is central to republican government. "It is *essential* to such a government that it will be derived from the great body of the society, not from an inconsiderable proportion or a favored class of it. . . ." [180] To him it is the House of Representatives which best expresses the equality of the proposed new Constitution:

> Who are to be the electors of the federal representatives? Not the rich, more than the poor; not the learned, more than the ignorant; not the haughty heirs of distinguished names, more than the humble sons of obscure and unpropitious fortune. The

177 *Second Treatise,* sec. 4, p. 4.
178 *Ibid.,* sec. 54, pp. 31–32.
179 *Ibid.,* sec. 59, p. 34.
180 *The Federalist Papers,* No. 39, p. 241.

> electors are to be the great body of the people of the United States. They are to be the same who exercise the right in every State of electing the corresponding branch of the legislature of the State.
>
> Who are to be the objects of popular choice? Every citizen whose merit may recommend him to the esteem and confidence of his country. No qualification of wealth, of birth, of religious faith, or of civil profession is permitted to fetter the judgment or disappoint the inclination of the people. [181]

Yet Madison, as a Lockean liberal, does not favor social or economic equality but only a political equality. He believes that for liberty to exist one must accept inequality in property,[182] but no republican regime could exist without political equality.

Is there any evidence that Americans believe in political equality? Although data is limited regarding the value itself (though conflicts are well documented in Chapter VI), there is indication of support. A poll in 1958 showed that three-quarter believed that the Declaration was right in saying that all were equal.[183] Another study – although done only in a single neighborhood – gives a further indication of what I believe is the general feeling of Americans regarding the abstract ideal of equality: 98 percent of those questioned believed that all people should be treated as equals in the eyes of the law; 98 percent believed that everyone should have an equal opportunity to get ahead; 91 percent believed that all should have an equal right to hold public office.[184] This is certainly an impressive unanimity.

Apparently, then, Americans believe in equality – but do they believe in the restricted political equality of the tradition, or do they value a full and complete social equality? [185] In his depth-psychological study, Lane found that the American common man fears equality. He fears that equality where men are occupationally, socially, and economically equal. Although

181 *Ibid.,* No. 57, p. 351.

182 *Ibid.,* No. 10, p. 78.

183 Hazel Erskine, "The Polls," *Public Opinion Quarterly* (Spring, 1969), p. 149. Also see Table V.8.

184 Frank R. Westie, "The American Dilemma: An Empirical Test," *American Sociological Review* (August, 1965), pp. 531–532.

185 Boorstin, p. 176.

Lane does not investigate political equality of the Lockean variety per se, his findings do suggest that Americans believe in this kind of equality.[186] In the United States, belief in equality seems to center around belief in equality *before the law* and, in general, equality in any dealings with government. However, economic equality and social equality are rejected. The Lane study suggests that the American common man supports a Lockean political equality but not full social or economic equality.

Table IV.50 suggests that Americans believe that they will be treated equally before the law. Expectations are also shown for four other countries. The United States and the United Kingdom both have essentially the same expectations of equal treatment. In both more than eight out of ten nationals believe that they will get equal treatment. Perceptions of equal treatment are much lower for the other countries studied. In the American political culture the belief in equality seems reflected in an expectation that members actually will be treated equally. It is clear that such a culture should reject any legal means to create a privileged aristocratic class. Table IV.51 suggests that almost all reject titles of nobility for the United States. Attitudes of political equality should also be noticeable when people view symbols of aristocratic inequality. When the British king and queen visited in 1939, although most Americans approved of the visit, most rejected the attendent symbols of inequality. Two-thirds of the women said they would shake hands rather than curtsey and seven out of ten men said they would shake hands rather than bow.[187]

All of these empirical observations suggest that the limited view of equality expressed by the tradition is the type of equality which actually is in the American political culture. Data presented below on property and achievement beliefs will further substantiate this conclusion. Religion beliefs will also explain a type of equality held by Americans at the level of morality. For the present, it will be maintained that a "fear of equality" exists when equality is defined as full economic and social equality. Equality in the United States (other than the

186 Lane, ch. 4 and p. 61. Also see Stanislaw Ossowski, *Class Structure in the Social Consciousness* (New York: Free Press, 1963), p. 107.

187 AIPO, 1939; in Cantril, p. 391.

TABLE IV.50 *Equality before the law*[a]

	U.S.		*U.K.*		*Germany*		*Italy*		*Mexico*	
	Bureaucracy	*Police*	*Bureaucracy*	*Police*	*Bureaucracy*	*Police*	*Bureaucracy*	*Police*	*Bureaucracy*	*Police*
Expect equal treatment	83%	85%	83%	89%	65%	72%	53%	56%	42%	32%
Do not expect equal treatment	9	8	7	6	9	5	13	10	50	57
Depends and other	4	5	6	4	19	15	23	21	5	5
Don't know	4	2	2	0	7	8	11	13	3	5

Source: Almond and Verba, p. 70.

[a]Actual questions: "Suppose there were some questions that you had to take to a government office – for example, a tax or housing regulation. Do you think you would be given equal treatment – I mean, would you be treated as well as anyone else?" "If you had some trouble with the police – a traffic violation maybe, or were accused of a minor offense – do you think you would be given equal treatment? That is, would you be treated as well as anyone else?" (1960)

TABLE IV.51 *Create an American aristocracy?*

Question: Do you think it would be a good idea if we had titles like Lord, Duke and Sir in this country the way they have in England? (OPOR, 1942)

Yes	0%
No	98
DK	2

Source: Cantril, p. 868.

moral equality of religion) is limited to political equality. Americans believe in equal political rights as they relate to the major political rule institutions of the polity, viz., the right to vote, the right to representation, and the right to be heard by officials. They also believe in equal treatment before the law, equal opportunity, and equal access to the political world – without regard to class considerations.

Tocqueville is often interpreted as saying that he found total social equality in the United States in the early ninteenth century. However, Tocqueville's social equality merely means that no hereditary class exists.[188] He specifically states that he does "not mean that there is any deficiency of wealthy individuals in the United States." [189] Indeed, he says that he knows of no country "where a profounder contempt is expressed for the theory of the permanent equality of property." [190] Coming from an aristocratic background he was naturally most impressed with the lack of aristocracy. Americans in the twentieth century tend to see equality from the same perspective. They believe in political equality but not in economic nor total social equality. They simply do not believe in hereditary, closed classes and the attendant forms of servility which follow from a rigid class society.

Though the poll data regarding American support for equality is limited, the data which is available does suggest a commitment to equality. The national poll data and the intensive study by Lane seem to support the hypothesis conten-

[188] Tocqueville, p. 37.
[189] *Ibid.*, p. 36.
[190] *Ibid.*

tion that political equality has cultural support within the United States.[191]

PROPERTY

The third generally recognized belief of the Lockean liberal tradition is the value of property. Certainly this is a central value for Locke. Indeed, it is the reason why civil society is created:

> The great and chief end therefore, of man's uniting into commonwealths, and putting themselves under government, is the preservation of their property; to which in the state of nature there are many things wanting.[192]

In the state of nature all men are fully free and morally equal. However, every individual's private property is "very unsafe, very insecure. This makes him willing to quit this condition which, however free, is full of fears and continual dangers." But what does Locke mean by private property? Some have restricted the definition to land only. Locke, however, is clear that property is more than land: "By property I must be understood here, as in other places, to mean that property which men have in their persons as well as goods." [193] Property is everything that relates to the individual. Locke's view of property also has a special relationship to the person's work and achievement. In the state of nature the rights of property are directly related to the effort expended by the individual:

> Every man has a *property* in his own *person.* This nobody has any right to but himself. The labor of his body and the *work* of his hands, we may say are properly his. Whatsoever, then he removes out of the state that nature hath provided and left it in, he hath mixed his labor with it, and joined to it something that is his own, and thereby makes it his property.[194]

Property, then, is everything that makes man a human being. His work, his job, his personal and real property – everything he has control over is his property in the state of nature. But

[191] Of course, belief in equality does not mean Americans will necessarily support it in specific situations. This problem is considered in Chapter VI.

[192] *Second Treatise,* sec. 124, p. 3; also see sec. 3.

[193] *Ibid.,* sec. 123, p. 173; sec. 173, p. 102.

[194] *Ibid.,* sec. 27, p. 17.

the chaotic condition in nature is too unsafe for man's property and he chooses to contract for civil society. In civil society, however, he does not give up his right to property. No one may lawfully take his property. Government serves to protect personal property from confiscation by others – including government itself.[195] Because of the central position given to this value, for the hypothesis to be sustained there must be substantial support for this value in the United States.

The Federalist Papers are so arranged that they emphasize the preservation of property (i.e., "prosperity") as the main justification for the new Constitution. All the early chapters focus attention on this value. After the introduction, the next twelve papers discuss ways the Constitution will help protect Americans from threats to property. Madison certainly is in the tradition of Locke with his justification for the institution of government. Government exists to protect the individual from the anarchy of the state of nature. Indeed, the creation of the American government itself is justified because the conditions in the American states could easily lead to anarchy "as in a state of nature." [196]

In one sense the property which must be protected in civil society is a source of dissension for that society, Madison maintains that property is "the most common and durable source of faction." [197] On the other hand the unequal distribution of property, though the cause of faction conflict, is necessary if man is not to be pressed into an illiberal uniformity.

> The diversity in the faculties of men, from which the rights of property originate, is not less an unsuperable obstacle to a uniformity of interests. The protection of these faculties is the first object of government. From the protection of different and unequal faculties of acquiring property, the possession of different degrees and kinds of property immediately results.[198]

Property is thus necessary because its protection insures that individual liberty and achievement possibilities survive. In-

[195] *Ibid.*, Note, then, that the property value is related to the liberty value.

[196] *The Federalist Papers*, No. 1, p. 36; and No. 2, p. 37; No. 52, pp. 324–325.

[197] *Ibid.*, No. 10, p. 79.

[198] *Ibid.*, p. 78.

deed Madison states that the protection of individual faculties is the first object of government. Property is a basic liberal value because its protection allows the individual to be free and secure.

Can the liberal tradition hypothesis be sustained for the property value? After the impact of the Great Depression of the 1930's there certainly could be some doubt. In order to investigate this problem two analytical categories will be used. The first describes attitudes toward government ownership of property and the second considers property in its relationship to limitations on individual achievement faculties.

There is probably little dispute over the amount of support for the value of *personal* property. However, another belief is more questionable. How much support does the value of private ownership of the *means of production* have in the United states?[199] To study support for the property value this study will focus upon a potentially less supported attitude regarding property — the attitude toward private ownership and control of business. Table IV.52 provides an attempt to measure support for this value. The question used there is especially valuable because it deals with a situation with which the individual has immediate interest, personal knowledge, and direct concern. The respondents were asked whether they would prefer governmental or private control of the business they worked for. More than eight out of ten supported the private property value. Other attitudinal data also show support for private ownership of production. Even as measured during the late stages of the depression, majorities would prefer to work for private concerns.[200] Majorities also support private business as generally more efficient than government.[201]

199 An interesting empirical study of the essentially one-industry city of Akron, Ohio, analyzed this very question a few months after the Goodyear rubber strike riot in 1937 (the "sit down" strike was first used here). Even in this type of setting there was high support for personal property: 76 percent of the scores on a property value scale were moderately favorable. Support for large corporate property was even moderately high, although corporate property limitations were accepted "in the interest of human welfare." See Alfred Winslow Jones, *Life, Liberty and Property* (Philadelphia: Lippincott, 1941), esp. pp. 322–335, 378, and table 8, p. 383. Also see Chapter VI.

200 FOR, 1940; in Cantril, p. 345.

201 FOR, 1946; in *ibid.,* p. 351. Belief on governmental and private control of production is somewhat complicated because of a wording diffi-

TABLE IV.52 *Governmental vs. private ownership*

Question: Do you think that you would be better off if the concern you worked for were taken over and operated by the federal government, or would you prefer to leave it under the present business management? (AIPO, 1941)

Prefer government operation	7%
Prefer present business management	81
No opinion	12

Source: *Public Opinion Quarterly* (Fall, 1941), p. 495.

The rule value of the Constitution prescribed that certain functions should be performed by government. It is not expected, therefore, that Americans would necessarily believe in private ownership of constitutionally mandated business institutions. Certain other functions are undoubtedly marginally constitutionally prescribed and would probably also be accepted as government owned. The hypothesis merely maintains that for the overwhelming majority of productive situations, Americans should believe in private ownership and control of property. Table IV.53 shows that in the late 1930's the public supported government operation of postal services. This constitutionally prescribed business is not necessarily bound to the value tenets of property because of the legitimization given to it by the specific assignment of postal function to the government by the Constitution. As such it is a unique production institution within the American system.[202]

culty in a question asked regarding government ownership. Instead of using the word *government,* the question asked whether there should be "public" or private ownership. The word *public* is misleading and suggests such differences as, for example, private as opposed to public ownership of voting stock in a corporation. The use of *government,* or *federal government,* is much more specific and easily comprehended. This can be demonstrated, to some degree, by the incompatibility of the results obtained by the use of *public* in a survey taken in July, 1936, and the word *government* in one taken only four months later. The use of *public* is also inconsistent with results of all other polls taken on this question. Fortunately, the word was rarely used by poll takers; see Cantril, p. 694, questions 1–3. Because of this confusion, the July, 1936, poll is disregarded, as is any other that uses the term *public.*

202 However, 41 percent would favor setting up the "Federal Post Office Department as an independent agency not directly operated by the Federal Government," 35 percent opposed so doing, and 24 percent had no opinion; see *Gallup Opinion Index* (April, 1970), pp. 4–5.

TABLE IV.53 *Domain of government ownership: postal service*

Question: On the following things we have found that people differ as to the degree to which government should function. Do you think that the government should operate all, some, or none of the postal and parcel-post services? (FOR, 1939)

Government operate all	87%
Some	5
None	4
Don't know, depends	4

Source: Cantril, p. 345.

In the more usual productive situations, Americans believe in private ownership of property. Ten economic areas which have been investigated with public opinion data show little support for government ownership. Table IV.54 shows that in no case tested was there a majority in favor of governmental ownership or operation of private business. Only during the depression was there even any substantial support for public ownership of the means of production. The situation of the banks in that period shows the least support for the value. Even this case may overstate the support for governmental control. Note the 1942 sample taken by *Fortune*. This sample has the only question on banks which allows a choice between government regulation and government ownership. When allowed a choice between regulation and ownership, less than 10 percent seem to want outright government ownership. This was probably the situation throughout the depression; that is, little support for governmental ownership or control but substantial demand for governmental regulation of certain businesses. When the stress created by the depression and World War II recede,[203] even these levels of support for government ownership decrease to where, by 1946, two-thirds of the population believe that banking should be left in the private domain. In the other nine areas there is support for regulation and in some areas support for some governmental ownership (especially of some natural resources), but in general there is little support for governmental ownership of the means of

[203] See Chapter VI for a further example of depression-property interaction.

production. Support for total removal of government from the economy is usually high enough to be considered cultural. The more remote the effects of the depression, the greater this support. For all property areas Americans support private ownership, with some regulation in periods of stress.

More insights into American beliefs in the value of property can be seen in Table IV.55. This table presents more current poll data on the general question of governmental-private property relationships. Approximately twice as many persons would leave policy problems like electric power and housing to the private domain.[204] The question is unfortunately very broad and it could be interpreted in a much narrower, non-cultural sense but it does suggest that current general attitudes are similar to the more detailed opinions measured in the 1930's and 1940's shown in Table IV.54. If the large "no opinion" group is removed from Table IV.55, *two-thirds of those with an opinion favor private control.* A comparative perspective can also help explain the American commitment to private property. Great Britain and the United States have many common political culture characteristics.[205] Yet, as Table IV.56 clearly shows, the Lockean liberal tradition is much more evident in the United States. In fact, there is a completely reverse set of attitudes for the two countries on the property value. While few in Britain disapprove of government ownership, few in the United States approve of it.

The second aspect of property suggested by the tradition, and widely noted by observers of the American political culture, is the *emphasis upon achievement.*[206] Hartz states that

[204] It should be noted, however, that most drastic wording of the question can produce different results. If asked whether the government should "own *some* power plants," a plurality would agree that the government should. This is a much more extreme question (could the Atomic Energy Commission own its own plant? The Army? Dubuque?). The surprising point is that almost four out of ten would not agree to even this modest government involvement; see SRC, 1964; in ICPR.

[205] Gabriel A. Almond, "Comparative Political Systems," *Journal of Politics* (May, 1956), pp. 391–409; and Robert R. Alford, *Party and Society* (Chicago: Rand McNally, 1963), ch. 1.

[206] For example, see Almond, *The American People and Foreign Policy,* p. 39; John Gillin, "National and Regional Cultural Values in the United States," *Social Forces* (December, 1955), pp. 107–113; Fred I. Greenstein, "A New Light on Changing American Values," *Social Forces* (May, 1964), p. 449; Seymour Martin Lipset, "A Changing American Character?" in

TABLE IV.54 *Domain of private ownership*

Type of business	Percentage who believe government should: Own and operate (own all)[a] business	Regulate (own some)[a] business	Leave alone (own no)[a] business	Don't know; depends
1. Factories producing essentials				
FOR, 1939	7%	14%	70%	8%
2. Grocery stores				
FOR, 1942	2	25	62	11
3. Automobile companies				
FOR, 1942	3	24	59	14
FOR, February, 1943	10	23	56	11
FOR, April, 1943	9	18	57	17
4. Telephone and telegraph				
FOR, 1939[a]	15	14	58	13
FOR, 1942	7	32	47	14
5. Coal mines				
AIPO, 1936	27	–	64	9
AIPO, 1945	29	–	59	12
AIPO, 1947	33	–	61	6
AIPO, 1949	25	–	58	17
AIPO, 1953	15	–	73	12
6. Railroads				
FOR, 1939	14	18	49	19

FOR, 1942	11	37	39	14
AIPO, 1946	26	—	67	7
AIPO, 1949	22	—	60	18
AIPO, 1953	14	—	75	11
7. Banks				
AIPO, November, 1936	36	—	56	8
AIPO, December, 1936	37	—	49	14
AIPO, July, 1937	46	—	39	15
AIPO, August, 1937	40	—	42	18
FOR, 1942	8	57	25	10
AIPO, May, 1945	27	—	61	12
AIPO, September, 1945	25	—	64	11
AIPO, 1946	26	—	66	8
AIPO, 1949	22	—	63	15
AIPO, 1953	14	—	76	10
8. Electric utilities				
FOR, 1939	15	20	48	17
9. Meat packing				
FOR, 1933	9	30	46	15
10. Natural resources				
FOR, 1939[a]	21	34	33	12

Source: Since this table summarizes thirty separate tables, the questions have not been listed. See the following pages of Cantril for each question: (1) p. 345; (2) p. 349, (3) p. 349, (4) p. 695, (5) AIPO release of October 10, 1953, (6) pp. 728–729 and AIPO release, (7) p. 272 and AIPO release, (8) p. 695, (9) p. 349, (10) p. 345.

[a]The wording was slightly altered. For the exact wording, consult the source.

TABLE IV.55 *Government vs. private allocation*

Question: The government should leave things like electric power and housing for private businessmen to handle. (SRC)

	1956	*1958*	*1960*
Disagree	23%	23%	22%
Agree	42	43	48
DK, depends, etc.	36	34	30

Source: ICPR.

TABLE IV.56 *Government ownership in Britain and the United States*

Question: Do you think the government should own the railroads in this country? (AIPO, BIPO, 1945)

	Great Britain	*United States*
Approve	70%	23%
Disapprove	30	77

Source: *Public Opinion Quarterly* (Fall, 1945), p. 365.

the United States is a country "where virtually everyone, including the nascent industrial worker, has the mentality of an independent entepreneur." As Lane's detailed study documents (though data is only applicable to the working class), Americans do not want economic equality. They strongly support the value of achievement. As Lane's interviews show, the common man fears that social equality would result in a dreary, disorganized society without incentives and without economic progress.[207]

Poll data suggest that Americans place a great emphasis upon the importance of their jobs.[208] Indeed, it may be called

---

Seymour Martin Lipset and Leo Lowenthal, eds., *Culture and Social Character* (New York: Free Press, 1961), pp. 136–171; and Morris, pp. 45–61. The Morris comparative analysis of values suggests that achievement ("way" 12) ranks fourth for the United States, seventh for Norway, eighth for Nationalist China, eleventh for India, and thirteenth for Japan (for percent of first choices out of thirteen) among college students. Also see David C. McClelland, *The Achieving Society* (Princeton: Van Nostrand, 1961).

[207] Hartz, p. 89; Lane, ch. 4, and p. 78.

[208] Cantril, pp. 893–897.

TABLE IV.57 *Achievement and jobs in Britain and the United States*

Question: Which of the following statements do you most agree with? (A) The most important job for the government is to make it certain that there are good opportunities for each person to get ahead on his own. (B) The most important job for the government is to guarantee each person a decent and steady job.

	Great Britain (BIPO, 1947)	*United States (FOR, 1947)*	*United States (SRC, 1968)*[a]
A Achievement value	40%	50%	51%
B Job value	55	43	34
No opinion	5	7	15

Source: *Public Opinion Quarterly* (Fall, 1947), p. 478; and ICPR.
[a]In 1968, the wording was slightly altered. For exact wordings, consult the source.

the common man's most valued property. However, the value of achievement, which is also based on property, seems to be more widely valued than the guarantee of a job per se.[209] Table IV.57 shows that a majority of Americans support the achievement value over the job value. (It also shows that achievement seems to be a more important belief in the United States than in Great Britain.) Although the differences between the two property values are not large enough for a culture belief (because there is a value conflict), Table IV.57 does show the comparative strength of the achievement value.

A more direct test of the relative strength of the achievement value is provided in Table IV.58. Here achievement is directly compared to non-achievement. Six out of ten respondents would rather choose a job with prospects for advancement and little security than one which is secure with little opportunity. Support for the achievement value is cultural. Support can also be approached from another perspective. Data are available which show support for achievement-related values. Americans appear to value work and its rewards.[210]

[209] If achievement values are removed, people would as soon work for the government as private industry; see AIPO news release of August 30, 1947.

[210] Almost six out of ten believe that those who do not get ahead "just don't have enough will power"; SRC, 1956, in ICPR; *Gallup Opinion*

TABLE IV.58 *Achievement vs. steady job*

Question: If you had your choice, which would you prefer? (FOR, 1940)

A job that pays a high wage but with a 50–50 chance of getting promoted or fired	61%
A steady job earning just enough to get by on, but with no prospect of advancement	33
Depends, don't know	6

Source: *Public Opinion Quarterly* (June, 1940), p. 355. Also see Cantril for similar data in 1942 and 1943, p. 530.

In a 1945 survey, "training" was chosen as the most important value which people believed children should receive from school. In 1948, "hard work" received the most support from the sample as to what would get "a young person ahead the fastest." Again in 1964, Free and Cantril found that the three qualities which Americans thought would make a person successful were: good education and training (71 percent); initiative, effort, and hard work (66 percent): and character and will power (59 percent).[211] In all three surveys, items such as good luck, fine art appreciation, moral character, good citizenship, and knowing the right people were de-emphasized. Table IV.59 directly compares the values of ability and luck. Substantial majorities support the value more congenial to achievement. Consider another attitude related to achievement. If people thought that it was of no use to try, an important incentive to achieve would be removed. Table IV.60 shows that the overwhelming majority of Americans do not feel that it is of no use to try to get anywhere in life.

Index (July, 1967), p. 17. In 1939, over 80 percent believed the able on relief should work; AIPO, 1937; in Cantril, p. 897. In 1964, 79 percent believed that "we should rely more on individual initiative and not so much on good welfare programs" (Free and Cantril, p. 191).

211 Training: FOR, 1945; in Cantril, p. 186. Hard work: FOR, 1948; in *Public Opinion Quarterly* (Spring, 1949), p. 174. Free and Cantril, p. 193.

TABLE IV.59 *Ability vs. luck as determinant of success*[a]

	1939 (AIPO)[b]	*1942 (OPOR)*[c]	*1970 (AIPO)*[b]
Luck	15%	12%	8%
Ability	79	70	86
Pull	–	23	–
No opinion	6	5	6

Source: AIPO, 1939; in Cantril, p. 829; OPOR, 1942; in *ibid.*, p. 830; AIPO, 1970; in *Gallup Opinion Index* (May, 1970), p. 14.

[a]More than one response was given by some respondents.

[b]Actual question: "Do you think people who are successful get ahead largely because of their luck or largely because of their ability?"

[c]Actual question: "Do you think success is dependent mostly on luck, on ability, or on pull?"

TABLE IV.60 *Powerlessness and achievement*

Question: I have often had the feeling that it's no use to try to get anywhere in this life. (SRC, 1956)

Agree	11%
Disagree	87
Don't know, etc.	2

Source: ICPR.

The importance of the value of achievement is also very strongly shown in Table IV.61. Three different questions in different years showed that more than six out of ten would not limit the rewards of achievement. However, people are somewhat less supportive of the wealth which does not appear to be earned. In two Gallup polls, one in 1936 and one in 1937, only slight pluralities (rather than cultural majorities) believed that government should not limit large private fortunes.[212] Opinion was evenly divided on whether a person with a fortune of one million dollars should be able to keep it.[213] Consider the following mass public support for property. When asked if the government should take from those who have much and give to those who have little, one-third of the

212 AIPO, 1936, 1937; in Cantril, p. 1040, pp. 3 and 5.

213 FOR, 1935; in *ibid.*, p. 5, q.1.

TABLE IV.61 *Limit achievement rewards?*

	Limit inheritance (AIPO, 1937)[a]	*Confiscate above needs (FOR, 1939)*[b]	*Limit amount earned (FOR, 1940)*[c]
Yes	30%	16%	24%
No	62	76	70
Don't know	8	8	6

Source: AIPO, 1937; in Cantril, p. 1041; FOR, 1939; in *ibid.*, p. 1041; FOR, 1940; in *Public Opinion Quarterly* (June, 1940), p. 354.

[a]Actual question: "Should a limit be placed on the amount of money any person can inherit?"

[b]Actual question: "Do you think our government should confiscate all wealth over and above what people need to live on decently and use it for the public good?"

[c]Actual question: "Do you think there should be a law limiting the amount of money any individual is allowed to earn in a year?"

population agreed without qualifications. Yet another fifth would do it within some bounds and 41 percent would *not* agree to the principle of progressive taxation at all! [214] In summary, as suggested in Table IV.61, Americans overwhelmingly believe that a person should be able to keep the property to which he believes his achievement drive has entitled him.[215]

After considering the very large amount of poll data (much of it representing stressful periods) available on the value of property, one must be impressed with the commitment of the American people to it. Either from the perspective of private ownership of the means of production or the perspective of the value of achievement, this central measure appears to support the liberal tradition hypothesis. Cultural consensus exists on the belief in: private superiority in business management; in private ownership of factories, grocery stores, auto companies, coal mines, telephone and telegraph, utilities, meat packing companies, natural resources (to a lesser extent), railroads, and (though again to a more limted extent during the depression) banks; achievement as a value; and in not limiting achievement rewards. Support appears substantial enough to

[214] FOR, 1937; in *ibid.,* p. 1041, p. 7. Two years later a majority was opposed to redistributing wealth by "heavy taxes on the rich" (FOR, 1939; in *ibid.,* p. 8.

[215] V. O. Key, Jr., *Public Opinion and American Democracy* (New York: Knopf, 1961), p. 91.

accept the property value as a belief of the American political culture.

RELIGION

The value of religion is the final important belief value hypothesized as being in the American political culture. It is a belief which is not always thought of as a part of a Lockean liberal tradition,[216] but one which does appear central to both Locke and Madison. Locke is very clear in the *Second Treatise* on the religious base of man's dependence. Indeed, man is property in his relationship to God.

> For men being all the workmanship of one omnipotent, and infinitely wise maker; all the servants of one sovereign master, sent into the world by his order, and about his business; they are his property, whose workmanship they are. . . . [217]

In other sections of the *Treatise,* Locke emphasizes the tradition belief in God, especially his role as Creator. Locke's state of nature is sometimes thought to be a place where man is totally free. But according to Locke, man is still subject to law. "The natural liberty of man is to be free from any superior power on earth. . . ." [218] Man, as a religiously based individual, is always responsible to God's natural law.

From the moral equality of man before God, Locke deduces a natural-law precept demanding not only moral equality but also requiring religiously inspired altruism in man's relations

[216] Locke has been interpreted as having a deviousness which actually hides a hedonist philosophy rather than a religious one; see Leo Strauss, *Natural Right and History* (Chicago: University of Chicago Press, 1965), pp. 202–251, esp. pp. 249–251. I believe this rests upon a fundamental misunderstanding of Locke's theory, which is carefully analyzed by Richard Ashcraft, "Locke's State of Nature: Historical Fact or Moral Fiction?" *American Political Science Review* (September, 1968), pp. 898–915; also see Sterling Lamprecht, *The Moral and Political Philosophy of John Locke* (New York: Russell and Russell, 1914; reissued 1962). For present purposes, I do not, however, have to resolve the question. My hypothesis concerns the interpretation of Locke that has been believed by Americans. The "theory of deviousness" was not developed until the second part of the twentieth century. Presumably all were "fooled" until then. Also see John Dunn, *The Political Thought of John Locke* (Cambridge: At the University Press, 1969).

[217] *Second Treatise,* sec. 6, pp. 5–6.

[218] *Ibid.,* secs. 77, 241; sec. 22, p. 15.

with his fellow man. Being subject to God's law and being his property,

> and being furnished with like faculties, sharing all in one community of nature, there cannot be supposed any such subordination among us, that may authorize us to destroy one another, as if we were made for one another's uses, as the inferior creatures are for ours. Every one, as he is bound to preserve himself . . . ought he as much as he can to preserve the rest of mankind . . . and not . . . take away, or impair the life, or what tends to be the preservation of the life, the liberty, health, limb or goods of another.[219]

The value of religion, thus, has two aspects for the tradition. The first is a belief in and attachment to God and His precepts and the second a prescription for a moral responsibility to the welfare of the other members of the community. Both must be supported for the hypothesis to be accepted.

As has been noted, Locke represents the values and Madison more their application. Since Madison is less interested in values and since religion is only tangentally related to governmental application, one would not expect much mention of this value in *The Federalist Papers.* Religion is mentioned at a few points, however. In fact, at one point Madison practically attributes the adoption of the Constitution and the successes of the Revolution to the direct intervention of God. In Federalist No. 37, Madison reflects on the compromises which were necessary to frame the Constitution and on the number of conflicting interests which underlay them.

> The real wonder is that so many difficulties should have been surmounted, and surmounted with a unanimity almost as unprecedented as it must have been unexpected. It is impossible for any man of candor to reflect on this circumstance without partaking of the astonishment. It is impossible for the man of pious reflection not to perceive in it a finger of that Almighty hand which has been so frequently and signally extended to our relief in the critical stages of the revolution.[220]

At another point, Madison uses the importance of religious rights to highlight the importance of civil rights. Religious freedom is also used as one of the proofs that equality will

219 *Ibid.,* sec. 6, p. 6.
220 *The Federalist Papers,* No. 37, pp. 230–231.

reign under the new Constitution. Although religion was to be separate from government, it was viewed by Madison as an important value.[221]

An idea of the importance of this value to the American belief system can be obtained from Lane's in-depth study of the American common man. Lane found religion to be central to the common man's ideas of political values. "The concept that is most closely associated with the idea of freedom, mentioned first and most frequently, is the *freedom of religion.*" According to Tocqueville, this value has been important for some time. The early French observer noted, "There is no country in the world where the Christian religion retains a greater influence over the souls of men." [222] Today, the Census reports that 71 percent of the people in the United States identify themselves as Protestants, 23 percent as Catholics, and 3 percent as Jews. That is, 97 percent of the American people identify themselves with one of these three religions. About six out of ten actually are recorded by the churches as members, and about seven out of ten believe they belong to a church.[223] It appears that between 40 and 50 percent actually attend services in a normal week.[224]

> About 70 to 75 percent of the American people, it may be safely estimated, regard themselves as members of churches, another 20 to 25 percent locate themselves in one or another religious community without a consciousness of actual church membership — they constitute a "fringe of sympathetic bystanders" so to speak.[225]

221 *Ibid.*, No. 51, p. 324; No. 57, p. 351; *Letters and Other Writings of James Madison*, 1:14.

222 Lane, p. 22; Tocqueville, I:294.

223 Reported in Will Herberg, *Protestant, Catholic, Jew* (Garden City, N.Y.: Doubleday, 1960), pp. 65, 48–49; U.S. Bureau of the Census, *Statistical Abstract of the United States, 1970* (1970), pp. 41–42.

224 *Gallup Opinion Index* (January, 1967), p. 10; 1955–1966 data. For SRC data, see ICPR election codebook, 1952–1964. Also see Cantril, pp. 699–700. It should also be noted that attendance at church has declined over the period studied by public opinion polls. Between 1955 and 1969 attendance declined from 49 percent in an average week to 42 percent; see *Gallup Opinion Index* (January, 1970), p. 5.

225 Herberg, p. 49. Also see Bernard Lazerwitz, "Religion and Social Structure in the United States," in Louis Schneider, ed., *Religion, Culture and Society* (New York: Wiley, 1964), pp. 426–439; and W. Seward Salisbury, *Religion in American Culture* (Homewood, Ill.: Dorsey Press, 1964), esp. ch. 4.

TABLE IV.62 *Do you personally believe in God?*

	1944	*1948*	*1952*	*1954*	*1968*
Yes	96%	94%	97%	96%	98%
No	1	3	–	–	2
Undecided	3	3	–	–	0

Source: AIPO, 1944; in Cantril, p. 744; AIPO, 1948; in *Public Opinion Quarterly* (Spring, 1948), p. 173; *Catholic Digest* Poll, 1952; in Herberg, p. 91; AIPO, 1954; in *ibid*; AIPO, 1968; in "Special Report on Religion," *Gallup Opinion Index* (February, 1969), p. 15.

Table IV.62 presents evidence as to the commitment of Americans to a belief in a Supreme Being. As is shown, belief in God is virtually unanimous in the United States. It should not be assumed, however, that these figures would be found in any religious or Christian nation. Table IV.63 shows that belief can vary substantially between countries even within the same general religious tradition. Whereas more than nine out of ten believe in God in Brazil, Greece, Australia, Canada, and the United States, only about six or seven out of ten believe in France (basically because only 17 percent of the people who identify as Communists believe in God) and in Sweden.

As the table shows there is also a wide range of opinions on whether there is a life after death. In every case, substantially fewer people believe that there is a life after death than believe there is a God. Table IV.64 reviews the situation in the United States over a longer time period. Approximately seven out of ten believe there is a life after death.[226] The smaller support for specific qualifications of religious belief like this one suggest that American religious belief may not be especially orthodox.[227] For example, while in 1968 almost everyone believes in God, 85 percent believe in heaven (up from 72 percent in the 1950's); 65 percent in hell (from 58 percent), and

[226] Two other polls support this observation: FOR, 1942, poll of high school students; in Cantril, p. 311; FOR, 1948; in *Public Opinion Quarterly* (Spring, 1949), p. 172.

[227] Herberg, ch. 5; Yoshio Fukuyama, "The American Conscience," in Roger L. Shinn, ed., *The Search for Identity* (New York: Institute for Religious and Social Studies, Harper and Row, 1964), p. 174. On the depth of commitment: 27 percent feel deeply religious, 63 percent somewhat religious, 7 percent hardly religious at all, 1 percent not religious at all, and 2 percent not sure; see Harris, 1965; in *Polls* (Winter, 1967), p. 75.

TABLE IV.63 *Belief in God and life after death, fifteen countries*

	Do you believe in God?						*Do you believe in life after death?*					
	1948			*1968*			*1948*			*1968*		
	Yes	*No*	*Unc.*	*Yes*	*No*	*DK*	*Yes*	*No*	*Unc.*	*Yes*	*No*	*N.O.*
Brazil	96%	3%	1%	—	—	—	78%	18%	4%	—	—	—
Greece	—	—	—	96	2	2	—	—	—	57	28	15
Australia	95	5	—	—	—	—	63	20	17	—	—	—
Canada	95	2	3	—	—	—	78	9	13	—	—	—
United States	94	3	3	98	2	0	68	13	19	73	19	8
Austria	—	—	—	85	10	5	—	—	—	38	56	6
Switzerland	—	—	—	84	11	5	—	—	—	50	41	9
Norway	84	7	9	73	12	15	71	15	14	54	25	21
Finland	83	5	12	83	7	10	69	11	20	55	23	22
West Germany	—	—	—	81	10	9	—	—	—	41	45	14
Netherlands	80	14	6	79	13	8	68	26	6	50	35	15
Denmark	80	9	11	—	—	—	55	27	18	—	—	—
Great Britain	—	—	—	77	11	12	—	—	—	38	35	27
Sweden	80	8	12	60	26	14	49	17	34	38	47	15
France	66	20	14	73	21	6	58	22	20	35	53	12

Source: Gallup International Surveys, 1948; in *Public Opinion Quarterly* (Spring, 1948), pp. 154, 173; 1968 data from "Special Report on Religion," pp. 15, 18.

TABLE IV.64 *Do you believe in life after death?*

	1936	*1944*	*1948*	*1953*	*1965*	*1968*
Yes	64%	76%	68%	77%	72%	73%
No	36	33	13	7	28	19
DK	–	11	19	16	–	8

Source: AIPO, 1936 and 1944; in Cantril, p. 310; AIPO, 1948; in *Public Opinion Quarterly* (Spring, 1948), p. 173; Catholic Digest Poll, 1953; in Herberg, p. 91; Harris, 1965; in *Polls* (Winter, 1967), p. 75; AIPO, 1968; in "Special Report on Religion," p. 18.

60 percent in the devil.[228] During the depression a slightly higher percentage of people even said that greater economic security was more needed than an increase in religious commitment to help solve current problems, though that was undoubtedly strongly influenced by the depression crisis.[229] Herberg is impressed with the datum that 54 percent of those who said that religion was very important replied that the beliefs did not have any effect on their "ideas of politics and business." [230] Yet with these qualifications in mind, 89 percent of Americans believe in the Trinity, 80 percent believe that Christ is divine, 90 percent pray to God (56 percent frequently), 86 percent believe the Bible is divinely inspired, 72 percent want their children raised as church members, 98 percent want their children to receive religious instruction, and eight out of ten go to church sometime during the year.[231] Of eleven countries surveyed in 1968, Gallup found the highest rate of church attendance was in the United States.[232] Indeed, by every measure the United States is comparatively quite committed to the religion value.[233]

In addition to support for the basic religion value, the liberal tradition also required support for the value of aiding other members of the community. That is, as Locke said, man

[228] "Special Report on Religion," *Gallup Opinion Index* (February, 1969), pp. 16–17, 20; also see Herberg, p. 91.

[229] FOR, 1938; in Cantril, p. 742.

[230] Herberg, p. 73. Of course, there is no reason to suppose that the mass is aware of the shaping influence of political culture.

[231] *Ibid.*, pp. 91–92; also see "Special Report on Religion," p. 12; and 1968 SRC data reported in the Appendix.

[232] "Special Report on Religion," p. 3.

[233] Also see Norval Glenn, "Negro Religion and Negro Status in the United States," in Louis Schneider, ed., *Religion, Culture, and Society* (New York: Wiley, 1964), p. 626.

should believe that he is bound to do "as much as he can to preserve the rest of mankind." According to the tradition, there should be some feeling that people should help others, especially when "the life, the liberty, health, limb or goods of another" are threatened. This desire to help can be referred to as altruism.

The cooperation necessary for any meaningful altruism has been shown (see Table III.6) to be relatively strong in the United States. Free and Cantril's data indicate that only 6 percent of Americans would say that human nature is basically bad and "you can't be too careful" concerning people whereas 67 percent say that human nature is basically good (25 percent found it mixed and 2 percent said "don't know"). The overwhelming majority believe, according to Stouffer that no decent person would "ever think of hurting a close friend or relative." The more positive side of this belief can be shown by two further examples. Hyman and Sheatsley report that two-thirds of those questioned at the end of World War II said they would be willing to stay on the rationing system if it would help the suffering people of Europe. A Roper study in 1947 found that about two-thirds of those questioned believed that foreign aid was best justified by altruistic arguments. In a 1965 Harris poll, 84 percent said they felt bad that people in the United States go hungry and two-thirds worried about people abroad who did not have enough to eat. Similarly large majorities sympathized with people who live in slums and felt sorry for the way the Negro has been treated.[234]

It is sometimes thought that a large minority do not believe in humanitarian welfare for the needy. To some extent this observation is based on legitimate public opinion data. Most surveys show a substantial portion of the public is opposed to federal welfare programs to some degree. For example, in 1965 a Gallup survey indicated that only about 40 percent of the public generally had favorable feelings toward welfare and relief programs. A similar percentage had some reservations about the programs and 6 percent wanted to do away with the programs completely. However, much of the opposition measured in questions like this relate to other

[234] Free and Cantril, p. 187; Stouffer population sample, 1954; in ICPR; Hyman and Sheatsley, p. 46; NORC, 1947; in *Opinion News* (January 1, 1948), p. 6; Harris, 1965; in *Polls* (Winter, 1967), p. 73.

TABLE IV.65 *Support for welfare altruism*

Question: Welfare relief programs are needed at the local level so that no one will be without the necessities of life.

Agree	86%
Disagree	6
Don't know	8

Source: Free and Cantril, 1968, p. 192.

dimensions existing in the survey instrument and not to welfare or altruism directly. Many Americans are opposed to *federal* welfare programs without being opposed to welfare itself. This is rather clearly demonstrated in a question asked by Free and Cantril to emphasize *local* relief programs and thereby removing the local-federal control issue (Table IV.65). Polls have consistently shown 67–80 percent of Americans supporting government help to those without any other means of subsistence.[235]

Indications of support for altruism may also be ascertained by data on charitable contributions. One major study estimated that, in the 1950's, as much as $3 billion was voluntarily contributed by individuals for charitable purposes.[236] A study of income tax deductions indicates that the great bulk of these contributions, both in number and in total amount, was in small contributions by a very large number of contributors.[237] Willingness to devote time to altruistic pursuits may even be a better indicator of support for altruism than contribution of funds.[238] Here commitment appears especially strong. In 1959 Gallup asked if individuals would be willing to devote time to serve their community. As Table IV.66 shows, almost six out of ten gave the more altruistic re-

[235] Michael E. Schiltz, *Public Attitudes Toward Social Security: 1935–1965* (Washington, D.C.: U.S. Government Printing Office, 1970), p. 111. Also see AIPO, 1965; in *Polls* (Summer, 1965), p. 72; and Schiltz, p. 152.

[236] Alfred de Grazia and Ted Gurr, *American Welfare* (New York: New York University Press, 1961), p. 12.

[237] F. Emerson Andrews, *Philanthropic Giving* (New York: Russell Sage Foundation, 1950), p. 59.

[238] Actually, giving money seems to be rather low on the list of more altruistic activities; see Pitirim A. Sorokin, *Altruistic Love* (Boston: Beacon Press, 1950), p. 14.

TABLE IV.66 *Support for community service altruism*

Question: Suppose you were asked to serve on a committee which would deal with some local problems – such as housing, recreation, traffic, health, and the like – would you accept? (AIPO, 1969)

Support	59%
Not support	40
Don't know	1

Source: *Gallup Opinion Index* (April, 1969), p. 19.

TABLE IV.67 *Government supervise religion?*

Question: Do you think that our government should or should not supervise all religious observances by establishing a national church? (FOR, 1939)

Should	4%
Should not	92
DK, depends	4

Source: Cantril, p. 104.

sponse. The same survey also found that 50 million Americans have, at one time or another, actually performed formal, voluntary charitable work.[239]

The liberal tradition has also been highly antagonistic to governmental control of religion. Table IV.67 indicates that more than nine out of ten people do not believe in the establishment of a national church and the supervision of religious observances. Yet informal alliances between religion and the political system are accepted, as is suggested by data from a Gallup poll in the early 1950's. When asked if God should be acknowledged in the pledge of allegiance to the flag, about seven out of ten agreed that it should (Table IV.68).

AMERICAN BELIEF SYSTEM

The generally recognized Lockean values of liberty, equality, and property, and a fourth value, religion, have been identified as the belief values of the liberal tradition. Each of the values seems to be represented by a culture-like distribu-

[239] AIPO, 1969; in *Gallup Opinion Index* (April, 1969), p. 18.

tion in the United States. For all four measures substantial majorities of the public were found to express support. Certain subsidiary values were also identified. Education as a subsidiary value of liberty, achievement as a subsidiary value of property, and altruism as a subsidiary value of religion were found to have culture-level support. All seven values are considered components of the political culture of the United States.

Table IV.41 suggests that Americans tend to identify liberty as the most valued belief within the subsystem. This is true only in a comparative sense, however. In terms of absolute commitment the other components seem well accepted. More than eight out of ten choose liberty over economic security and seven out of ten choose individual freedom over duty to the state. Seven out of ten support freedom of the press while more than 95 percent support freedom of speech. The subsidiary value of education is accepted by eight out of ten. For the major belief of political equality, support is given by at least three-fourths of the population. Data indicate that 81 percent prefer private over public ownership of property (and culture support exists in many other areas of private property). Over six out of ten support achievement more than job security. More than six out of ten would not limit achievement. Finally, there is virtual unanimity of belief in God and seven out of ten accept the somewhat more demanding religious belief in life after death. Eighty-six percent support religious altruism. Evidently, liberty, equality, property, and religion can be considered the accepted values of the belief subsystem of the American political culture.

DEVELOPED AMERICAN POLITICAL CULTURE

This chapter has attempted to present as much attitudinal information as is available as a test of the liberal tradition hypothesis. Public opinion poll data have been reviewed over a period of three decades. Almost two hundred individual survey questions were used to try to test the hypothesis. The rule and belief systems have been systematically investigated using language directly from the *Second Treatise* and *The Federalist Papers* to provide the specific research hypotheses to be tested. Data was then introduced to test these statements. In

TABLE IV.68 *"Under God" added to pledge of allegiance?*

Question: It has been suggested that the words "Under God" should be added to the oath of allegiance to the flag. . . . Would you favor or oppose this change? (AIPO, 1953)

Favor change	69%
Oppose	21
No opinion	10

Source: AIPO news release of May 9, 1953.

every case the attitudes of Americans seem to fulfill the expectations of the liberal tradition hypothesis.

The question naturally arises whether support for the values of the tradition has changed over time. Unfortunately, it is difficult to test the hypothesis by era with the limited data available. If the questions used in this chapter were asked in the same form at set time intervals, I believe that the data would show some variation in support for individual values of the tradition between eras, together with consensus support for each value and a generally stable support for all of the values considered together as a political culture. Table IV.69 presents a very rough attempt to summarize support for the major developed values of the liberal tradition for which there are data over time. Although the summary is only very approximate and the absolute levels shown should not be taken as precise, it is probably an acceptable approximation. The data suggest that support for the values vary but that support remains consensual over time. At least six of ten support all values in every era. When looking at individual values, the differences in support across time do not appear great. There are differences, yet the differences are within a consensual limit. The mean support for all the values presented in this table (which may be viewed as a crude approximation of total political culture support), suggests that total support varies even less than does support for individual values. In each era roughly three-quarters of the members support the full political culture. Both an analysis of this data and an inspection of the individual data sources cited throughout the preceding two chapters give the impression of a consensual support for the liberal tradition over time.

TABLE IV.69 *Liberal tradition support by era*

Value	*Mean support for developed political culture values* 1936–1945	1946–1955	1956–1970
1. Popular rule and elections	–	86%	79%
2. Legislative predominance	71%	62	78
3. Federalism	65	–	65
4. Decentralized parties	62	69	73
5. Liberty	72	72	68
6. Equality	98	–	85
7. Property	60	66	66
8. Achievement	74	–	86
9. Belief in God	96	95	98
10. Religion	70	72	73
11. Altruism	–	–	73
Mean support for political culture by era[a]	74	75	77

Source, by value number: (1) Tables IV.4, IV.5, IV.6, and n. 40; (2) Table IV.22; (3) Tables IV.31, IV.32 ("No opinion" excluded), IV.29 dichotomized; (4) Tables IV.33, IV.34; (5) Tables IV.42, IV.43, IV.44, IV.45, IV.46; (6) Tables IV.51, V.8, IV.50, VI.22; (7) Tables IV.54 for coal, railroads, and banks, IV.55 ("Don't know" excluded); (8) Table IV.59; (9) Table IV.62; (10) Table IV.64; (11) Tables IV.65, IV.66.

[a]Mean support for the values for which data are available in all three time periods is as follows: 1936–1945, 72 percent; 1946–1955, 73 percent; 1956–1970, 76 percent.

In summary, in every research situation considered in the descriptive sections of this work, the data have shown consensual support for the values of the liberal tradition. Based upon the research done by those authors who have been cited throughout the last three chapters, on the historical materials considered, and on the attitudinal support data investigated here, I conclude that the liberal tradition is the core of the political culture of the United States. In my understanding, science only accepts the best description – not some absolute description. I believe that the present findings meet that criterion. The analysis does not prove in some absolute sense that the liberal tradition *is* the core of the political culture of the United States. It does seem to suggest, however, that the liberal tradition hypothesis explains the data quite well. In my opinion, no other integrated thesis could explain as much American opinion data as well.

CHAPTER V

Cultural Consensus and Cleavage

THE NATURE OF CONSENSUS SUPPORT IN LIBERAL REGIMES

CONSENSUS AND CONFLICT IN LIBERAL TRADITION REGIMES

An extensive analysis of public opinion and historical data seems to suggest that the liberal tradition is the political culture of the United States. Culture support was found for the major values associated with the tradition. Support for a sense of national identity and a sense of community identity appeared high. The associated Lockean value of trust was, comparatively speaking, strongly supported. These primordial relationships were called the identity system. The analysis done above suggests that this cultural subsystem was highly accepted when compared to other political systems. Support was also widespread for the major symbols related to the tradition. A liberal tradition society would be expected to focus upon symbols pertinent to the contract and the values of the tradition. The analysis in Chapter III suggests that the United States has a high attachment to these symbols. Both of these subsystems of political culture seem to show a strong consensus on the basic beliefs associated with a Lockean political culture.

The developed subsystems also show consensus support. The Constitution and other liberal rule institutions are supported rather strongly. Even the liberal tradition preference for legislative predominance seems supported by substantial majorities regardless of substantive policy area. Support for limited popular rule, for the three branches of the national government, for federalism, and even for decentralized parties is substantial — in fact, a culture consensus. Americans also seem to have attachment to a relatively coherent system of liberal beliefs. Liberty, equality, property, and religion all seem accepted by substantial proportions of the American people. For these developed subsystems, it was deemed necessary to relate the available attitudinal data closely to specific expressions of the tradition in the *Second Treatise* and *the Federalist Papers.* Even though this more demanding method of hypothesis testing was used, a substantial consensus on liberal tradition values was empirically described.[1]

Support for political values does appear, then, to be consensual. But what is the nature of this support? How widespread must support be to allow one to accept the descriptive hypothesis? Some further conceptual distinctions must be made before the necessary extent of support can be fully evaluated. It has been specified that support is measured by the percentage of agreement on values by the ordinary members. However, support is not viewed as residing on a completely undifferentiated mass. Analytically speaking, a political system is conceived as existing at different object-levels and as composed of different structural units. Although the study is limited to mass support there are differentiated aspects of mass support. To study and evaluate meaningfully the extent to which member support must be widespread, one must refine the idea of consensual support to consider these other important perspectives.

From one perspective, it may even appear that an incompatibility exists between the values of the liberal tradition and a too widespread support. Must there not be a conflict between ideas and interests for a liberal regime to be said to be liberal? Did not Madison himself maintain that conflict between interests was necessary in a free society? [2] With a con-

[1] Compare the values identified here as the core values with those identified in two other works: Willmoore Kendall and George W. Carey, *The Basic Symbols of the American Political Tradition* (Baton Rouge: Louisiana State University Press, 1970); and Robert A. Dahl, "The American Oppositions: Affirmation and Denial," in Robert A. Dahl, *Political Oppositions in Western Democracies* (New Haven: Yale University Press, 1966), pp. 63, 35–39.

[2] *The Federalist Papers* (New York: New American Library, 1961), No. 10.

sensus on values, can liberal liberty even be preserved?[3] Is the idea of consensus not inconsistent with the norms of democratic regimes? The empirical finding of consensual support for political values may mean that the American regime cannot be a liberal tradition regime.

On the other hand, although democratic theorists have held that conflict is necessary in liberal regimes, they have also stressed what they have believed to be the undesirable consequences of too intense conflict.

> Inherent in all democratic systems is the constant threat that the group conflicts which are democracy's life-blood may solidify to the point where they threaten to disintegrate the society. Hence conditions which serve to moderate the intensity of partisan battle are among the key requisites of democratic government.[4]

In a desire to preserve the very democratic regime itself, then, democratic theorists have also sought means to limit conflict. One of the major ways this has been accomplished has been through an offsetting consensus on values.[5] This view has tended to deemphasize the value of conflict (while pointing up the benefits that can flow from the plurality of interests associated with a regime based upon liberty).[6] In this view, a too intense cleavage would destroy a democratic regime.[7]

[3] V. O. Key, Jr., *Public Opinion and American Democracy* (New York: Knopf, 1961), p. 41.

[4] Seymour Martin Lipset, *Political Man* (Garden City, N.Y.: Doubleday, 1963), pp. 70–71.

[5] On this point see Key, pp. 51–53, and the discussion in the present work on pages 33–34, 363–366.

[6] David B. Truman, *The Governmental Process* (New York: Knopf, 1951), pp. 517, 507–516; Robert A. Dahl, *Pluralist Democracy in the United States: Conflict and Consent* (Chicago: Rand McNally, 1967), p. 270; Arend Lijphart, *The Politics of Accommodation* (Berkeley: University of California Press, 1968), esp. ch. 1; Robert A. Dahl, *Who Governs?* (New Haven: Yale University Press, 1961), ch. 7, 24, 28; Nelson W. Polsby, *Community Power and Political Theory* (New Haven: Yale University Press, 1963), ch. 6; Gabriel A. Almond, *The American People and Foreign Policy* (New York: Praeger, 1960; originally published 1950), ch. 7; V. O. Key, Jr., *Parties, Politics and Pressure Groups,* 4th ed. (New York: Crowell, 1958), pp. 13–14 and ch. 6; and Robert A. Dahl, *A Preface to Democratic Theory* (Chicago: University of Chicago Press, 1956).

[7] *The Federalist Papers,* No. 10; Truman, pp. 520–524; Dahl, *Pluralist Democracy, Democracy in the United States,* chs. 11, 12.

In general, democratic theorists have attempted to resolve the paradox between consensus and conflict by holding that democratic regimes should embrace both values. A democratic regime is viewed as balancing consensus and cleavage.[8] One major study of democratic values finds that balancing is the basic function of a "civic culture" which distinguishes it from other cultures.[9] A balance must be achieved because "too much consensus would be deadening and restrictive of liberty; too much cleavage would be destructive of society as a whole." [10]

THE AMERICAN REGIME AND SYSTEM LEVELS

It seems to me that the idea of balance is not a wholly satisfactory explanation of the nature of consensus as it acts in a liberal tradition regime. It does not appear differentiated enough. A more useful explanation seems to be deducible from our system model. Two aspects of that model seem especially relevant. We have used the analogy of system and how it functions. We have also viewed the analogy as acting differently at various levels. Environmental stimuli act upon systems and cause stress. But stress acts differently depending upon the level – community, regime, or authority – being affected. For example, conflict at the authority level may not influence the regime:

> Not all disagreements or cleavages in these . . . orientations [to authority-level conflicts over party, candidates, or issues] need necessarily constitute a threat to the survival of a system. Whether or not it does will depend upon the subject matter or object with respect to which the disagreement prevails. For example, it is obvious that democratic systems manage to survive even in the face of deep cleavages with respect to all these orientations.[11]

[8] Bernard R. Berelson, Paul L. Lazarsfeld, and William M. McPhee, *Voting* (Chicago: University of Chicago Press, 1954), p. 318.

[9] Gabriel A. Almond and Sidney Verba, *The Civic Culture* (Boston: Little, Brown, 1965; originally published 1963); pp. 356–360.

[10] Berelson, Lazarsfeld, and McPhee; also see Talcott Parsons, " 'Voting' and the Equilibrium of the American Political System," in Eugene Burdick and Arthur J. Brodbeck, eds., *American Voting Behavior* (Glencoe, Ill.: Free Press, 1959), pp. 100–114.

[11] David Easton and Robert D. Hess, "Youth and the Political System,"

Policy-level conflict apparently is quite widespread. Historically, it appears that most conflict has taken place at the authority, then at the regime, and most infrequently at the community level.[12] It appears that stress at one of these levels is different from and not necessarily related to stress at another. Again, the historically large amount of stress at the authority level has not seemed necessarily to lead to stress at other levels.[13]

It appears that democratic theory implicitly calls for conflict *but it does not specify the level at which conflict is to take place.* I would suggest that not specifying levels of conflict is a central analytic problem of normative democratic theory. If the level at which stress operates is critical in understanding political systems (as I believe it is), then for any theory of regime to be useful, it must specify its expectations in terms of these levels. Perhaps normative democratic theory could be more analytically precise if it could more carefully consider the problem of conflict from the perspective of Easton's three levels of political analysis.

From the perspective of these levels of analysis, there appear to be four ways to deal with conflict: (1) eliminate conflict (especially through coercion) within the community; (2) secure agreement on specific policy; (3) allow completely unregulated conflict; or (4) secure consensus on a process which provides rules within which conflict takes place.[14] The first way to deal with conflict would be inconsistent with the liberal tradition's identification of faction conflict as a necessary by-product of liberty.[15] In the United States this solution has not been accepted. There is simply too much evidence of value conflict in the American system for this to be the means used or accepted.[16]

---

in Seymour Martin Lipset and Leo Lowenthal, eds., *Culture and Social Character* (New York: Free Press, 1961), p. 229.

12 *Ibid.*, p. 230.

13 *Ibid.*

14 Dahl, *Pluralist Democracy in the United States,* pp. 16–17.

15 *The Federalist Papers,* No. 10, p. 78.

16 For example, see Lloyd A. Free and Hadley Cantril, *The Political Beliefs of Americans* (New Brunswick, N.J.: Rutgers University Press, 1967).

The second means also appears inconsistent with the tenets of the liberal tradition. To secure agreement on all policies would assume that there were not significant interests in the society that had real opposing interests. The tradition assumes that there are conflicting interests.[17] Although a liberal regime could attempt to get this agreement, it could not rely upon it. Key's important study of American opinion, suggests, however, that this is the means used in the United States. Yet a close examination of Key's data indicates that when he is talking about the policy level he is actually often dealing with more fundamental values – relating to the liberal tradition – which are actually at least partially at the regime level. Key reports data on support of rule institutions, support for education, and support for national defense as examples of policy agreement. The present analysis suggests that these questions are actually surrogates for rule, education, and vertical identity values of the political culture. He also presents data on (1) refusing to fire governmental personnel accused of being Communist without proof, and on (2) giving fair treatment to Negroes if they are not receiving it. These issues, though not directly tradition values, do seem to be closely related to the liberal tradition values of Constitutional and judicial rules, and to altruism and equality. All of these examples, then, appear to have culture elements. In each situation the concurrence was at a more fundamental level than specific policy.

Key additionally mentions support for old-age pensions for needy persons and federal aid to education[18] as examples of how the American system uses agreement on specific policy to limit conflict in society. Although at first glance his data suggest historic support for federal aid to education, the most detailed study of public opinion regarding education suggests that support is not deep-seated.[19] Nonetheless, majorities have appeared to support federal aid ever since 1938.[20] However,

[17] *The Federalist Papers,* No. 10, pp. 77–80.

[18] Key, *Public Opinion and American Democracy,* pp. 31; 38–39; 30; and 34.

[19] Frank J. Munger and Richard F. Fenno, *National Politics and Federal Aid to Education* (Syracuse, N.Y.: Syracuse University Press, 1962), p. 96.

[20] *Ibid.,* p. 92.

in 1964 the Survey Research Center changed the question. Instead of asking whether the individual supported federal aid to education, a choice was presented between federal, and state or local aid only. As shown above in Table IV.34, a plurality was actually opposed to federal aid when the question was asked this way (similar results had actually been obtained as early as 1943 by the National Opinion Research Center). I would conclude that Americans consensually support aid to education (it is, after all, a culture value), but that *federal* aid to education presents an authority level conflict.

Support for old-age pensions is Key's only example of specific policy support as a way to limit social conflict which is not directly influenced by a liberal tradition value. Of course, this policy is shaped by the liberal tradition value of altruism. Yet support for federal old-age pensions under Social Security is not simply a surrogate for a tradition value nor a direct result of shaping. It appears essentially to be the result of environmental stress. For specific policy support to have the major role in limiting conflict for the political system, however, there must be support for most, if not all, policies raised in the political system. Support for one policy cannot fulfill this function unless it has a general shaping ability. Of course, if support for any one policy did have this characteristic, it would just indicate that it was a culture value. Even if support for Social Security were cultural, this fact alone would not disprove my hypothesis. Although the liberal tradition is viewed as the core of the political culture of the United States, there are undoubtedly other values that affect its political system. My contention is that these other values are not nearly as explanatory as those related to the liberal tradition.

Nevertheless, support for Social Security does not appear cultural. It is political, and support is widespread. It has also been stable over a relatively long period of time (but, of course, for a much shorter period of time than the Lockean values identified here). Social Security support fails the test because it does not shape. It does not have the generality necessary to be considered a cultural value. Support is related to a very specific policy and not to a class of phenomena. It appears that support for Social Security was not even transferred very strongly to the closely related policy of medical care under social security.[21]

There is undoubtedly agreement on specific policy in the United States, and provision of old-age pensions is one example. However, support for specific policies does not seem general enough to be the major means of handling conflict. It not only seems to be inconsistent with liberal tradition values but it does not seem to be the means used in the United States. The data show too much conflict over policy attitudes.[22] Do liberal tradition regimes then call for unregulated conflict of interests? Madison provides an unambiguous reply: "The regulation of . . . interests forms the principle task of modern legislation." [23] The principle function of a liberal regime is to regulate conflict.

To approach the paradox of the necessity for both conflict and consensus more rigorously, it appears necessary to differentiate types of support further. In regard to conflict actually two terms have been used – *cleavage* and *conflict.* In considering the common sense conceptions of the words, *cleavage* seems to be the more drastic of the two terms and implies more of a division than simple conflict. It may be useful, then, using the concept of levels of analysis, to define *cleavage* as *difference at the regime or community level* and reserve the term *conflict* to *disagreement at the authority level.* Likewise, *consensus* can be defined as *agreement at the regime or community level.* The term *agreement* can be limited to *concurrence at the authority level.* Both consensus and agreement on the one hand and cleavage and conflict on the other, however, would be empirically the same. For example, concensus and agreement would both appear in the form of what Key calls a "J curve" distribution of opinion.[24] This is an empirical distribution of opinion which shows many in the population con-

[21] In July, 1962, Gallup asked how medical costs should be met for the aged. The social security option was chosen by 44 percent, the private insurance option by 40 percent, and don't know was 16 percent. Also see Michael E. Schiltz, *Public Attitudes Toward Social Security: 1935–1965* (Washington, D.C.: U.S. Government Printing Office, 1970), p. 140.

[22] Free and Cantril.

[23] *The Federalist Papers,* No. 10, p. 79.

[24] Key, *Public Opinion and American Democracy,* p. 28.

curring on some proposition with few people in opposition and few undecided. The preceding analyses of the identity, symbol, rule, and belief systems of the political culture screen suggest that the opinions measured take the form of J curves. Authority-level opinion may or may not be in the form of J Curves. It is the assumption regarding the level of the system at which opinion is assumed to exist, its stability over time, and its shaping force, which determine whether one is considering agreement or consensus. A J curve which describes attitudes in the political culture is a consensus curve. A J curve of authority-level, policy opinion attitudes is an agreement curve.

In the United States there are J curves of *agreement* on policy at the authority level. The data of the previous two chapters suggest that there is also very considerable *consensus* on community and regime values. The consensus evident in the available data is on the values of the liberal tradition. But, once again, is not this very consensus on liberal tradition values incompatible with these values? The paradox is resolved by applying the conceptual distinctions regarding levels of system and the concept of culture values. A regime, being value-defined, can value conflict only if the conflict does not threaten the full regime itself (unless conflict is essentially the only value defining the regime). A liberal tradition regime would not force total policy agreement but there is no reason why it could not accept consensus to sustain its regime values. Too much active community cleavage would create antagonistic groups which could not conflict in the restricted manner required by the tradition. Major active cleavage on symbols and sense of national identity would make civil society communication almost impossible. Without community trust, associative cooperation would seem to have little chance for success. Without some consensus on rules and beliefs, the conflict valued by the tradition probably could not take place without destroying the regime; consequently, the third means of handling conflict must be ruled out. It appears, then, that a liberal tradition deals with conflict through the final means mentioned above – it secures consent on a process which provides rules within which conflict can take place.

Making the conceptual distinction between object-levels of the political system and between different types of support at the two major levels — that is, between consensus and agreement, and between cleavage and conflict — allows one to specify more clearly what democratic theorists mean when they say they want to balance consensus and cleavage. If we assume that they want enough stability for the community and democratic regime to maintain themselves and yet have enough disagreement to prevent stagnation and repression, it can be maintained that most democratic theorists actually prescribe *consensus at the regime and community levels and conflict at the authority level.*

Since the liberal tradition is a democratic theory of the regime, one of the answers to the question as to how widespread support for a liberal tradition regime must be can now be specified. It must be a consensus support at the community and regime levels. The Madisonian expectation for conflict is fulfilled by allowing it to take place over policy. The excesses of intense cleavage are protected against by an overarching protective consensus on regime rules. It is assumed that the conflict at the authority level can take place *because of* the consensus that exists at the more fundamental levels. In a liberal tradition regime the paradox of valuing conflict and needing consensus must be met by allowing policy conflict within a protecting consensus on regime and community values.

AUTHORITY LEVEL AND SYSTEM

Our model emphasizes levels of analysis but its primary focus is upon the analogy of system. Systems act to reduce stress. Presumably they do this at all levels — even at the authority level. It also seems reasonable to suppose that there is some interrelation between levels. What happens at one level can affect the others:

> As long as no powerful group challenges the form of the regime or the unity of the given political community, severe conflict at the governmental level with regard to which party is to hold power in any period offers very little threat to the rest of the political system. If no powerful group questions the nature of the regime, the consensus with regard to basic orientations at this level leaves the way free for even deep and protracted pluralistic struggles over control of public policy. To be sure, if

> these governmental or policy conflicts persisted over a long period, the cleavage of this level would filter through to other levels. . . . At this point, cleavage at the lower, that is, the governmental level begins to act upon the erode consensus at the higher level.[25]

The system analogy thus suggests that democracies (or any type of political system) may act to moderate even authority-level conflict so as to protect the community and the regime from future cleavage.

In a liberal tradition regime, groups conflict over policy isues. Most issues are resolved within the regime and forgotten.[26] However, some issues are so important to a group that the conflict can cause stress on the regime. This is especially likely if one group or set of groups is constantly arranged against another. If disputes range along important structural lines and one group tends to win time after time, the losing group may come to reject the regime. This happened in the United States in the middle of the nineteenth century and led to the Civil War.[27] Without moderate conflict, without cross-cutting rather than cumulative divisions, without multiple group memberships for individuals,[28] and without time to ease severe stress,[29] conflict can easily degenerate into active cleavage. In trying to avoid active cleavage, then, while the norms of democracy demand conflict at the authority level, prudent democratic systems probably moderate it — if they value relative system stability.

Moderate conflict within a consensus seems to be the general pattern in the United States. Apparently, the American regime attempts to balance policy interests so that although no group is completely satisfied, none (or at least very few) are completely alienated.[30] Indeed, it appears to attempt, within

[25] Easton and Hess, p. 230.

[26] Some are resolved through the political system but most outside it.

[27] Dahl, *Pluralist Democracy in the United States,* pp. 288–289; also see his *Preface to Democratic Theory,* pp. 104–105.

[28] Truman, pp. 508–510.

[29] Lipset, pp. 70–79.

[30] *The Federalist Papers,* Nos. 10, 51. On the minority status of the weak in the American system, by a critic of the consensus-limited conflict nature of the American regime, see Michael Parenti, "Power and Pluralism: A View from the Bottom," *Journal of Politics* (August, 1970),

the limits imposed by its values, to achieve a large degree of agreement even at the authority level.

> Perhaps if Americans converged on a single principle it would be this: Unanimity, though unattainable, is best; institutions must therefore be so contrived that they will compel a constant search for the highest attainable degree of consent.[31]

The United States appears to meet the expectations of a liberal tradition regime. It seems to have both regime consensus and a relatively higher, but restricted, degree of authority-level conflict. The empirical description of American political culture in the two previous chapters suggests that there is consensus on the values of the regime. Decisions by authorities,[32] attitude differences on policy between conflicting sets of leaders,[33] and between attentive publics [34] suggests that there is conflict at the authority level over policy.

Yet the conflict at the authority-level in the United States is not unrestricted. Indeed, the general pattern appears to be to attempt to please as many as possible. Certainly the literature on Congress [35] – even the unfriendly literature – suggests that agreement is characteristic of that central institution. Some of the more recent analyses of the presidency suggest

---

p. 528. Apparently even weak groups are *listened to,* but their problems are not necessarily *solved;* see Harold L. Wolman and Norman C. Thomas, "Black Interests, Black Groups and Black Influence in the Federal Policy Process," *Journal of Politics* (November, 1970), pp. 894–895. These authors note that one of the reasons they are not solved is that no one is sure what the solutions are; see p. 829.

[31] Dahl, *Pluralist Democracy in the United States,* p. 329.

[32] In Congress, for example, see David R. Mayhew, *Party Loyalty Among Congressmen: The Difference Between Democrats and Republicans, 1947–1962* (Cambridge: Harvard University Press, 1966).

[33] Herbert McClosky, Paul J. Hoffman, and Rosemary O'Hara, "Issue Conflict and Consensus Among Party Leaders and Followers," *American Political Science Review* (June, 1960), pp. 406–427.

[34] Donald J. Devine, *The Attentive Public: Polyarchical Democracy* (Chicago: Rand McNally, 1970), ch. 6.

[35] A few examples are Richard F. Fenno, Jr., *The Power of the Purse* (Boston: Little, Brown, 1966); J. Leiper Freeman, *The Political Process* (New York: Random House, 1955); Bertram Gross, *The Legislative Struggle* (New York: McGraw-Hill, 1953); George Galloway, *The Legislative Process in Congress* (New York: Crowell, 1953); and Nelson W. Polsby, *Congress and the Presidency* (Englewood Cliffs, N.J.: Prentice-Hall, 1964). A more hostile analysis is James Macgregor Burns, *The Deadlock of Democracy* (Englewood Cliffs, N.J.: Prentice-Hall, 1963).

that even this core institution acts more consensually than some had previously thought.[36] I do not wish to imply as one author has accused other consensus analysists of implying,[37] that America is a "happy land" of "flat" attitudes and consensus.[38] Although it appears that the system acts to moderate conflict even at the authority level, much conflict in attitudes does exist at that level – some of it relatively severe.[39] The general picture, nonetheless, appears to be moderate conflict within a broad consensus.

SUPPORT AND SOCIAL STRUCTURE

For one to accept the descriptive hypothesis of widespread support for the liberal tradition, then, support must be consensual. The empirical description of the identity, symbol, rule, and belief subsystems of the political culture in Chapter III and Chapter IV allow the conclusion that a broad consensus on regime-level values does exist. From this perspective, positive member support for tradition values, especially on rules, can be interpreted as agreement on a process regulating conflict. If all agree to the process, conflict can take place within boundaries. The acceptance of these boundaries is an agreement to limit conflict to these boundaries. A value support such as this has an obvious relationship to widespread political culture support. From the perspective of the political system, a consensus on regime values which acts as a limit on policy-level conflict can be seen as an input structure which acts as a screen on system processes. This is precisely how the maintenance process was explained in Chapter I for a political system with a shaping political culture.

This perspective, however, does not encompass all the meaningful dimensions of mass value support. Indeed, it only provides a background for the detailed analysis of the extent of American value support to be performed in this chapter. It

36 Richard Neustadt, *Presidential Power* (New York: Wiley, 1960).

37 John Higham, "The Cult of Consensus," in Edward Handler, ed., *The American Political Experience* (Lexington, Mass.: Heath, 1968), p. 42.

38 Indeed, at the authority level, there are important differences in such social structures as region, education, occupation, and religion; see Norval D. Glenn, "Massification Versus Differentiation," *Social Forces* (December, 1967), pp. 172–180.

39 Dahl, "The American Oppositions," pp. 48–53.

was important to distinguish between agreement and consensus support and how they differently relate to the political system, but they have mainly been analyzed to aid in the more detailed analysis of consensus support. For regime-level support to aid in maintenance, it must be widespread. However, consensus support is only one measure of this quality. A second is provided by the sociological perspective. This view suggests the importance of support within the important social divisions in the society. To be considered widespread, value support must also extend through relevant social structure.

Any differentiated society experiences some severe conflict which is potentially of cleavage proportion. In a diffuse society much affected by the forces of specialization like the United States,[40] this potential is very high. Many of the American value differences between groups are "the hallmark of a society with marked internal social differentiation." [41] One might expect that such a society would be highly subject to cleavages between subcultures with different life styles. However, the potential appears greater than it actually is, for two reasons. First, in the United States, groups allow their members considerable latitude in following norms.[42] They do not demand total acceptance. In the second place, many groups do not interact to any great degree with alien subcultures. Contacts are few and those few are in areas of overlapping interest. Since there are few interactions, actual social stress is minimized.[43]

For many potential cleavage situations, then, the cleavage is simply ignored. Therefore, although social differentiation may lead to active cleavage, it does not necessarily do so. One may have noted that *cleavage* has appeared in this section modified by two terms, suggesting that a further conceptual refinement of the term is necessary. Systemic cleavage has been viewed as existing in two states – potential and active. *Potential cleavage* exists as a latent division along

[40] See Fred W. Riggs, *Administration in Developing Countries* (Boston: Houghton Mifflin, 1964), pp. 23–38; and David Apter, *The Politics of Modernization* (Chicago: University of Chicago Press, 1965), p. 254.

[41] Robin M. Williams, Jr., *American Society*, 2d ed. rev. (New York: Knopf, 1960), p. 375.

[42] *Ibid.*, pp. 377–395.

[43] *Ibid.*, pp. 375–377.

the social structure divisions existing in any differentiated and specialized society. Potential cleavage is merely structural and may or may not match actual divisions relevant to the political community or regime. *Active cleavage* develops when the potential cleavages coincide with value differences which are relevant to the political regime (or its political culture). Active cleavage exists in political culture when relevant social structure is divided on community or regime values and when this division also represents a rejection of the liberal tradition values by an important group in the social structure. Even active cleavage, however, does not necessarily imply an imminent threat to regime maintenance. It must still be translated from attitude to behavior. Yet, in the long run, active cleavage exists as a threat to maintenance.

What is the current situation in the United States? Is liberal tradition value support widespread across potential cleavages? The empirical investigation of some of the major social structures will be the task of the remainder of this chapter. But what social structure should be investigated? Obviously, any society can be analytically differentiated in an almost infinite number of ways. Dahl suggests that the major ways are regional, socio-economic, subjective class, ethnic and religious, and ideological. Burnham maintains that the important differentiations are sectional, community-society, and ethno-cultural.[44] The present study will use the more traditional academic classifications of political, economic, ecological, and societal. These not only have the benefit of familiarity but they also include the categories suggested by the two authorities.

POTENTIAL CLEAVAGE AND CULTURE SUPPORT

AMERICAN POLITICAL CULTURE AND CLEAVAGE

Based upon the foregoing discussion, on the concept of culture itself, and on the description of the subsystems of political culture, one should hypothesize that American political culture support would be widespread across all relevant social

[44] Dahl, *Pluralist Democracy in the United States,* ch. 14; Walter Dean Burnham, "Party Systems and the Political Process," in W. N. Chambers and W. D. Burnham, *The American Party Systems* (New York: Oxford University Press, 1967), pp. 282–287.

structures in the society. Of course, there are likely to be differences but one would not expect marked antagonism to the liberal tradition in any major group. Unfortunately, the data available do not allow any real measure of *degree* of antagonism. All that can be done is to show how *many* deviate from the tradition norm. It is possible to focus upon those who actually reject values or upon the larger ambivalent groups, but the present study has focused upon positive support. It is here hypothesized (in the more restrictive sense) that *positive support* for the tradition as a whole should exist within all relevant social structure. Operationally, we should expect, first, that there should not be any cumulative active cleavages across all major values of political culture for any substantial *group*. Second, majorities should support the *values* of the tradition in all important subgroups. Finally, even if majorities support the values there should be few large differences between social subdivisions. Arbitrarily, 10 percent will be considered a large difference.[45]

Unfortunately, since the resources available for the present study are limited and are based exclusively upon secondary analysis, it is able to investigate neither all the relevant social structures for each subsystem nor all of the important measures of each subsystem. In some cases the operational measures of the subsystem which were used are not the most representative of the subsystem, but they were used because they were the only ones available for detailed analysis.[46] Obviously, this restricts the usefulness of the analysis. However, the analyst even under optimal conditions always lacks some operational precision if for no other reason than that operationalizations are necessarily only approximations of the analytic precision of concepts. Here, the operationalizations are additionally restricted because of the constraints of the research design. None-

[45] This may be an appropriate point to note a methodological problem for subgroup analysis on samples. As a subgroup's size decreases, the amount of survey error increases. In evaluating the following subgroup data, one should be aware of this limitation. For Gallup approximations, see practically any edition of the *Gallup Opinion Index;* on the general problem, see Samuel Stouffer, *Communism, Conformity, and Civil Liberties* (New York: Wiley, 1966), pp. 270–272.

[46] Basically the researcher was limited to data available through the Inter-University Consortium for Political Research at the University of Michigan. Their kind cooperation is gratefully acknowledged.

theless, I believe they are sufficiently representative to merit their use.

One measure of each major identity system is available for secondary analysis. National identity will be tested with the data of Table III.3.[47] Unfortunately, the data for the early period are limited to only two cleavages. This necessitates using other data beyond our time frame. The horizontal dimension of identity will be tested with the community trust variable (Table III.6). Although the operationalization is weak, it does measure a central aspect of sense of community identity.[48] There was only one measure of symbol support available, and data for it were available only for economic, ecological, and certain societal differentiations. Fortunately the question is central to the symbol system, though, since it is the one on the Constitution (Table III.8, 1940). Much more data is available for the developed subsystems. The rule system uses the following measures: importance of popular rule (page 148, n. 40), belief in legislative predominance (Table IV.14), support for federalism (Table IV.29), and support for decentralized parties (Table IV.35). Belief system operationalizations are belief in individual freedom (Table IV.43), expectation of equality before the law (i.e., police treatment, Table IV.50), private property (Table IV.54, railroads, 1946), and religion in regard to a life after death (Table IV.64, 1968). Even though four questions are used for each developed subsystem they are not, again, necessarily the most appropriate. They are, however, the ones most appropriate considering the available data.

Realizing the constraints of the research design and the resources available it is surprising how representative the selected measures of the full political culture are. Trust is not only a central measure of identity but it also represents the

[47] The data breakdown can be found in William Buchanan and Hadley Cantril, *How Nations See Each Other* (Urbana: University of Illinois Press, 1953), p. 213.

[48] It is weak in the sense that we are analyzing the weaker operationalization of trust rather than that of Table III.5, which is unavailable. As stated on page 99, the second response category to the Table III.6 operationalization appears to gauge two dimensions of the Lockean value system rather than the one required for efficient operationalization. This may account for the fact that this is the tradition measure which is least fulfilled. Undoubtedly a better operationalization would show fewer cases of active cleavage — e.g., see Table V.12.

dimension of identity which is particularly related to the Lockean tradition. Vertical identity is well represented with a question on attitudinal attachment to national identity. For the symbol system, certainly few symbols could be more important than the one available to represent it — the Constitution. For the rule system, all of the major subcategories of rules are represented. The question used on freedom is not the same as belief in liberty, but since support is much higher for other measures of liberty, the freedom value is probably a conservative estimate of liberty. Again, expectation of equal treatment is not exactly belief in equality. However, it is probably a close enough approximation at least to suggest some relevant social differentiation. Although the property and religion [49] questions seem to represent their values clearly, the cleavage analysis is restricted to description of the groups analyzed by the original investigators.

There also appears to be a rather good distribution by two other criteria. Five different poll organizations are represented. More importantly, a rather long time period is also represented. Three measures are from the 1940's, two in the 1950's, four in the single year 1960, and two others in the late 1960's. Politically, at least one year during the term of every president from Franklin Roosevelt to Lyndon Johnson is represented. The data overrepresent more recent attitudes but this has the benefit of concentrating upon what many assume to be the most likely period of value change.

With these limitations and advantages in mind, it is proper to proceed to a description of the major cleavages. In accordance with these considerations, I will assume that the eleven measures used are representative of all of the values of the tradition. Of course, they only approximate the true complexity of this value system; yet they are, I believe, reasonable surrogates of the tradition. The operational questions are: (1) Are these values accepted by a majority in each subgroup? and (2) Is there a large difference in support between subgroups?

[49] The belief in God question (Table IV.62) was considered for use. Belief in God was so unanimously accepted, however, that there were no cleavages of any substance for any social grouping. The question used is thus a conservative estimate of support for the religion value.

POLITICAL CLEAVAGES

Most American history, with its emphasis upon party struggle, mass movements, and ideological conflict, suggests that the significant active cleavages in the United States are political. Yet the concept of culture and the results of the foregoing empirical investigation of the American political culture have already suggested that this will probably not be so. As already mentioned, even in the case of the political party and its contrived symbols of conflict, consensus is the normal situation in the United States.[50] However, in order further to substantiate empirically the depth of attachment to the liberal tradition, it is necessary to explore each potential cleavage in some detail.

Political Party. The textbook conception of political party suggests that the people are divided in a rather basic way between the two contending national parties. Typically the focus of party conflict is upon a sharp contest between great masses of believing partisans in each party. The voting studies have erased to some extent this rather overly rationalistic conception of party identifiers as extremely ideological and highly program oriented. Yet these studies have suggested that psychological party identification is rather important in showing relevant party differences in regard to elections.[51] Other data suggest that there is also some connection between party identification and conflict on policy issues.[52] Of course, both of these party conflicts are at the authority level.

Is party an active cleavage for the political culture? Table V.1 shows that most people do not perceive party differences to be very severe—even during a period when intellectuals perceived substantial domestic differences.[53] When asked if parties differ on authority-level policies only about a third of those who know what the government is doing believe such

[50] Mitchell, ch. 9.

[51] Angus Campbell *et al., The American Voter* (New York: Wiley 1964), ch. 5.

[52] Angus Campbell, Gerald Gurin, and Warren E. Miller, "Political Issues and the Vote," *American Political Science Review* (June, 1953), pp. 359–385.

[53] See Key, *Public Opinion and American Democracy,* pp. 83–85.

TABLE V.1 *Perceived differences between parties*

	1942 (AIPO)[a]	*1946 (AIPO)*[a]	*1952 (SRC)*[a]
Great deal	25%	30%	21%
Little	42	43	41
None	26	21	32
No opinion	7	6	6

	1944 (NORC)[b]	*1960 (SRC)*[b]
Differences	41%	47%
No differences	48	40
Don't know	11	13

Source: Hadley Cantril, *Public Opinion, 1935–1946* (Princeton: Princeton University Press, 1951), pp. 579–580; SRC, ICPR.

[a]Actual question: "Do you think it makes a great deal of difference or only a little difference which party runs the country?"

[b]Actual question: "In general do you think there are any real differences between the Democrat and Republican parties?"

differences exist.[54] Almond and Verba report that party adherents in the United States tend to view their opposites in the other party in generally favorable terms. More than six in ten Democrats do not think that the Republican Party would endanger the country and a like proportion of Republicans think the same of the Democratic Party.[55] This type of trust between parties does not exist in all countries.[56] Almond and Verba further note that marriage across party lines is viewed indifferently in the United States.[57] Table V.2 presents the most recent data on interparty perception. Although party identifiers perceive members of their own party more favorably, a majority feel positively disposed toward members of the other party. Data from the same source (not reported in the table) indicates that only 10 percent of the Democrats and 18 percent of the Republicans view the other party "highly unfavorably."

54 Campbell *et al., The American Voter,* p. 106.

55 AV, 1960; in ICPR.

56 Almond and Verba, pp. 86–94. Note especially the Italian and Mexican party cleavages.

57 *Ibid.,* p. 97.

TABLE V.2 *Perception of other party*

Question: How would you rate the Republican (Democratic) Party?

	Party of respondent	
Perceive favorably	*Republican*	*Democratic*
Own party member	97%	92%
Other party member	55	57

Source: Derived from *Gallup Opinion Index* (September, 1966), pp. 28–29.

The above data, however, do not relate directly to the importance of political cleavage but to the general importance individuals consciously assign to party difference. Table V.3 presents a breakdown of ten measures of the political culture.[58] The data seem to show some tendency for Republicans to support the liberal tradition values more than Democrats. However, most of these differences are small, and some can probably be assigned to the time factor. In the one case where Democrats support the tradition more, it seems reasonable to assume that at least part of the reason for supporting legislative predominance at the time was because there was a Republican president and a Democratic Congress. Partisan considerations may also help explain Republican support data on the federalism value. No doubt some of the Republicans who supported the state government prerogative did so because at that time the states were the main stronghold of Republican Party strength. Yet it is important to note that in both cases, majorities of the other party supported the liberal tradition value even though it was against their own immediate interest.

Still other differences in support can be related to immediate, personal, short-term interests. The higher socio-economic status of Republicans can account for some of the higher support given by them to the property value. The greater number of higher status Republicans could also explain the differences in trust. Higher concentration of Midwesterners may help explain religion differences. Other social characteristics undoubtedly could explain more. The most obvious pattern of

[58] Unfortunately no control for party was taken by the data source for the other value.

Table V.3, though, is the pattern of consensus. This pattern is also apparent among those without specific identification to a party. Independents seem to act as a middle group when compared to party identifiers at the authority level. However, on cultural matters they seem to provide as high a general support as the party identifiers.

Although somewhat less reliable (because of the very small size of the sample group), the data suggest that the "apoliticals," i.e., those who do not identify with either party or consider themselves political independents, represent the only significant active cleavage within the political-party social differentiation. This group isolates itself from the political realm — politics does not exist for them. They have no interest in or knowledge of or feeling that politics is meaningfully related to them.[59] These apoliticals represent only 5 percent of the population when meaningfully operationalized. It is questionable if they represent in any sense a relevant public for the political system.

With the exception of the apoliticals, the political party division of the population does not seem to offer an example of active cleavage within the political culture. Substantial majorities of Democrats, Republicans, and Independents support all the measures of the liberal tradition.

Political Publics. Much scholarly literature suggests that there is a conflict between the values of the elite and the masses which are led by them.[60] Regime values may be supported by elites and not by the masses and vice versa.[61] A political system is probably especially dependent on the sup-

[59] Robert E. Lane, *Political Life* (New York: Free Press, 1959), p. 153; also see Edgar Litt, "Political Cynicism and Political Futility," *Journal of Politics* (May, 1963), pp. 312–323; and especially Lester W. Milbrath, *Political Participation* (Chicago: Rand McNally, 1965), ch. 3.

[60] Two classic examples from very different perspectives are Karl Marx and Frederick Engels, "Manifesto of the Communist Party," in Carl Cohen, ed., *Communism, Fascism, and Democracy* (New York: Random House, 1962), pp. 89–110; and G. Mosca, *The Ruling Class,* trans. Hannah D. Kohn, Arthur Livingston, ed. (New York: McGraw-Hill, 1930). For a more modern interpretation of a similar thesis, see C. Wright Mills, *The Power Elite* (London: Oxford University Press, 1956), pp. 275–276.

[61] Easton, pp. 166–167.

TABLE V.3 *Political cleavages in the American political culture: percentage of support for the liberal tradition value*

	Identity system		*Symbol system*	*Rule system*				*Belief system*			
Political grouping	*National identity (AIPO, 1971)*	*Community trust (AV, 1960)*	*Constitution (FOR, 1940)*	*Popular rule (SRC, 1960)*	*Legislative predominance (SRC, 1958)*	*Federalism*[a] *(AIPO, 1967)*	*Decentralized parties (SRC, 1958)*	*Individual freedom (AV, 1960)*	*Expect equality (AV, 1960)*	*Property*[b] *(AIPO, 1953)*	*Religion (AIPO, 1968)*
Political party											
Democrat	87%	54%	NA	90%	66%	69%	55%	67%	85%	71%	70%
Republican	88	62	NA	95	56	85	57	69	90	83	80
Independent	83	60	NA	92	67	70	64	68	86	75	73
Apolitical	NA	49[c]	NA	60	25	NA	41	69[c]	81[c]	NA	NA
Political public											
Attentive	NA	62	NA	97	72	NA	68	71	88	NA	NA
Non-Attentive	NA	51	NA	87	59	NA	56	66	83	NA	NA

[a]In this and all subsequent tables, replies for this question are dichotomized between those choosing the state and those choosing national government values; the "neither" and "no opinion" responses are ignored (see Table IV.29).

[b]The 1953 data are used since support for Independents does not appear in the 1946 data. In 1946, Democrats showed 61 percent support and the Republicans 77 percent.

[c]The form of the question makes this category much larger than it is for the SRC data. This, of course, makes the classification "apolitical" somewhat suspect in these cases.

port of the powerful.[62] Unfortunately, little reliable data on American leadership attitudes regarding political culture are available. The massive Stouffer mass-elite study (which included attitude data from community leaders) did show that leaders were more committed to some elements of the tradition than was the public as a whole.[63] The present work, however, cannot deal with the higher-level elite. Gabriel Almond suggests a much lower level elite, though, that can be investigated with our available resources. Almond maintains that "any social form of action involves division of labor and division of influence." Specialization of labor takes place within politics also. Almond views the people not as one mass public but as a hierarchy of publics which is based upon *attention given to politics.* The amount of attention given by individuals ranges from a complete absence of attention on some people's part to the complete attention given by a professional lobbyist or a congressman. Even though we are forced to ignore the highest levels of attention given by the decision-makers and the other specialized elites, the concept of attention does allow some investigation of elites. Almond maintains that at any level of attentiveness, those who are attentive are more important for policy-making. In a previous work I have presented some empirical evidence that at the authority level those who pay attention in a moderate way, called the Attentive Public (by paying attention to politics in the newspaper, in conversations with friends, etc.), do have attitudes which are closer to the outputs actually adopted than does the mass public.[64]

Opinions for this low level Attentive Public as contrasted with the Non-Attentive Public are presented in Table V.3.[65]

[62] *Ibid.,* pp. 198, 166–167.

[63] I do not mean to imply that the elite necessarily support the liberal tradition. Undoubtedly they support some of it but they may reject some aspects. The present study is of mass support. Lack of support by the elite would, of course, be a significant active cleavage, but this is a matter for future empirical testing. For the data suggesting greater elite support of regime values, see Stouffer, esp. ch. 2; also see Herbert McClosky, "Consensus and Ideology in American Politics," *American Political Science Review* (June, 1964), pp. 361–382; James W. Prothro and Charles M. Grigg, "Fundamental Principles of Democracy," *Journal of Politics* (Spring, 1960), pp. 276–294; and Key, *Public Opinion and American Democracy,* p. 537.

[64] Almond, p. 137–139; Devine, *The Attentive Public,* chs. 5 and 7.

[65] The Attentive Public is the one-third to one-quarter of the public

As Key had predicted, there are political culture (as well as authority level) differences between these publics.[66] For each political culture measure for which this differentiation could be made, the Attentives seem to support the liberal tradition values more than those less attuned to politics. In a study in Iowa, support for the legislative value was also found to be greater among higher level Attentives.[67] It is most important to note, though, that the Non-Attentives also support the tradition with majorities in each tested situation. Therefore, while there are differences between these publics at the culture level, there is nothing that could be described as an active cultural cleavage. Support for the liberal tradition culture is more akin to a consensus for both publics. It may not be important for a political system if the Non-Attentive public withdraws at the authority level. However, the concept of political culture concerns the whole population. Withholding of support by any substantial group could negate the hypothesis. The data reported here suggest high support from Attentives. However, even those who do not pay attention at the authority level support the community and regime political culture valucs.

Political Ideology. One does not have to be very familiar with American politics to know that one of the favorite means of explaining political conflict has been to explain it as a struggle between "liberals" and "conservatives." What would journalists do without being able to label politicians and political proposals conservative or libcral? Indeed, could even the average Attentive be able to follow the confusing multiplicity of politics without these labels? Certain recent empirical work suggests that these labels have limited usefulness to the mass public. Only slightly over half of the population has even a rough idea of what liberal and conservative mean in

---

most attentive to politics; the Non-Attentive is the remainder of the population. For the exact operationalization, see Devine, *The Attentive Public,* chs. 4 and 5.

66 Key, *Public Opinion and American Democracy,* p. 51. Also see Devine, *The Attentive Public,* p. 119.

67 G. R. Boynton, Samuel C. Patterson, and Ronald D. Hedlund, "The Missing Links in Legislative Politics: Attentive Constituents," *Journal of Politics* (August, 1969), pp. 717–721.

relation to party conflict (with only 17 percent having a reasonably clear conception).[68] Twenty-nine percent can give no meaning to any relationship between party and ideology.[69] Consistency on opinions is low (within two years 13 out of 20 change their position on issues) [70] and constraint or intercorrelation between issues is low.[71]

"Liberals" and "conservatives" are not only difficult to find empirically, they are also difficult to define. McClosky tries to define conservatives *attitudinally* and he finds that they are illiberal. Schoenberger defines them *behaviorally* and his data suggests that conservatives are quite libertarian – more so than a sample of registered Republicans. Indeed, confusion is so rife that Free and Cantril are forced to call Lockean liberalism "conservatism." I believe that there is so much confusion because a comparative perspective is not taken. Comparatively, the Hartz thesis maintains that both American "liberals" and "conservatives" are liberal. Each is a tendency within Lockean liberalism. Hartz believes that the essential difference between the two strains is an economic one: those who identify with and are supported by the wealthy he calls Whig; the other group identifies with and is supported by the poor, and is called petit-bourgeois.[72] I would argue that the difference is somewhat more subtle than this. For example, this view cannot explain why in the 1964 election the "liberal" Lyndon Johnson received most of his financial support from large

68 Philip E. Converse, "The Nature of Belief Systems in Mass Publics," in David E. Apter, ed., *Ideology and Discontent* (New York: Free Press, 1964), pp. 220–221.

69 *Ibid.*

70 *Ibid.*, p. 239.

71 *Ibid.*, p. 229; and Norman R. Luttbeg, "The Structure of Beliefs Among Leaders and the Public," *Public Opinion Quarterly* (Fall, 1968), pp. 398–409. On the relevance of ideology, also see Dahl, *Pluralist Democracy in the United States*, pp. 356–370.

72 Herbert McClosky, "Conservatism and Personality," *American Political Science Review* (March, 1958), pp. 27–45; Robert A. Schoenberger, "Conservatism, Personality and Political Extremism," *American Political Science Review* (September, 1968), pp. 868–877; Lloyd A. Free and Hadley Cantril, *The Political Beliefs of Americans* (New Brunswick, N.J.: Rutgers University Press, 1967), ch. 1; for a more explicit distinction see Allen Guttman, *The Conservative Tradition in America* (New York: Oxford University Press, 1967), ch. 6; Louis Hartz, *The Liberal Tradition in America* (New York: Harcourt, Brace and World, 1955), pp. 92, 15.

contributors and the "conservative" Barry Goldwater received much of his from small contributors.[73]

The argument in this work is that although liberals and conservatives disagree — even with substantial conflict — this conflict is on the authority level. In any political system the political culture will not be interpreted in precisely the same way by any two groups. American Lockean liberals and conservatives may emphasize different values of the tradition but they both support all of the Lockean values of the community and the regime [74] — certainly at the mass level. The difference between liberals and conservatives, I would suggest, is that they emphasize different aspects of the tradition and they appraise environmental stress differently. For example, liberals probably tend to be less interested in the value of property than are conservatives. Yet even this difference is questionable. Free and Cantril found that policy-defined conservatives actually opposed big business more than did liberals. (Perhaps conservatives are narrowly bourgeois in their property belief — supporting small rather than big business.) Again it would seem that liberals could be defined as stressing equality more than conservatives, who stress liberty. There is some data to support this.[75] Yet these tendencies seem to be just that — tendencies within liberalism. American liberals would probably support the tradition value of property more than would European socialists, and American conservatives would support equality more than would European conservatives.

It appears to me that Schoenberger's behavioral measure of conservatives tends to define the essential difference between American liberals and conservatives. Although ideology undoubtedly is meaningful to only about one-fifth of the public, Professor Schoenberger seemed to find it among registrants with a third party. He allowed the behavioral action of

[73] In 1964, Democrats received 69 percent of their funds from contributions of $500 or more, and Republicans received only 28 percent of their funds from large contributors; see *Congressional Quarterly Almanac* (Washington, D.C.: Congressional Quarterly Service, 1965), p. 157.

[74] Easton, pp. 103, 337–339.

[75] Free and Cantril, pp. 132, 120–129. Huntington suggests a difference between equality and development on the one hand and liberty and tradition on the other. Samuel P. Huntington, *Political Order in Changing Societies* (New Haven: Yale University Press, 1968), p. 160.

registering with the New York State Conservative Party to define a conservative. When comparing these conservatives to a group of Republicans, he found that the only significant difference between the groups was a greater fear of central government on the part of conservatives.[76] It may be that the central value difference between American ideologies rests upon the amount of environmental stress each group is willing to allow to modify the culture values. Liberal Lockeans would tend to emphasize the strength of the environmental stress and the need for the political culture to adjust to changing developmental conditions. Conservative Lockeans would tend to emphasize values in the organism and the need to preserve these values in the face of change.[77] Yet these are not invariant positions. Both recognize stress and both recognize the importance of values. The difference is one of emphasis.

The main area of disagreement comparatively, then, is upon the role of the central government. Liberals believe that the role of Washington should be expanded to meet modern stress conditions; conservatives believe that its scope should be restricted in accordance with the tradition values of liberty, property, and federalism. One question which is available to the present study for manipulation can be used to measure this basic difference. Table IV.55 presented a distribution of beliefs which was identified as a property value. However, the question could also be used as a crude operationalization of *liberal* and *conservative.* The question asks whether the government should leave things like electric power and housing to private businessmen. Unfortunately, the question only poorly approximates a general liberal-conservative dimension. For example, those from the mountain states tend to say no — probably because of the large power projects in those states and not because of sentiment on the question of housing.[78] Nonetheless, the question does appear to measure the dimension to some degree and is the best one available.

Table V.4 shows the support for the liberal tradition by

[76] Schoenberger, p. 875.

[77] For a similar observation see Talcott Parsons, *Politics and Social Structure* (New York: Free Press, 1969), p. 181. The observation suggests that American liberals and conservatives may more properly be called modern liberals and traditional liberals.

[78] Devine, *The Attentive Public,* p. 104.

TABLE V.4 *Ideological political cleavages in the American political culture: percentage of support for the liberal tradition rule value*

	Legislative predominance	*Decentralized parties*	*Popular rule*
Liberal	69%	59%	93%
Conservative	67	59	96

Source: SRC, 1958, 1960; ICPR.

liberals and conservatives for three of the rule system measures. The similarity of support for the liberal tradition is amazing. To all intents and purposes there is no difference between opposing ideologists at the mass level regarding these cultural rule values. This is confirmed in 1968 by a better measure, directly asking if the central government is getting too powerful (see page 374). Again, there is probably more support for rule system values than for beliefs, but there appears to be a general consensus on all tradition values. Of course, there would be substantial disagreement on application of the values to specific policy situations (see Chapter VI), but general support of the tradition values appears high. Therefore, regardless of the situation at the authority level,[79] it appears that there is a substantial regime-level consensus upon liberal tradition values by both "conservatives" and "liberals" in the United States.

[79] Free and Cantril find rather substantial differences at the authority level. Calling differences at this level "operational" ideology, they find policy-level differences as well as differences on policy-related questions. Certain of these questions could be incorrectly interpreted as cultural, and so deserve some comment. Free and Cantril show disagreement on a trust question, but it is *political* trust, not trust in people (which is the tradition value); see Free and Cantril, p. 117. For substantiation of this difference between types of trust, see Joel D. Aberbach, "Alienation and Political Behavior," *American Political Science Review* (March, 1969), pp. 86–99. Free and Cantril also show differences between "operational" liberals and conservatives on confidence in the three branches of government. Although the scores differ, only two cases show scores below a 0.5 favorable response, and both of these were for the very small "completely conservative" group; see Free and Cantril, p. 118. It should also be again mentioned that the scale used cannot be simply interpreted. A 0.4 score may actually indicate a majority support. Their definition of liberal and conservative also rests upon some questionable operationalizations – e.g., the federal-aid-to-education question probably measures the education value rather than federal aid.

There is even some fragmentary evidence that ideologically sensitive groups thought to be antagonistic to the basic values of the regime also support the tradition. The lack of support by intellectuals for the property value is supposedly a simple datum of American politics. However, a random probability sample of a large border state university showed that a majority of the faculty thought that "nationalization of the basic industries would lead to an intolerable degree of governmental control." [80] Even more suprisingly, a 1968 sample of midwestern Kennedy and McCarthy student campaigners *who also identified with the "New Left"* found that a plurality even of this group identified with the property value. Forty-seven percent agreed that "democracy depends fundamentally on the existence of free enterprise." [81] Of course, these groups accept these values to a much lower degree than other groups, but the interesting finding is that support is high even among relatively alienated ideological groups. Undoubtedly there is total alienation in some groups but the little data available suggests that it is limited to a very small segment of the population.[82] For the overwhelming majority, authority-level ideology supports regime Lockean liberalism.

In summary, none of the potential political cleavages seem to be active political culture cleavages. The evident differences between political parties, publics, and ideologies appear to exist at the authority level. The liberal tradition values are supported by all of the relevant political groupings in the political system.

CLASS CLEAVAGES

Most sociological political scientists probably maintain that potential cleavage along class lines is the cleavage most liable to become active. The concept of class has little meaning for

[80] Sample of the University of Maryland faculty, April, 1968 by the author: 56 percent agree, 7 percent undecided, 37 percent disagree.

[81] David H. Everson, "The Kennedy and McCarthy Workers in Indiana," *Public Affairs Bulletin* (September–October, 1968), p. 3: 47 percent agree, 16 percent neutral, 36 percent disagree.

[82] The CBS survey of youth finds that only those classified as "revolutionaries" reject the American regime. They represent about 1 percent of the *youth* population and 3 percent of the college-educated youth population. *Generations Apart* (New York: CBS News, 1969), pp. 69, 39. Also see pages 272–276.

many Americans, however, and it is difficult to define operationally.[83] On the other hand much of the empirical analysis which has been done in American politics suggests policy relevance for operational variables which could be considered class related. Income, occupation, socio-economic status, and education all can be considered operational measures of class and each has been among the variables with the strongest relationship to political explanation at the authority level.[84]

Apparently there are authority-level differences between class groups. However, the present study is interested in whether there are active cleavages which can threaten the political culture. Do classes approximate the active cleavage suggested by Marx at the "class conscious" stage? Table V.5 presents data regarding differences in class support for the liberal tradition values. Both income and education are used as operationalizations of class since, I assume, they both tend to express differences which approximate the idea of class in the American setting.[85] Except for the measures of national identity, support of the states in the federal system, and belief in a life after death, the differences between classes meet the criteria set for significant differences. In many cases, differences as great as 20 percent exist between class subcultures. Classes do differ at the culture level.

The data of Table V.5 even suggest some relationship between class measures and culture values. Correlations were computed between degree of culture support and the income operationalization. These represent one of only two cleavage situations where associations are of a substantial nature. These are reported (for the data available for secondary analysis) in Table V.6. In most cases there is a statistically significant relationship at the alpha .05 level. There is a some-

[83] Key, *Public Opinion and American Democracy,* pp. 138–149.

[84] The studies are numerous. Four of the more sophisticated are Campbell et al., *The American Voter,* p. 90, 251–255; Richard Centers, *The Psychology of Social Classes* (New York: Russell and Russell, 1961), ch. 10; Robert R. Alford, *Party and Society* (Chicago: Rand McNally, 1963); and Norval D. Glenn and Jon P. Alston, "Cultural Distance Among Occupational Categories," *American Sociological Review* (June, 1968), pp. 365–386.

[85] Data for interviewer-rated socio-economic status and objective occupational status were also analyzed. Results were not reported since the general findings were similar to those for income.

TABLE V.5 *Class cleavages in the American political culture: percentage of support for the liberal tradition value*

	Identity system		*Symbol system*	*Rule system*				*Belief system*			
Class grouping	*National identity (BC, 1948)*	*Community trust (AV, 1960)*	*Constitution (FOR, 1940)*	*Popular rule (SRC, 1960)*	*Legislative predominance (SRC, 1958)*	*Federalism (AIPO, 1967)*	*Decentralized party (SRC, 1958)*	*Individual freedom (AV, 1960)*	*Expect equality (AV, 1960)*	*Property (AIPO, 1946)*	*Religion (AIPO, 1968)*
Income level											
$0–$2999	92%[a]	40%	58%[b]	78%	56%	75%	54%	56%	80%	–	78%
$3000–$4999	95[a]	53	64[b]	88	64	75	57	68	82	61%[d]	79
$5000–$7499	96[a]	64	71[b]	95	68	74[c]	62	71	89	68[d]	70
$7500 plus	97[a]	70	80[a]	97	73	74[c]	64	81	93	71[d]	71
Education											
0–7 years	89[e]	35	NA	68	43	69[e]	52	52	80	62[e]	75%[e]
8 years	–	44	NA	92	62	–	57	61	83	–	–
9–11 years	–	57	NA	90	57	–	53	66	80	–	–
12 years	87[e]	66	NA	95	69	75[e]	59	78	91	72[e]	73[e]
Some college	80[e]	69	NA	97	68	75[e]	60	78	89	74[e]	71[e]

[a]These categories were very poor, below average, average, and wealthy.
[b]These categories are not strictly comparable. The categories were poor, lower middle class, upper middle class, prosperous.
[c]The AIPO income categories differed at the highest categories, i.e., $5000–$6999 and $7000 and above.
[d]Income distributions were not available, so the following occupational categories were used: manual workers, white collar workers, and business and professional workers.
[e]Only gross categories of grammar school or less, high school, and at least some college were available. For education, the AIPO data are used for national identity.

TABLE V.6 *Relationship between measures of the political culture and income*

Income	*Trust*	*Legislative predominance*	*Decentralized parties*	*Individual freedom*	*Expect equality*	*Popular rule*
Gamma	.361	.151	.076	.278	.293	.555
Chi-square, $p <$	.001	.001	NS	.001	.05	.001

what pronounced tendency for those with lower incomes, as opposed to more middle and upper groups, to trust others in the community less, and a tendency for the poor to support popular rule less, individual freedom less, legislative predominance less, and to expect equal treatment less. What is true for the income measure is, to a lesser extent, also true for the education measure.

Even with these differences, however, it is important to note that *all class levels have a majority opinion in favor of the liberal tradition values for most of the culture measures.* It is only in the case of the trust and legislative predominance values for the very lowest class that there is less then a majority sustaining the culture beliefs. Both of these exceptions, however, are important. Each of them is a peculiar value of the liberal tradition. The trust measure of the identity subsystem is the only one which has less than majority support for both income and education. It can, thus, be easily related to the concept of class. The low class (approximately one-fifth of the population) rejects the trust value by a substantial margin. Only about four in ten support this important Lockean value. By our criteria an active cleavage exists for this measure of the tradition. However, as has been noted, the more satisfactory trust measure could show support. Even with the present operationalization, however, one cannot conclude that there is a low-class active cleavage. Table V.5 suggests that the general picture is support for the tradition by the low class. Even within one subsystem, low support levels may be compensated for by high support from other elements in the same subsystem. The very high support for national identity may well offset the low community support and the low class may not even lack tradition support for a full political culture subsystem. Even so, lower levels of support for trust from the low class does represent a partial exception to the generalization of majority support for all tradition values.

Legislative predominance is accepted by the low class as defined by income, but it is rejected by those with less than eight years education [86] (about 16 percent of the population). Of course, education may not even be an adequate measure of class. It may be used to operationalize socialization or even sophistication.[87] Table IV.13 has shown increasing support for legislative predominance as an individual progresses through school. The lowest educated group may not have attended school long enough to be socialized, or they may not have the capacity to value the more complex idea of collegial (as opposed to personal rule.[88] It is interesting to note that it only takes completion of grade school to push support to 62 percent.

The data suggest that there are class-related active cleavages in the political culture. Yet Table III.4 has shown that, comparatively, the United States is not very class conscious.[89] The analysis of this section shows only two active class cleavages, and that these are both offset by high support even within the same subsystems (legislative predominance is balanced by three other rule values). I conclude that there is, in general, support for the tradition. Table V.5 shows uniformly high support for the tradition by the middle and upper class groups. It also shows that for 85 percent of the tests the liberal tradition is accepted by the low class group.[90]

[86] Also see Prothro and Grigg.

[87] For example, see John Dewey, *Democracy and Education* (New York: Macmillan, 1916).

[88] David Easton and Jack Dennis, *Children in the Political System* (New York: McGraw-Hill, 1969), p. 117. Note the majority support on page 374.

[89] Also see David R. Segal, "Social Structure Bases of Political Partisanship in West Germany and the United States," in William J. Crotty, ed., *Public Opinion and Politics* (New York: Holt, Rinehart and Winston, 1970), pp. 216–235. One comparative study suggests that there is no necessary relationship between class cleavage and regime maintenance. See Morris Janowitz and David R. Segal, "Social Cleavage and Party Affiliation: Germany, Great Britain and the United States," *American Journal of Sociology* (May, 1967), pp. 601–618.

[90] Many class theories expect that property support should be weak for the low class. Table V.5, however, shows substantial support. Other data show that labor union members also oppose government ownership — 62

The norm is potential rather than actual cleavage. All class groups tend to support the liberal tradition as a whole.

ECOLOGICAL CLEAVAGES

In a regime so oriented to geographical representation not only in its legislature but in the essence of its federal structure, ecological cleavages could easily be thought to be crucial. Two ecological cleavages seem to have been prominent in the literature in regard to the United States and will be investigated here.

Regionalism. The importance of this variable in explaining American politics has been thought to be waning since the appearance of class politics during the 1930's.[91] However, regional interpretations have been popular throughout this period.[92] There is even some evidence that these authority-level differences have been increasing in the very recent past.[93] Value research has suggested that ecological subcultures do present different views of the world and different beliefs of what is good in that world.[94] In a nation the size of the

---

percent support the property value; see AIPO release of January 29, 1947. This is again confirmed in 1953 when support by manual workers increases to 70 percent; see AIPO release of October 10, 1953.

[91] Key, *Politics, Parties and Pressure Groups,* pp. 270–272.

[92] A few of the better examples are Alford, p. 83; Thomas A. Flinn, "Continuity and Change in Ohio Politics," *Journal of Politics* (August, 1964), pp. 521–544; V. O. Key, Jr., and Frank Munger, "Social Determinism in Electoral Decision," in Burdick and Brodbeck, eds., *American Voting Behavior* (New York: Free Press, 1959), pp. 281–299; Ira Sharkansky, "Economic Development, Regionalism and State Political Systems," *Midwest Journal of Political Science* (February, 1968), pp. 41–61; and Daniel J. Elazar, *American Federalism* (New York: Crowell Company, 1966).

[93] Walter Dean Burnham, "American Voting Behavior and the 1964 Election," *Midwest Journal of Political Science* (February, 1968), pp. 1–40; Norval D. Glenn and J. L. Simmons, "Are Regional Differences Diminishing?" *Public Opinion Quarterly* (Summer, 1967), pp. 176–193; Donald J. Devine and Ronald F. Weber, "Social Structure, Electoral Structure, and the 1968 Presidential Election," unpublished manuscript.

[94] John Gillin, "National and Regional Cultural Values in the United States," *Social Forces* (December, 1955), pp. 107–113; Dean Jaros, Herbert Hirsch, Frederich J. Fleron, Jr., "The Malevolent Leader," *American Political Science Review* (June, 1968), pp. 564–575; William A. Scott, "Empirical Assessment of Values and Ideologies," *American Sociological Review* (June, 1959), pp. 299–309; Evon Z. Vogt and Thomas F. O'Dea, "A Comparative Study of Values in Social Action in Two Southwestern Com-

United States it would be surprising if there were not some regional differences. The present study is interested in whether these differences represent active culture cleavage.

One authority has suggested that American regional cleavage is best represented by "the ongoing conflict between the Northeastern metropole on the one hand and the Southern and Western regions, which stood in a quasi-colonial relationship to the metropole, on the other." [95] Table V.7 does provide some substantiation that the East differs from the other regions. For the five of the eleven measures for which data is available and differences are observable, the East seems to be differentiated from the other regions. It is noticeably higher in trust and symbol acceptance, and lower in support for federalism, decentralized parties, and religion than the "colonial" regions. It and the Midwest are also notably less supportative of the freedom value. The West seems most distinguishable in terms of higher support for individual freedom.

Other analysts choose the South as the most atypical of the American regions because of the importance of the race issue. Table V.7 shows that, except for its high support for the religion value, the South's other two noticeable deviations can be related to the Negro. The states have provided a means for southern protection against the civil rights incursions of the federal government and southerners consequently support federalism to a much higher degree than do the other regions. Southerners also are noticeably less supportative of the equality value. On the other hand, this value is only a crude approximation of the true tradition equality value. Table V.8 reports data more directly related to this tradition belief. As can be seen, there is a substantial difference between southern opinion and the amount of support for the value outside of the South.

In looking at these regional differences, however, one must

---

munities," in Edward E. Sampson, ed., *Approaches, Contexts, and Problems of Social Psychology* (Englewood Cliffs, N.J.: Prentice-Hall, 1964), pp. 474–484; and Jonathan P. Lane, "Isolation and Public Opinion in Rural Northeast Brazil," *Public Opinion Quarterly* (Spring, 1969), pp. 55–68.

[95] Walter Dean Burnham, "Party Systems and the Political Process," in William Nisbet Chambers and Walter Dean Burnham, eds., *The American Party System* (New York: Oxford University Press, 1967), p. 284.

TABLE V.7 *Ecological cleavages in the American political system: percentage of support for the liberal tradition value*

	Identity system		*Symbol system*	*Rule system*				*Belief system*			
Ecological grouping	*National identity (AIPO, 1971)*	*Community trust (AV, 1960)*	*Constitution (FOR, 1940)*	*Popular rule (SRC, 1960)*	*Legislative predominance (SRC, 1958)*	*Federalism (AIPO, 1967)*	*Decentralized parties (SRC, 1958)*	*Individual freedom (AV, 1960)*	*Expect equality (AV, 1960)*	*Property[a] (AIPO, 1953)*	*Religion (AIPO, 1968)*
Region											
East	86%	62%	71%	91%	58%	58%	51%	65%	88%	72%	60%
Midwest	87	54	NA	94	61	77	57	65	88	77	78
West	85	53	55	92	62	74	64	80	90	79	70
South	85	53	NA	88	65	85	58	69	80	73	85
Size of community											
Less than 5,000	88	55	NA	NA	61[b]	82[c]	57[b]	68	85	NA	NA
5,000–20,000	88	51	NA	NA	60[b]	83[c]	52[b]	62	85	NA	NA
20,000–100,000	85	61	NA	NA	61[b]	71[c]	63[b]	73	92	NA	NA
Over 100,000	87	55	NA	NA	61[b]	64[c]	54[b]	69	84	NA	NA

[a]Data are taken from 1953 since no ecological data are available for 1946.
[b]Size of local community for this group actually is: rural, 2,500–10,000; suburban, 10,000–100,000; urban, 100,000 and over.
[c]Size of local community for this category actually is: under 2,500; 2,500–49,999; 50,000–499,999; 500,000 and over.

TABLE V.8 *Regional cleavages and belief in equality*

Question: Do you think that the Declaration of Independence was right, only half right, or not right at all when it stated that "all men are created equal"?

	Percent "right"
Northerners	82%
Southerners	67

Source: Ben Gaffin and Associates, 1958; in Hazel Erskine, "The Polls," *Public Opinion Quarterly* (Spring, 1969), p. 149.

again note that although the range of support varies considerably within regions, in no case does less than a majority of the population support any tradition value. The East does have lower levels of support for more of the values than does any other region; but it also has higher support for some of the tradition values. The South is lower in support for equality than the other regions but cultural majorities do support the value. Indeed, each of the regions supports all measures of the tradition values.

Local Community Size. Two theories of the influence of community size upon cleavage have been thought important in the United States. The more traditional theory has suggested that small communities tend to develop the important civic virtues because the people have a close relationship to familiar and proximate government. This view sees the large cities as dehumanizing agencies which tend to destroy community trust and communal patterns of association.[96] The second view sees the large cities as centers of culture which tend to liberate the individual from the social restraints of pastoral intimateness. In the newer urban setting the individual is free and able to pursue the higher virtues without the intense demands for uniformity sought by local communities.[97]

Table V.7 suggests that there is little tendency for either the small community nor the large city to have more support

[96] For example see Lewis Wirth, "Urbanism as a Way of Life," *American Journal of Sociology* (July, 1938), pp. 1–23.

[97] For example, see Lewis Mumford, *The Culture of Cities* (New York: Harcourt, Brace, 1938).

for the liberal tradition. Indeed, except for the federalism value, there seems to be some tendency for the political culture values to be accepted most in the medium sized city.[98] However, again the most descriptive aspect of the data is the uniformity of the acceptance of the values for all size classifications.

In summary, ecological differentiation suggests some interesting hypotheses regarding different levels of culture support, but little evidence of important active cleavages exists. Some differences undoubtedly reflect authority-level conflict which may provide areas of potential cleavage. For both region and community size, however, all important groupings support the values of the tradition.

SOCIETAL CLEAVAGES

Of the innumerable ways to differentiate a population, four final societal divisions have been thought important in American social analysis. Sex, religion, age, and race differentiations will be used to further test whether the tradition is supported by all relevant social groups.

Sex. In the United States the two sexes normally perform very different societal roles. In the family specialization of labor, the male is oriented to the outside world, the female to the home. Not surprizingly these different role orientations affect political attitudes. Since politics is part of the outside world, research has shown that the man is more politically aware and that he participates in politics to a greater extent than does the woman.[99] But are there significant support differences between the sexes for the political culture?

Table V.9 suggests that the differences relating to sex characteristics do not seem to be important for political culture. Two measures are at the 10 percent difference suggested above. Yet these, and the other differences which occur, actually relate to the greater percentage of "no opinion" responses from

[98] This suggests that Dahl's optimum city size of between 50,000 and 200,000 may be the most appropriate for culture support; see Robert A. Dahl, "The City in the Future of Democracy," *American Political Science Review* (December, 1967), pp. 953–970.

[99] Lane, pp. 209–216; Fred I. Greenstein, "Sex Related Political Differences in Childhood," *Journal of Politics* (May, 1961), pp. 353–371.

TABLE V.9 *Societal cleavages in the American political culture: percentage of support for the liberal tradition value*

	Identity system		*Symbol system*	*Rule system*				*Belief system*			
Societal grouping	*National identity (BC, 1948)*	*Community trust (AV, 1960)*	*Constitution (FOR, 1940)*	*Popular rule (SRC, 1960)*	*Legislative predominance (SRC, 1958)*	*Federalism (AIPO, 1967)*	*Decentralized parties (SRC, 1958)*	*Individual freedom (AV, 1960)*	*Expect equality (AV, 1960)*	*Property[a] (AIPO, 1953)*	*Religion (AIPO, 1968)*
Sex											
Male	95%	55%	NA	91%	67%	74%	62%	68%	82%	NA	70%
Female	96	56	NA	89	56	75	52	69	88	NA	76
Religion											
Protestant	87[e]	55	NA	NA	63	78	58	70	86	NA	79
Catholic	84[e]	58	NA	NA	56	65	53	64	85	NA	75
Jewish	NA	65	NA	NA	71	NA	58	65	88	NA	NA
Other	NA	71	NA	NA	53	NA	53	79	86	NA	NA
None	NA	38	NA	NA	50	NA	63	63	78	NA	48[b]
Age[c]											
Young	96[f]	56	NA	95	60	68	62	84	87	76	71
Middle	94[f]	63	NA	93	63	74	58	69	85	76	73
Old	97[f]	48	NA	84	60	77	53	60	85	73	75
Race											
White	86[e]	59	64%	92	62	NA	56	70	88	NA	81[d]
Non-white	83[e]	29	50	74	52	NA	60	51	61	NA	81

[a]No data was available for 1946, so 1953 was used.
[b]This data is for "non-church goers."
[c]Young is 18–29 years; middle, 30–49; old, over 50.
[d]This question was worded, "Death is not tragic for the person who dies, only for the survivors," See Erskine, p. 155.
[e]The 1971 AIPO data are used.
[f]The 1971 figures are 78 percent, 86 percent, and 91 percent.

women (which probably is cultural but is not considered here as *political* culture) and not differences in negative support. For each measure reported both men and women are highly supportive of the liberal tradition.

Religious Affiliation. Religious cleavages have been historic dividers of states — perhaps the most divisive of any cleavage. The data of Table V.9 indicate that religious differences regarding fundamental regime values are substantial. In every case for which data is available our criteria for significant differences between groups is met. This confirms other empirical work which suggests differences between religious groups.[100] On the other hand, there is not a statistically significant relationship (for X^2, alpha = .05) between religious affiliation and most of the culture measures. Since Protestantism is the dominant religious subculture, it is not surprising that Protestants support the tradition more than does the second largest religion on six out of the seven measures. When the other religions are considered, however, the smaller religious groupings are often higher in support than are Protestants. Protestants actually support the tradition most highly on only two of the measures — a record matched by Jews and those identified as "other" religionists.

For all the major religious groupings, Table V.9 shows majority support for the tradition values. However, those with no religious affiliation (about 4 percent of the population as operationalized here) are the lowest in support on five values and are below majority support on two of these measures. One of these measures is the religion value — which, of course, is to be expected. The other is the trust value.[101] Yet this group does have a majority support for four of the six measures for which we have data.

[100] Gerhard Lenski, *The Religious Factor* (Garden City, N.Y.: Doubleday, 1960); Lawrence H. Fuchs, *Political Behavior of American Jews* (Glencoe, Ill.: Free Press, 1956); Seymour Martin Lipset, "Religion and Politics in the American Past and Present," in Robert Lee and Martin E. Marty, eds., *Religion and Social Conflict* (New York: Oxford University Press, 1964), pp. 69–120; and Harry M. Scoble and Leon D. Epstein, "Religion and Wisconsin Voting in 1960," *Journal of Politics* (May, 1964), pp. 381–396.

[101] Again, the operational problems with trust should be kept in mind. See note 48.

Although there are relatively large differences in support between religious groups, all of those with a religious affiliation support all of the tradition values. Even the group which has no religious affiliation supports most of the measures of the tradition.

Age. Another potential source of cleavage for the political system could be the breakdown of the socialization process. If the values of the culture are not transferred from generation to generation, political culture cannot be the historical process envisioned here. Each generation must accept the values or the tradition cannot be maintained. To some extent, no doubt, each generation questions the values of the preceding one. Some values are accepted, others are discarded. Of course, it is the argument of the present work that each succeeding generation has accepted the values of the liberal tradition. Many values have been discarded through the evolution from colony to industrial society but the core values of the political tradition must have been maintained if the hypothesis is to be accepted. If a socialization hiatus existed to the extent that one generation rejected the political values of the preceding one, the liberal tradition would not be the value system of the political culture for long. Is there any evidence to suggest that there is a break in the socialization process for political culture values?

Table V.9 suggests that the only significant differences between age groups appear for the values of trust and freedom. A rather remarkably larger proportion of the young believe in individual freedom as opposed to service to the state.[102] There is also a difference between the middle group and the oldest on this value. Indeed there is a moderate, significant statistical relationship between these two variables ($G = .272$; $X^2 = 48.8$; $p < .001$). In the other case of difference in attitudes, it is the old who most reject the tradition value of community trust — the value whose support has been most subject to variation throughout this analysis. This lack of

[102] One psychologist maintains that an increase in "privatism" — focus on the self rather than institutions — is the major difference between the young and previous generations, see Jeffrey K. Hadden, "The Private Generation," *Psychology Today* (October, 1969), p. 32.

support for trust by the old is the only age case in Table V.9 where support falls below the majority level. However, a further age breakdown shows that for the age group 18–25 only 48 percent support the value. The closeness of this support to majority support suggests that the operationalization of trust could be crucial for both the old and the very young.

The overall picture, in any event, appears to be one of consensus on tradition values. In the overwhelming number of situations in Table V.9, the values are supported by substantial majorities of each age cohort in the population. There does not seem to be an active cleavage between the young and the old. Actually, the young appear to be more supportive of the tradition than the old. While the old are least supportive in five situations, the young give the least support of the three generations in only two situations (the religion and federalism values) and the differences there are so small that one cannot have much assurance that any real difference exists in the population. For five values, the young are even more supportive than the middle-aged group. Even where a majority of the young (18–25 years) did not support the tradition trust value, it could be more a case of maturing than of rejecting a value of the tradition. Trust may take longer to inculcate through the socialization process than the other values of the tradition.

Conventional wisdom would take issue with these findings. The young are supposed to reject the values of their parents. Is there not a "generation gap" between the young and old? Three major national surveys have recently attempted to test whether there is a generation gap in values in the United States.[103] The first, a national study of both college and non-college youth sponsored by the Columbia Broadcasting System, concluded that large minorities of American youth are "sharply critical of our economic and political systems." These minorities were especially said to exist on the college campuses. On the other hand, this study also said that the findings

[103] Because of the indefinite boundaries of the youth and student populations, the three studies are less than ideal random samples. They appear to be quota samples and undoubtedly overrepresent the active student population. Since the active student is more opposed to tradition values, these samples are actually a more rigorous test than the true population would be.

TABLE V.10 *Generation, education, and political culture support: percentage of support for the liberal tradition values*

	Total adult	*Total youth*	*Non-college youth*	*College youth*
American democracy can respond to needs	NA	89%	91%	87%
American way of life can solve problems	89%	94	95	93
Hard work will always pay off	83	74	79	56
Right to private property is sacred	91	85	87	75
Religion very or fairly important	NA	84	87	69

Source: CBS News study *Generations Apart*, pp. 24, 28, 5, 5, 10.

of the study could be interpreted in another way — "with emphasis on the large majority of youth who retain orthodox and traditional values." In other words, a study which has been used to document the "generation gap" actually finds that the large majority of youth accept the traditional values of their polity.

The second survey, a Gallup study of college students, shows a substantial difference between the values of the young college student and traditional values relating to sex mores and college life.[104] In these areas there does appear to be a generation gap.[105] However, the Gallup data do not show political differences. The third study is directly related to the political attitudes of college students. This survey, commissioned by *Psychology Today,* found no rebellion among college students. The chief investigator concluded that today's college youth strongly conformed to the stated ideals of their parents and teachers.[106]

A close analysis of these data confirms that there is no active cleavage for the young, and that there is no generation gap for political culture values. This is apparent even in the CBS study *Generations Apart* (*sic*). Questions which relate to support of liberal tradition values show high levels of support. As is shown in Table V.10, youth are almost as supportive of

104 *Generations Apart,* pp. 4, 3; *Gallup Opinion Index* (June, 1969), pp. 19–37.

105 Also see *Generations Apart,* p. 17. But perhaps not as large as these seem to imply; see the Harris data on these points, "Change, Yes — Upheaval, No," *Life* (January 8, 1971), pp. 22–30.

106 Hadden, p. 79.

the political culture values as are their parents. In each case support came from more than a simple majority. The largest differences found in the study are actually between college educated and non-college educated – both for youth and their parents.[107] In all cases, though, all age generations support tradition values. Only those few who call themselves revolutionaries (1 percent of the youth) do not support the Lockean tradition.[108] The *Psychology Today* study strongly affirms this conclusion. Only 14 percent reject solution of conflict through traditional rule institutions.[109] Much current literature suggests that property, especially the capitalist system, is rejected by the young. Yet 61 percent of the college students supported the following statement: "The free-enterprise system is the single economic system compatible with the requirements of personal freedom and constitutional government."[110] Fifty-four percent supported the achievement value.[111]

In summary, there does not appear to be a generation gap nor an active cleavage for the young. Indeed, they appear somewhat more supportive of the political culture than the old. The only possible exception to this generalization may be the trust value. This may be a value of middle age (though it appears with a majority support as early as the age group 25 to 29).[112] In every other situation, whether an American is

[107] Also see Philip E. Converse, Warren F. Miller, Jerrold G. Rusk, Arthur C. Wolfe, "Continuity and Change in American Politics," *American Political Science Review* (December, 1969), pp. 1104–1105.

[108] *Generations Apart,* p. 69. The regime is rejected by 76 percent of this group.

[109] Hadden, p. 34.

[110] *Ibid.,* p. 35.

[111] *Ibid.,* p. 68. For the religion value, see Vernon Jones, "Attitudes of College Students and Their Changes: A 37-Year Study," *Genetic Psychology Monographs* (February, 1970), esp. p. 78.

[112] Compare with Easton and Dennis, pp. 300–303, where support for police rises from 53 percent for the age group 21–25 to 61 percent for the 26–30 group. However, see their very different interpretation of youth support (pp. 54–57 and ch. 14). Data from the 1968 SRC survey find an interesting division in attitude on the Chicago protests. Only the college educated under thirty supported the protest, but interestingly, the 21–25 age group is less favorable than the 26–29 age group; see John P. Robinson, "Public Reactions to Political Protest: Chicago, 1968," *Public Opinion Quarterly* (Spring, 1970), pp. 6–7. Other data even show that there is not a gap on one authority-level issue where "everyone" expected it – on Vietnam. Indeed, the young are somewhat more "hawkish" than the old;

young, middle aged, or old, he tends to support the values of the liberal tradition.

Race. Few divisions appear to have more potential for active cleavage in the United States than the one involving race. The only time the regime was seriously threatened, it stemmed partly from this cleavage. In one sense the racial problem seems to be an extension of the ethnic cleavage situations of the immigration periods in American history. Yet in another it seems to be so much more intractible that many have suggested that it is qualitatively different. Much of this thought seems to me to be based upon the simplistic "melting pot" analogy and its applicability to ethnic groups. Much of the opinion which finds the Negro cleavage qualitatively different assumes that the Negro has not been assimilated and that other ethnic groups have.

It would appear, however, that a pure melting pot assimilation never took place in the United States.[113] One authority maintains that America has not had either a melting pot or cultural pluralism but rather a structured pluralism within the major religious groups and within occupational categories.[114] Sociological theory seems to suggest that there are four major ethnic groups: white Protestants, Catholics, Jews, and Negroes.[115] Yet, although these observations seem to have some validity, there also seems to be some validity to the melting pot thesis. To me, Michael Parenti, following upon the work of Erich Rosenthal, makes the conceptual breakthrough which explains both observations. Parenti maintains that there are actually two dimensions to the integration of societies. He uses

---

see Milton J. Rosenberg *et al., Vietnam and the Silent Majority* (New York: Harper and Row, 1970), p. 65. Two other studies (both beyond our 1935–1970 time frame) also strongly reinforce the conclusions reached here. See *Gallup Opinion Index* (March, 1971) and "Change, Yes — Upheaval, No." Also see Seymour Martin Lipset, "Polls and Protest," *Foreign Affairs* (April, 1971), pp. 548–555.

113 Nathan Glazer and Daniel Patrick Moynihan, *Beyond the Melting Pot* (Cambridge: M.I.T. Press, 1963), p. 13.

114 Milton Gordon, *Assimilation in American Life* (New York: Oxford University Press, 1964), p. 159.

115 Glazer and Moynihan, p. 314. Also see Will Herberg, *Protestant, Catholic, Jew* (Garden City, N.Y.: Doubleday, 1960), pp. 19–23 and *passim.*

the Parsonian separation between cultural and social systems to make the crucial distinction.[116] This distinction is relevant to the theory used in the present analysis.

Parenti conceives the first dimension of social integration to be acculturation. This is a process whereby an ethnic group accepts the beliefs, values, and other norms and practices of the society. The second dimension is called assimilation and relates to the interrelations and associations which actually exist among individuals.[117] If the "melting pot" means that both these processes have taken place, then the United States never had a melting pot. The American system has been able to acculturate in every important sphere of social life almost all of the people who have been introduced into the community. The thesis of the present work is that it has done so politically. However, sociological research in the United States does not suggest that there has been any widespread social assimilation. Parenti maintains that "in the face of widespread acculturation, the minority still maintained a social substructure encompassing primary and secondary group relations composed of fellow ethnics." [118] In the United States, in other words, many ethnic groups have undergone acculturation but not assimilation. He maintains that his thesis applies even in the supposedly homogeneous suburbs.[119]

If the Negro is compared to the other ethnic groups, then, the comparison must be made on cultural patterns, not on assimilation criteria. The test must be made on cultural values. Both observation and empirical data suggest that current Negro militancy, while certainly not trivial, is not a real revolution – it is not fundamentally directed at the regime. It can more properly be called a rebellion or an insurrec-

[116] Michael Parenti, "Ethnic Politics and the Persistence of Ethnic Identification," *American Political Science Review* (September, 1967), p. 718; Erich Rosenthal, "Acculturation Without Assimilation? The Jewish Community of Chicago, Illinois," *American Journal of Sociology* (November, 1960), pp. 275–288.

[117] Parenti, p. 718.

[118] *Ibid.*, p. 719.

[119] *Ibid.*, pp. 719–722. Also see the evidence on the persistence of ethnic psychological identification, pp. 722–724; and Rosenthal, pp. 275–288. Notice how this differs radically from Easton's and Dennis's attempt to deal with socialization as it relates to the melting pot; see Easton and Dennis, p. 40.

tion.[120] Although protests are relatively severe and sustained, they do not indicate loss of support for the political culture values. While dramatic differences in attitudes at the authority level do exist,[121] these do not necessarily directly affect the political culture. Even lower rates of participation [122] do not necessarily show less regime acceptance but more probably relate to lower rates shown by comparable class groups. Higher levels of psychological alienation and authoritarianism [123] seem to have similar bases. Beyond these considerations, it should be emphasized that most of the differences cited above are related to other social systems and not primarily to the *political* system. For political culture the situation could still be one of consensus – even if economic and social values were evidenced by substantial divergences.

Dwaine Marvick maintains, however, that the racial situation is an active cleavage in the American political culture:

> Negro Americans in many ways are excluded from the dominant political culture of their community and nation, and are denied its rewards. Norms and roles are learned in a special Negro subculture. . . . [124]

Marvick holds that there is a separate black subculture with different political values.[125]

[120] Aaron Wildavsky, "The Empty-head Blues: Black Rebellion and White Reaction," *Public Interest* (Spring, 1968), pp. 3–16; and David O. Sears, "Black Attitudes Toward the Political System in the Aftermath of the Watts Insurrection," *Midwest Journal of Political Science* (November, 1969), pp. 515–544.

[121] William Brink and Louis Harris, *Black and White* (New York: Simon and Schuster, 1966), appendix D.

[122] Sidney Verba, "Democratic Participation," *The Annals* (September, 1967), p. 70.

[123] Hazel Erskine, "The Polls: Negro Philosophies of Life," *Public Opinion Quarterly* (Spring, 1969), pp. 151–152, 157.

[124] Dwaine Marvick, "The Political Socialization of the American Negro," *The Annals* (September, 1965), p. 113.

[125] In my opinion his data shows differences only in perceptions of performance, not values; see *ibid.*, pp. 119–120. Another study also suggests racial value difference: Leonard Broom and Norval D. Glenn, "Negro-White Difference in Reported Attitude and Behavior," *Sociology and Social Research* (January, 1966), pp. 187–200. However, this study reports an "Index of Dissimilarity" between the races of only 12.7 percent while the differences between Southern and Northern whites is 14.3 percent and the differences between high and low educated whites is 15.3

TABLE V.11 *Relationship between measures of the political culture and race*

Race	*Trust*	*Legislative predominance*	*Decentralized parties*	*Individual freedom*	*Expect equality*	*Popular rule*
Gamma	.502	.195	.111	.352	.646	.561
Chi-square, $p <$	.001	.05	.05	.001	.001	.001

Is there a distinguishable Negro *political* culture? Does this important segment of the community actually reject the liberal tradition? Table V.9 certainly shows substantial differences between whites and blacks regarding the values of the tradition. In seven out of the nine situations reported there, fewer blacks support the tradition values than do whites. Table V.11 shows that there is a statistically significant relationship between race (i.e., white/non-white) and the values of the tradition. It is relatively strong for popular rule, trust, and expectation of equal treatment (each of these values is related negatively to non-white racial characteristics).

In spite of these differences, however, Table V.9 shows that both races accept the tradition values. In only one situation, on the trust value, does less than a majority support the measure of the value. Even in that situation the findings are subject to question. It will be remembered that the operationalization of the trust value has been suspect through our whole analysis. A better operationalization is presented in Table V.12, which makes the data of Table V.9 very suspect. Table V.12 shows that almost six out of ten Negroes trust other people. It may even be, then, that in the single cultural subsystem for one element of which Table V.9 shows less than majority black support, a substantial majority of blacks feel a sense of community. For the other dimension of the subsystem, sense of national identity, Negroes strongly support political culture – over 80 percent believe their country is worth fighting for in a war.[126]

Negro support for rule institutions also seems high. In 1966,

percent. That is, the race difference is smaller than the regional or educational ones.

[126] Brink and Harris, p. 274.

TABLE V.12 *Race cleavage and identity system trust*

Question: Do you think most people can be trusted? (NORC, 1964)?

	National	Negroes
Yes	77%	59%
No	21	40
Don't know	2	1

Source: Erskine, p. 153.

TABLE V.13 *Race cleavage and belief in equality: percentage who agree Declaration of Independence was right on "all men are created equal"*

Southern black	83%
Northern white	82
Northern black	75
Southern white	67

Source: Erskine, p. 149; see Table V.8.

60 percent thought Congress had been helpful to Negro rights; 65 percent thought the Supreme Court had been, and majorities of those with an opinion even thought state and local institutions were helpful.[127] In one study of a neighborhood in Washington, D.C., about two-thirds supported popular rule.[128] Data on belief values also show substantial support for the tradition among Negroes. Table V.13 shows that blacks tend to believe in equality, though, as for whites, there are regional variations.

There are different patterns of value support by race on other beliefs. A slight majority of Negroes who had an opinion believed that the government should not own or closely control newspapers (although 84 percent of the whole population thought it should not).[129] On another property value,

[127] *Ibid.*, pp. 240–242.

[128] Study conducted by Professor Ernest A. Chaples in May, 1969, of the Fort Lincoln area of Washington. This study also showed high amounts of national identity support.

[129] FOR, 1939; in Hadley Cantril, *Public Opinion, 1935–1946* (Princeton: Princeton University Press, 1951), p. 417.

blacks and whites supported the value equally – of those with an opinion, six out of ten believe that things like electric power and housing should be left to private business.[130] In a third case, blacks supported the tradition more than whites. In 1964, 90 percent (as compared to 82 percent for whites) believed that "if you try hard enough you can usually get what you want." [131] On the other hand, even though he supports the value, the Negro does not believe that he has as good a chance to get ahead as "most people." [132] Yet he believes as strongly as whites that the political system responds to him.[133]

This review of racial attitudes suggests that blacks and whites both accept the liberal tradition as their political culture. For each of the tradition values, majority support was found within both races. The system model suggests, however, that there is a relationship between demands and support. The generally lower level of Negro support [134] suggests that the denial of the full rewards of the system may have had an effect on support. Certainly authority-level neglect of wants can influence support levels and could lead to an active cleavage. The data suggest, however, that at least for the present, race is not an active cleavage in the American political culture.

AMERICAN CONSENSUS AND CLEAVAGE

For the liberal tradition to be the political culture of the United States, its values must have consensus support by its

[130] SRC, 1958; in ICPR. On the other hand, of *Southern* Negroes, a majority would not leave these activities in the hands of private businessmen. Data courtesy of Donald R. Matthews and James W. Prothro (1961).

[131] Erskine, p. 151.

[132] *Ibid.*, p. 149.

[133] Angus Campbell and Howard Schuman, *Radical Attitudes in Fifteen American Cities* (Ann Arbor: Survey Research Center, University of Michigan, 1968), p. 41. He also has a more positive view of government; see Everett F. Cataldo, Richard M. Johnson, and Lyman A. Kellstedt, "Political Attitudes of Urban Blacks and Whites," in Steven V. Monsma and Jack R. Van Der Shik, eds., *American Politics* (New York: Holt, Rinehart and Winston, 1970), pp. 47–59.

[134] Actually, the Negro is much more supportive of the religion value than the white; the Negro feels more strongly about his religious beliefs and he practices them more; see Erskine, p. 155. Also see Norval D. Glenn, "Negro Religion and Negro Status in the United States," p. 624, but also p. 638.

system members. The data of the early chapters suggest widespread support for the values of the Lockean liberal tradition. It has been argued, however, that support cannot be said to be consensual unless it is diffused through the important groups in the society. The present chapter has investigated the extent to which the tradition is supported within the major potential societal cleavages. Eleven major values (sense of national identity, sense of community identity, liberal symbol support, popular rule, institutionalization – critically legislative predominance, federalism, decentralized parties, liberty, equality, property, and religion) and three subsidiary values (education, achievement, and altruism) have been used to operationalize the political culture of the United States. Of the eleven *major values,* only four have been found not to be supported by a majority in every group investigated. The religion value was not supported by the group not identifying with a formal religion. Decentralized parties were not accepted by the apolitical. Legislative predominance was not supported by the apoliticals and the least educated. Sense of community identity (at least with the operationalization available for secondary analysis) was not supported by the apolitical, the lowest income group, the least educated, those with no religious affiliation, the old, or the non-white.[135]

Data was not as available for the *subsidiary values.* One study suggests that education is differently evaluated depending upon class level.[136] Yet Table V.14 shows that although 91 percent of the prosperous recommend college education to their children, 68 percent of the poor also support education.[137] Data on achievement also show less support in the lower class.[138] One study shows that in a choice between a

[135] Again the operationalization of trust is somewhat suspect. Although the direction of support is undoubtedly in the direction indicated (i.e., less support among these groups) the absolute level is probably higher than indicated and perhaps supported by majorities in each group. The data of Table V.12 are again relevant.

[136] Herbert H. Hyman, "The Value System of Different Classes," in Richard Bendix and Seymour Martin Lipset, eds. *Class, Status, and Power,* 2d ed. (New York: Free Press, 1966), p. 490.

[137] It should also be noted that college education is not the same as education. College education may be less relevant for the low classes and material-gain education more important. On this point, see Ephram H. Mizuchi, *Success and Opportunity* (New York: Free Press, 1964), pp. 74–80.

[138] Hyman, pp. 492–494.

TABLE V.14 *Education support by class*

Question: After the war, if you had a son (daughter) graduating from high school would you prefer that he (she) go on to college, or would you rather have him (her) do something else, or wouldn't you care one way or the other? (Roper, 1945)

	Percent preferring college
Prosperous	91%
Upper middle	91
Lower middle	83
Poor	68

Source: Herbert H. Hyman, "The Value Systems of Different Classes," in Reinhard Bendix and Seymour Martin Lipset, eds., *Class, Status, and Power* 2d ed. (New York: Free Press, 1966), p. 491.

TABLE V.15 *Achievement support by class*

	Support achievement (good pay, 50–50 risk)	*Not support achievement (low income, security)*	*Don't know*
Executives	87%	8%	5%
Factory labor	53	45	2
Unemployed	51	45	4

Source: Cantril, p. 530. See Table IV.58 for question wording.

government that guarantees equal opportunities and one that guarantees a steady job, a majority in most occupation groups supported achievement but the semi- and unskilled labor groups did not.[139] However, Table V.15 shows support for achievement by three quite different occupational categories. Achievement, when operationalized in this form,[140] seems to be supported by labor. Indeed, it is supported by a majority of the unemployed! Data also suggest that support for altruism differs by class. Although seven out of ten of the highest class would work for a community service committee, less

139 Centers, p. 62. This study also shows that all classes support private ownership of business (p. 61).

140 However, for some other operationalizations in selected populations, achievement is not supported by the lower status occupation groups; see the 1942 and 1943 data in Cantril, p. 530. Also see data in the Appendix.

than a majority of the lower class group would.[141] Yet, the lower class may support the value and simply not have the time to work on committees. Data do show that well over half of the total amount of charitable contributions in the United States come from the lower class groups.[142]

Although there appear to be differences in group support for education, achievement, and altruism, the values do appear to be accepted by all major groupings. The low classes, especially, however, seem to support these values less – and this may be functional for the regime.[143] Yet a majority appear to support the tradition values even in this group. The data are limited for the secondary values, but they too seem supported by the relevant potential cleavages in the American system.

For the major values, 307 tests have been performed to find the extent to which Lockean values were supported by thirty-seven major potential cleavages. Only 3.5 percent of these tests did not show majority support for the tradition values. The data may also be viewed chronologically. If the values are divided into four time groups – 1940–1948, 1958, 1960, and 1967–1968 – support does not differ significantly for the four periods. That is, there does not seem to be any trend over time, at least with these indicators, in support for the tradition. Even if we look at the single value indicator which has the least widespread support – and again the problem may be only operational – the data show that it is supported by 80 percent of the groups. Since, overall, 96 percent of the tests showed that the values were supported, it must be concluded that the liberal tradition has the widespread support necessary for a consensual political culture. The data which has been presented suggest that the liberal tradition is strongly supported by the politically relevant groups.

Thirty-seven major social groups were identified. Only five groups (14 percent of the groups) showed support which fell below a majority for even one measure of the tradition. Apolit-

141 AIPO, 1969; in *Gallup Opinion Index* (April, 1969), p. 19.

142 F. Emerson Andrews, *Philanthropic Giving* (New York: Russell Sage Foundation, 1950), p. 59.

143 Mizuchi holds that the low classes limit their achievement expectations to what appear to be realistic in their own life situation. In this way non-attainment of achievement expectations does not become a source of stress for the regime; see pp. 74–75.

icals had less than majority support for most measures, the low income group for one measure, the least educated for two measures, the non-religion group also for two measures, and the aged for one measure. Even though these groups had less than majority support for some measures, no group – with one exception – had less than majority support for at least two-thirds of the measures. Indeed, only three values of the tradition are involved in the rejection and these are supported for the majority of cleavage situations. Even for the situations where majorities do not support, the problem may be operational.[144] But even the data reported here show that these four groups highly support most of the tradition values. One must conclude that, generally, all relevant social groups support the tradition.

Apoliticals constituted the only group that did not support the political culture. This is not a surprising finding. In the political specialization of labor the apoliticals have removed themselves from the political system. This 5 percent of the population has become irrelevant in terms of output and, thus, is not part of the cultural screen. It is not a meaningful exception to the general finding of tradition support by all relevant potential cleavages. This observation of weak support by those unrelated to specialization to politics also helps explain a pattern of support among other groups. Not only do Republicans, Democrats, and Independents support more highly than apoliticals, but attentives support more than non-attentives, upper and middle classes support more than the lower, more highly educated support more than less, middle-aged support more than either young or old, and whites support more than blacks. The groups more specialized toward politics are more supportive of the political culture. Again, this pattern does not mean that the less politically-oriented groups do not support the tradition. Every group (except the apoliticals) supports the liberal tradition as the political culture of the United States.

In sum, whether one considers political, economic, ecological, or other societal differentiation, there is a wide consensus

144 If the operationalization problem allows trust to be excluded, these four groups accept the tradition with only two exceptions – those with less than eight years of education for legislative predominance, and the non-religious for the religion value.

across potential cleavage. Of course, not every conceivable group in the American system positively supports the tradition. Some smaller groups are probably ambivalent and accept the regime only by sufferance. Groups made up of combinations of the less politically specialized potential cleavages may present such situations. The data do suggest additivity of these dimensions. For example, 90 percent of all American system members support popular control; while 68 percent of the low education group (0 to 7 years of schooling) support the value and 60 percent of the apoliticals do. The group which is both less educated and apolitical, however, shows support of only 44 percent.

Other groups undoubtedly actually reject the regime and its value system. It was noted above that 76 percent of college students who call themselves revolutionaries reject American regime values. It is very possible that other small groups also reject the tradition values. In considering groups smaller than the general potential social cleavages studied here, however, one is no longer dealing with mass support. Politically relevant groups composing less than 1 percent of the population (as in the example given above) should be considered as low-level elites. Actually, support by elite groups may be more important for regime maintenance than mass support.[145] Yet, elite support is beyond the scope of the present study.

At the mass level, for both the masses and their major social groupings, the liberal tradition is supported as a consensual political culture. Although one could never, in some absolute sense, demonstrate the existence of the liberal tradition as the political culture of the United States, I will argue that the widespread agreement on political values reported above and the depth of support shown here[146] represent a reasonable empirical demonstration of the descriptive hypothesis. Lockean liberalism may be used as an empirical description of the core of the member political culture of the United States.

[145] Almond, *The American People and Foreign Policy,* p. 142; and Key, *Public Opinion and American Democracy,* p. 537.

[146] It should be noted once again that only positive support has been reported for each potential cleavage grouping. In most cases the largest part of the remainder of support is *neutral* rather than *negative*. Since neutral support may not be negatively related to regime maintenance, the data reported here may be a very conservative estimate of support. See the Appendix for more data.

CHAPTER VI

Political Culture Maintenance

POLITICAL CULTURE VALUES UNDER STRESS

The political system has been assumed to have two major processes associated with its attempt to cope with stress. At its most basic level it attempts to persist as a system.[1] The concept of culture, however, adds another dimension. The political culture is thought of as a subsystem of the full political system. This subsystem copes by attempting to maintain its value composition. The full political system, then, exhibits both persistence and maintenance.

The concept of culture, however, leads one to an emphasis upon regime uniqueness. Though culture as a basic process is undoubtedly of interest to persistence theory, the very idea of culture suggests that polities differ in fundamental ways. It may even suggest that there is no one answer to the system question "How do political systems operate?" Perhaps culture is such a fundamental structure for social analysis that analysis can best proceed through the study of actual, empirical regimes – with specific compatible or non-compatible political cultures. Some common functions can undoubtedly be identified[2] but one can, perhaps, best study social systems when

[1] David Easton, *A Framework for Political Analysis* (Englewood Cliffs, N.J.: Prentice-Hall, 1955) pp. 79, 86–89. Also see Chapter 1.

[2] Gabriel A. Almond, "Introduction," in Gabriel A. Almond and James S. Coleman, eds., *The Politics of Developing Areas* (Princeton: Princeton University Press, 1960), pp. 3–66.

their functions are related to their own particular structure[3] – including their value structure.

To study specific regimes then, value maintenance may be a central empirical focus.[4] The repository of behaviorally important regime values is the political culture. As a political subsystem, the political culture has the properties of a system described by Easton.

> The peculiar characteristic that political systems share with other social systems, and even some mechanical and biological ones, is that they are not necessarily defenseless in the face of disturbances to which they may be subjected. The processes and structures of political systems are not freely buffeted about by the vagaries of chance. Because of the kind of system it is, the members of a political system have the opportunity, of which, however, they may not always take advantage, to respond to stress in such a way as to try to assure the persistence of some kind of system for making and executing binding decisions.[5]

At this level, systems seek simple persistence. Culture systems, however, additionally seek to *maintain* their value constellations. Indeed, Parsons finds that maintenance of the value system is a basic problem confronting social systems.[6] The present chapter will analyze this maintenance and shaping process. It will attempt to explain value support (or lack of support) in these terms. The culture is viewed as screening stimuli which may threaten its own survival as a value system. Weak stimuli which act on the culture may be ignored completely. Stronger stimuli are screened in terms of the value constellation. Even substantial stress can be shaped by the political culture as it seeks to maintain its value composition.

Stimuli stress on the value constellation of political culture can be viewed as operating in two basic ways. The first type of stress arises when environmental stimuli cause a conflict

[3] Fred W. Riggs, "Structure and Function: A Dialectical Approach," (a paper prepared for delivery at the 1967 annual meeting of the American Political Science Association), p. 6.

[4] It is important even if it is a "second order" problem, as it is to Easton (p. 86).

[5] *Ibid.*, pp. 77–78.

[6] Talcott Parsons, "Culture and the Social System: Introduction," in Talcott Parsons *et al.*, eds., *Theories of Society* (New York: Free Press, 1961), p. 964.

between two component values within the political culture. This can be called *intracultural stress.* In this case, since the stimuli are aided by valued beliefs, there is a greater possibility that the system will be influenced. The stability of the system is threatened and stress is great. When this situation arises there is a high probability that commitment to the affected value will be reduced. Using the system analogy, a research model can be outlined. At time one, support for political culture Value A is measured. At time two, a stimulus is introduced which is related to Value B and which threatens Value A. During the period of the confrontation between stimulus-Value B and Value A, stress is high for the political culture system. Support for Value A is lowered to accommodate Value B. However, Value A (as well as Value B, since the stimulus had to accommodate itself to it) has had its influence upon the stimulus. It has shaped the stimulus, to some degree, to itself. The stimulus has thus been shaped and tempered by political culture. After the stimulus recedes or is converted, the stress eases and, in the normal culture situation, Values A and B return to their former state. In cases of extreme stress, however, values may be altered.

The study of intracultural stress would be rather direct if there was a simple hierarchy to political culture values. Unfortunately culture systems do not appear to be this amenable to study. Myrdal has suggested that all Americans have the same values but weight them differently. Commager, on the other hand, has suggested that Americans tend to weight many values equally without any hierarchy. Very little research has approached this problem. Morris's empirical study has attempted to rank general social values, and Rokeach is intensively studying the relative strength of liberty and equality.[7] In general, however, there is little systematic evaluation of relative value support. Yet even with an almost total lack of data, our observations in Chapter IV do allow some observations

[7] Gunnar Myrdal, *An American Dilemma* (New York: Harper, 1944), 1: xliv; Henry Steele Commager, *The American Mind: An Interpretation of American Thought and Character Since the 1880's* (New Haven: Yale University Press, 1950), p. 39; Charles Morris, *Varieties of Human Value* (Chicago: University of Chicago Press, 1956); Milton Rokeach, "The Role of Values in Public Opinion Research," *Public Opinion Quarterly* (Winter, 1968–1969), pp. 547–559.

on tradition values. Apparently the legislative predominates over the executive (Tables IV.14 to IV.22) and, under certain circumstances, over the judiciary (Table IV.25). Liberty seems to be the belief most spontaneously identified with the benefits of the American governmental system (Table IV.41), although absolute levels of support could be higher for other values.[8] However, these are preliminary observations rather than a systematic attempt to build a belief hierarchy. That must be left to some other study. All that can be said is that some values appear more important to the American political culture than do others. Yet as value theorists have suggested and as empirical study has tended to confirm, the great majority of people cannot very extensively arrange their values.[9] Perhaps for the mass some few values do predominate over others, but it is improbable that all are arranged in some precise value hierarchy.

The second type of stress arises from direct interactions between environmental stimuli and a political culture value. This can be called *environmental stress.* In this case, the normal result would be the shaping of the stimuli in the direction of the value. However, this relationship also depends upon the strength of the environmental stimuli. When the stimuli reach a certain strength, stress can be so great that the culture values change. Even in this situation, though, the culture influences the stimuli. The stimuli are changed in the process of interaction. In any situation, the effects of interaction with the political culture should be evident in the outcome.

If there were accurate measures of the strength of stimuli and of support for the political culture values reported in the previous chapters, the resolution of stimuli-value interaction could be rather easily predicted. Unfortunately, such is not the case. All the system analogy and the concept of culture allow us to state is that systems can persist and that culture

[8] Rokeach suggests that his research has already been able to show that, for Americans, liberty is more important than equality. However, he is using social equality rather than the political equality of the tradition as used here; see Milton Rokeach, "A Theory of Organization and Change Within Value-Attitude Systems," *Journal of Social Issues* (1968), pp. 13–33.

[9] Commager, p. 39; Robin Williams, Jr., *American Society,* 2d ed. (New York: Knopf, 1965), p. 413; Philip E. Converse, "The Nature of Belief Systems in Mass Publics," in David Apter, ed., *Ideology and Discontent* (New York: Free Press, 1964), pp. 206–261. Also see Chapter II.

systems will seek to maintain themselves. We would, thus, predict that in most situations the value would be maintained. Fortunately, there is enough data to allow at least a crude test of this stress situation. The shaping hypothesis holds that the liberal tradition beliefs should be maintained over time. However, as in the situation of intracultural stress, support for the value may be moderated during the time of confrontation to accord with strong environmental stimuli. It shall be the task of the next few sections to demonstrate this process for the value conflict both within the political culture system and between the environment and culture values.

A CHANGING AMERICAN CHARACTER?

What have social scientists said about recent changes in values in the United States? Have our basic political values changed? [10] David Reisman has suggested that Americans have recently undergone a fundamental change in character. Although the shift from inner directed to other directed could take place without a change in values, a conclusion such as Reisman's could be damaging for the present hypothesis. Lipset maintains, however, that the values noted by Reisman have always been present in the American character.[11] They could, and have, existed together with the values identified here as the liberal tradition.

Lipset suggests that there has been a historical consensus on the values of social equality and achievement.[12] A more general consensus is found by a political scientist. In his review of American values, Almond found a "surprising consistency and agreement between the early and later observers" of American beliefs. Indeed, the major sociological studies mentioned above (especially Coleman, Gabriel, Gillin, McDiarmid, and Williams) and historical studies such as Brogan's all point in

[10] Seymour Martin Lipset, "A Changing American Character?" in Seymour Martin Lipset and Leo Lowenthal, eds., *Culture and Social Character* (New York: Free Press, 1961), pp. 136–171, raises the question which heads this section.

[11] David Reisman, *The Lonely Crowd* (New Haven: Yale University Press, 1950); Lipset, pp. 170–171.

[12] Lipset, *passim*. In terms of my analysis, I maintain that although there has been substantial social equality in the United States, there has not been a mass belief in it as a value. To the extent to which it is accepted, social equality is seen as a derivative of political equality and moral equality.

the direction of a continuity in values. Two more long-range empirical studies of values also seem to justify a conclusion of consistency through time. Using polls of school children between 1896 and 1944, Greenstein found that societal heroes have not changed from "production" to "consumption," as claimed by some, but that the types of leaders to be emulated have remained stable. Furstenberg, in an exhaustive review of accounts by foreign visitors to the United States between 1800–1850, finds that the basic social institution of the family seems to have retained essentially the same values after the substantial environmental effects of industrialization that they had held before. Even Kluckhohn, who suggests that a change has taken place, seems to give almost as much emphasis to the unchanging values as to the changed. Reisman himself, in a later article, seems to say that perhaps he has overemphasized the change he noted in *The Lonely Crowd*.[13]

Although the present hypothesis argues for a general continuity in values, it does not exclude change. Certain of the changes noted by the above analysts do represent change. The difference between the present hypothesis and their views is one of perspective. The present analysis uses the political system model, composed of social and political structures and values, considered at three major levels of analysis. From this perspective, there has been change but it has not been change in political culture values. Change in society

> is a function not only of its value system, but of its more detailed structure, in particular of the kinds and levels of *differentiation*

[13] Gabriel A. Almond, *The American People and Foreign Policy* (New York: Praeger, 1960), p. 39; D. W. Brogan, *The American Character* (New York: Knopf, 1944), esp. ch. 1; and Commager, especially in the first and last chapters; Fred I. Greenstein, "New Light on Changing American Values: A Forgotten Body of Survey Data," *Social Forces* (May, 1964), p. 448; Frank F. Fustenberg, Jr., "Industrialization and the American Family: A Look Backward," *American Sociological Review* (June, 1966), p. 337; Clyde Kluckhohn, "Have There Been Discernible Shifts in American Values During the Past Generation?" in Elting E. Morrison, *The American Style* (New York: Harper, 1958), pp. 147–217; David Reisman, "Some Questions About the Study of American National Character in the Twentieth Century," *The Annals* (March, 1967), pp. 45–47. Some maintain that there has been change, but they appear to reflect stress change or authority-level change. See Stuart Chase, "American Values: A Generation of Change," and Elmo Roper, "The Politics of Three Decades," in *Public Opinion Quarterly* (Fall, 1965), pp. 357–376.

> of that structure. Structural differentiation is, however, a function of several variables other than the value system, and can change independently of any change in the value-system itself. One of our principal theses about American society is that, whereas the paramount value-system has (as we believe) remained stable, the structure of the society has been undergoing a major process of differentiation. We believe that a major part of the phenomena that form the center of the analysis of Reisman, Kluckhohn, and others are results of these structural changes and can be analyzed in terms, not of the breakdown or disappearance of the component of normative order, or of a new one at the general-value level, but of new *specifications* of the general value system, in relation to new structural and situational conditions.[14]

As stress acts on the social system, the system attempts to cope with the disturbance. Adjustments can be made at various levels through social values, social structure, political values, or political structure. Some social values and structures (e.g., the family) are very resistant to change. Others, such as the work-organization in American society, are amenable to change. In some cases structures can change (the church may be an example) and values remain essentially the same. The precise resolution in any social situation depends upon the relative strengths of the stimuli, the values, and the structures, and the level at which stress operates.[15]

As all social scientists have noted, there has been substantial change in the American social system. But change has not taken place randomly through the system. Change has tended to focus first on social structure, then on social values, then on political structure, and last on fundamental political values. Because of the strength of the American political culture, most change has been deflected to the physical and social environ-

[14] Talcott Parsons and Winston White, "The Link Between Character and Society," in Lipset and Lowenthal, eds., p. 103; also see Williams, p. 449.

[15] An important difference between the Parsonian view and the present one is that the present view is less deterministic and does not believe that changing social differentiation of a long-range character has had a deterministic effect upon any *stable* weighting of political values. Of course, Parsons is talking about social values in general, not political values. For a similar view of some of the determinism in Parsons, see Harold Kaplan, "The Parsonian Image of Social Structure and Its Relevance to Political Science," *Journal of Politics* (November, 1968), p. 899.

ment. While social change has been substantial, fundamental political change has been limited. The argument of the present analysis is that the change which does take place politically takes place at the authority level. Neither the regime values nor the essential regime political structure have changed to any significant degree. There is an influence at the regime level, though. Because of the force of the stress, support for tradition values is reduced temporarily. When stress is reduced, however, the regime values become culturally supported once again.

MODELS AND EXPLANATION

A major argument of the present study is that there is behaviorally significant consensual member support for political values in the United States. Yet the major empirical value studies in political science tend to state otherwise. Much of the discussion on value consensus is ambiguous.[16] Key's findings on the subject have already been discussed. Two other major empirical studies, one by McClosky and the other by Prothro and Grigg, are also ambivalent in discussing the existence of and the importance of consensus on fundamental values.[17] The confusion, it seems to me, results from a basic misunderstanding of the content of the fundamentals which are actually believed in the United States and from the use of the less explanatory mechanical, rather than the biological-system analogy. Both of these lead to difficulties in evaluating member support for values.

To take the second observation first, the direct stimulus-response analogy leads one to expect that values should be uniformly and consistently supported. For example, both Mc-

[16] For example, see the "yes there is, no there isn't" discussion of agreement on fundamentals (yes on democracy but no on any content to it in a consistent creed) in Marian D. Irish and James W. Prothro, *The Politics of American Democracy,* 4th ed. (Englewood Cliffs, N.J.: Prentice-Hall, 1968), pp. 76, 75–78.

[17] V. O. Key, Jr., *Public Opinion and American Democracy* (New York: Knopf, 1961), pp. 41–50 (see also pages 34–36); Herbert McClosky, "Consensus and Ideology in American Politics," *American Political Science Review* (June, 1964), pp. 361–382; and James W. Prothro and Charles M. Grigg, "Fundamental Principles of Democracy," *Journal of Politics* (Spring, 1960), pp. 276–294.

Closky and Prothro-Grigg are surprised that commitment to values are reduced when specific circumstances are introduced which affect these values. Their assumption seems to be that a value should be supported regardless of the intervening circumstances. The system analogy, however, emphasizes the limitations of the use of the one-to-one relationship under all circumstances, and leads one to suspect that intervening circumstances may be crucial. The system analogy makes one more aware of the *ceteris paribus* qualification in any suggested relationship. McClosky is aware that values are "ordinarily encountered not in pure form or in isolation but in substantive contexts that are bound to influence the ways in which we react to them." Prothro and Grigg also note that there is more than one dimension of values and that they can conflict. They even mention, in a footnote, that this approach is "a fruitful type of investigation." [18] Yet both articles present the analysis of value-conflict as a secondary concern to their major one of reporting reduced value support in particular situations. It seems to me that this is the result of a concentration upon the less explanatory, mechanical, deterministic model of democracy. The more fruitful approach seems to be from the perspective of system.

The use of the mechanical analogy also has its consequences in the selection of the value system to be tested. The second analytical difficulty thus follows from the first. The two empirical works under investigation test for a consensus with an inappropriate value system. Both are aware of a distinction between the "textbook model of democracy" and "the reality"; [19] and McClosky and Prothro-Grigg do show rather conclusively that the "textbook model" is not the value system of the United States. The point is, however, that this does not at all suggest that there is no consensus on fundamentals. *It only shows that there is no consensus on the fundamentals of the "textbook"* (McClosky) *or the "democratic"* (Prothro-Grigg) *model.*[20] Consider some of the issues discussed in the theories

[18] McClosky, p. 376; Prothro and Grigg, p. 293.

[19] McClosky, p. 377; Prothro and Grigg, p. 293.

[20] Also see Fred H. Willhoite, "Political Order and Consensus," *Western Political Quarterly* (June, 1963), esp. p. 302; David H. Everson, "Consensus

tested. Prothro and Grigg identify the value system tested with two beliefs – majority rule and minority rights.[21] This may be an appropriate test for the "democratic model" but it is not at all a test for the liberal tradition model. The tradition does not hold a simple conceptualization of majority rule. Quite the contrary, the tradition expects a *limited* majority rule.[22] Or consider the McClosky article. He finds that Americans show no agreement on the value of equality.[23] However, he tests for social and economic equality (although he calls it political equality) which are not parts of the liberal tradition value system.[24] Indeed, McClosky's "equality" values are confronted with opposed beliefs which are in the tradition – property and achievement.

By using political culture within the system model, many of these problems are avoided. A simple stimulus-response model is not entertained. Following Ranney and Kendall, the present work must reject this model and its New England town meeting counterpart as inappropriate for a complex polity.[25] By directing attention away from the "textbook model," a hypothesis more oriented to the complexity of the real world is advanced for testing. The concept of political culture, especially when integrated into the system model, allows the researcher to view value data from another perspective. It allows one to discard the amorphous conception of "fundamental" for the scientific concept of culture. But it is the concept of system itself and its function of stress regulation that makes the present approach most useful. This analogy suggests the possibility of a conflict of values. Surprise at a reduction under stress in adherence to values turns to an *expectation* of societal conflict processes and situational value stress. One does not expect that support for the value of liberty will be unaffected if

---

and Democracy," *Public Affairs Bulletin* (January–February, 1970), esp. p. 4; and Steven R. Brown, "Consistency and the Persistence of Ideology: Some Experimental Results," *Public Opinion Quarterly* (Spring, 1970), pp. 60–68.

[21] Prothro and Grigg, pp. 282–283.

[22] See pages 144–145.

[23] McClosky, pp. 367, 369.

[24] See pages 201–206.

[25] Austin Ranney and Willmoore Kendall, *Democracy and the American Party System* (New York: Harcourt, Brace and World, 1956), part 1.

a particular use of liberty threatens other values which are just as important (or perhaps more important) in the political culture; or that free speech will be absolute if it is perceived as a threat to other tradition beliefs.[26]

McClosky and Prothro-Grigg performed a needed test of the textbook model. In both works significant culture and system problems were raised. For example, both questioned whether agreement on fundamentals was needed in a democracy. The present work, however, with its *explicit* use of the system model and the concept of political culture, is additionally forced to deal with the concept of value stress. Of course, the explicit use of this formal theory is not absolutely necessary for one to be implicitly guided by it. In his widely read work on civil liberties, Samuel Stauffer used an *implicit* stress model to explain American beliefs.[27] In my opinion the Stouffer analysis is more explanatory because of this emphasis upon value conflict. American opinions in the area of Stouffer's study are the result of conflicts between support of civil liberties and fear of communism. If the value of research results is directly related to the descriptive power of the analogy used (as I believe it is), Stouffer was successful because of his use of a systems analogy. The use of an explicit political culture system in the present study is intended to derive even more fully the benefits of this extremely heuristic analogy.

AMERICAN VALUES UNDER STRESS

It has been the plan of this book to elaborate theory by the introduction of relevant data. The present section will provide explanatory data on value clashes as culture-system related. Our interest at this point is to look at a large amount of opinion data in possible intracultural and environmental conflict situations to judge whether the system analogy and

[26] The two theoretical considerations noted above explain most of the differences between the present analysis and those of McClosky and Prothro-Grigg. The remainder, especially in McClosky, can be traced to wording difficulties — e.g., "Any person who hides behind the law when he is questioned about his activities doesn't deserve much consideration." What does "consideration" mean — moral consideration, legal consideration, physical consideration, etc.? See McClosky, table 2, p. 366.

[27] Samuel A. Stouffer, *Communism, Conformity and Civil Liberties* (New York: Wiley, 1966; originally published 1955). Also see Alfred Winslow Jones, *Life, Liberty and Property* (Philadelphia: Lippincott, 1941).

the concept of cultural shaping can explain the data more satisfactorily than the deterministic models. Only if the present perspective provides a reasonable interpretation in these preliminary data situations will detailed tests of the shaping hypothesis be performed. Unfortunately the limitations of resources and design have forced the use of less than ideal tests especially at this preliminary stage. However, there is enough useful attitudinal data to allow a survey of certain value clashes. Most empirical data is limited to one question at one point in time but there is enough information on a few situations that they approach attitudinal case studies. All of the situations reported should be viewed as attempts to establish an empirical base for the later more direct testing of the shaping hypothesis.

INTRACULTURAL STRESS

The analysis of values in conflict – even of tradition values – is not original to the present study. As mentioned, Lipset analyzes conflict between equality and achievement. Sabine views the clash between liberty and equality as the basic conflict in the American political arena. A central value clash in Locke's own analysis is the conflict between freedom on the one hand and the necessity to be subject to rules on the other. Madison holds that a good government is one which is able to balance the conflicting values of stability and liberty. Indeed, Lerner holds that values can only be explained as an interplay between value "polar opposites." [28] Each of these authors deals with values in conflict. The present perspective views these not necessarily in natural conflict but in conflict as environmental stress acts simultaneously with one value and against another. The present perspective empirically expects reduced support under stress and a return to consensus support when stress is reduced.

Press Freedom and Belief and Rule Stress. Chapter IV shows that about eight out of ten people express support for freedom

[28] Lipset, pp. 136–171; George H. Sabine, "The Two Democratic Traditions," *Philosophical Review* (October, 1952), pp. 451–474; John Locke, *Second Treatise on Government,* in Sir Ernest Barker, ed., *Social Contract* (New York: Oxford University Press, 1962), sec. 95, p. 56; *The Federalist Papers* (New York: New American Library, 1961), No. 37; Max Lerner, *America as a Civilization* (New York: Simon and Schuster, 1957), p. 73.

TABLE VI.1 *Means of censorship*

Question: By what means should newspapers and magazines be prevented [from writing on the questions mentioned in Table VI.2]? (FOR, 1939)[a]

Public opinion	34%	Government control or censors	8%
Editor's good taste	29	Others	1
Libel and obscenity laws	15	Don't know	12

Source: Cantril, p. 417.
[a]Asked of those who thought newspapers or magazines should be prevented from writing about some subjects.

of the press. This amount of support is quite remarkable because, relatively, the press is not highly regarded as an institution. Most do not think that the press has actually abused its freedom,[29] and, as has been mentioned, very few favor censorship. Twice during the depression, however, Americans were asked to decide which group abused their power more – bankers, press, radio, pulpit, or veterans. In both 1936 and 1938 the press was mentioned most often as being the least responsible with their power.[30] In another survey in 1969, 42 percent were found to think that television networks favored a particular side in presenting political news while 40 percent thought that they dealt fairly with all sides.[31] Many of those who criticize the press think that press freedom should be restricted. A majority, however, favor an amount of control which is consistent with the tradition – that is, they do not generally believe that the government should enforce press restrictions.[32] As Table VI.1 shows, 63 percent would employ only broad societal and personal pressures. Another 15 percent would strengthen libel and obscenity law. Only 8 percent would use direct government control.

Although Americans have reservations on the press they do

29 FOR, 1939; in Hadley Cantril, *Public Opinion, 1935–1946* (Princeton: Princeton University Press, 1951), p. 417. Also see Hazel Erskine, "The Polls: Opinion of the News Media," *Public Opinion Quarterly* (Winter, 1970–1971), pp. 630–643.

30 FOR, 1936, 1938; in Cantril, p. 416.

31 AIPO, 1969; in *Gallup Opinion Index* (January, 1970), p. 8. For newspapers, the figures were: 37 percent deal fairly; 45 percent favor one side (p. 9).

32 AIPO, 1941; in Cantril, p. 417. Also see p. 10.

support the freedom of the press value rather strongly, with culture majorities. Yet value conflict can affect this support. Table VI. 2 reports support data for the value in some rather specific support situations. For most instances reported, there is a consensus support for freedom of the press. The only situations where this does not take place are when that value is confronted with questions perceived to threaten other tradition values. All three cases where a majority would restrict the press relate to moral-religious values. Allowing the press to attack a public official without allowing his reply seems to be the most extreme case of value conflict. The moral-connected idea of fairness seems to outweigh freedom of the press with approximately eight out of ten Americans.

Even some of the intermediate reductions in the number of people committed to freedom of the press could be explained in terms of some of the values within the political culture. Attacking President Roosevelt threatens the executive rule-value; attacking the way a company operates its business and its labor policy threatens the property belief-value. In most situations, most of the American people support the value of freedom of the press. Yet Table VI.2 shows that support for a value depends upon the context in which it is raised. Consensus majorities support press freedom "even if it injures the sale of a product and lessens the income of the producer," but these majorities will be weakened if the press value threatens other more deeply held values such as, perhaps, moral values.[33]

Table VI.3 presents another example of value conflict within the political culture involving freedom of the press. In this situation the liberty value is chosen over the rule value by a substantial degree. However, the support for the abstract belief in freedom of the press is also reduced in this value conflict. Table IV.44 showed that about eight out of ten Americans support the abstract value of freedom of the press. When this value is in conflict with the judicial rule value, however, as shown in Table VI.3, support is reduced to six out of ten.

Freedom of the press is a central value of the liberal tradition. Much data suggests that Americans believe in it and that

[33] In 1969, 85 percent would like to have stricter state and local laws against obscene literature; see AIPO, 1969; in *Gallup Opinion Index* (July, 1969), p. 18.

TABLE VI.2 *Freedom of the press vs. various values*

Question: Even if it injures the sale of a product and lessens the income of the producer, do you think that newspapers and magazines should have the right of honest criticism about a book or movie, the quality of gasoline, the labor policy of an employer, the way a company operates its business? Do you believe that newspapers and magazines should be allowed to: (FOR, 1939)

	Support for press liberty	*Opposition to press liberty*	*Don't know*
Criticize a book or movie	82%	11%	7%
Describe methods of fighting syphilis	79	11	10
Show pictures of accidents and war	76	19	5
Attack gasoline quality	73	17	10
Attack the Communist Party	65	13	22
Attack the labor policy of an employer	60	28	12
Attack the way a company operates its business	58	31	12
Attack the Nazis	56	20	24
Attack President Roosevelt	54	38	8
Print divorce hearings	39	52	9
Print fine painting of a nude	37	53	10
Print a speech attacking a public official without printing his reply	12	80	8

Source: Cantril, pp. 416, 417.

TABLE VI.3 *Freedom of the press vs. judicial rule values*

Question: Which do you think is more important – that newspapers should be allowed to criticize the decisions of judges (courts), or that judges (courts) should be free from such criticism? (AIPO, 1940)

	More important value		
	Freedom of the press	*Judicial rule*	*No choice*
Judges question	64%	23%	13%
Courts question	63	22	15

Source: Cantril, p. 417.

it is an important part of the political culture. On the other hand, freedom of the press is not the only value in the culture. Other values are also strongly held within the tradition. When strongly held values conflict, support for each of them is reduced. It does not seem reasonable to argue that a decrease in support for freedom of the press in a conflict situation means that Americans do not support the value. It simply means that other values can influence support for press freedom. When these values appear threatened, support for the liberty value is reduced. When the conflict situation is resolved and the threat removed, support for the value apparently returns to a consensus level. Freedom of the press is a widely held value of the tradition, but it is not as widely held when it conflicts with other values in the tradition.

Other Stresses Between Tradition Values. Property, federalism, and altruism are highly supported tradition values. A simple model may expect that all values are held regardless of the situational milieu. The model used here, however, expects that certain values of the tradition could come in conflict with others and change the strength of support. When values associated with decentralization (federalism) are related to the property value, Americans support small businesses more than large ones.[34] When asked if "big business concerns are a good thing or a bad thing for the country," substantially fewer people will support this idea of property than will support the general value of property.[35]

Again, support for altruism leads Americans to accept the idea of welfare and relief but they do not accept it under all circumstances. Table IV.65 showed that 86 percent supported welfare as an ideal at the local level. However, support is much reduced for *federal* welfare programs. For example, a recent poll showed that most people had "mixed feelings" on welfare and relief programs – few were "favorable" but few rejected them completely.[36] The results are even more dramatic when achievement values conflict with altruism values. Over eight out of ten believe that the physically able should work while

[34] AIPO, 1937 and 1938; in Cantril, p. 335.
[35] AIPO, 1937; in *ibid.* Cf. Table IV.52.
[36] AIPO, 1965; in *Polls* (Summer, 1965), p. 72.

on relief.[37] Yet both values are accepted. In a study done in the 1930's in labor-troubled Akron, Ohio, even under extremely unfavorable circumstances the value of property was accepted.[38] Corporate property support was low. Even under this extreme stress, the value itself appeared to be supported; yet that support was mitigated by stronger support for the altruism value.[39] Within value systems, support for one value often conflicts with and influences support for others.[40]

Stress and the Judicial Value. The Supreme Court has frequently been involved in intracultural value clashes and, fortunately, enough data are available to more properly document this final example of intracultural value stress. Indeed, it may be taken as a case study in how this type of value conflict is viewed. The "court-packing" proposal of the 1930's is perhaps the most widely known public opinion value clash in recent American history. In this case two rule values were in conflict – the value of the executive and the value of the judiciary. Franklin Roosevelt sought to reduce the power of the Republican appointees on the Court who had invalidated crucial New Deal legislation. Shortly after his impressive electoral victory in 1936, Roosevelt requested that Congress give him the power to appoint additional justices for each judge over seventy years old on the Court. The Court and its adherents fought the change with judicial rule value support and Roosevelt fought back with executive rule values (of course, authority-level policy considerations were also involved, but for simplicity they will be ignored). Each participant in the conflict, thus, represented a central value of the tradition and each used the symbolism which would maximize the popular appeal of its position.[41]

Early in the controversy, Gallup asked the people if they

37 *Ibid.*

38 Jones, pp. 333–334.

39 *Ibid.*, pp. 351, 334. Also see the data from the depression era.

40 *Ibid.* Especially see Sabine.

41 Harry P. Stumpf, "The Political Efficacy of Judicial Symbolism," *Western Political Quarterly* (June, 1966), pp. 293–303; also see Robert A. Dahl, "Decision-Making in a Democracy: The Role of the Supreme Court as a National Policy-Maker," *Journal of Public Law* (Fall, 1957), pp. 279–295.

TABLE VI.4 *Executive vs. judicial values: FDR "court packing"*

Date of poll	*Favor FDR plan*	*Oppose FDR plan*
February 15, 1937[a]	47%	53%
March 10, 1937	48	52
March 15, 1937	49	51
April 5, 1937[b]	47	53
April 26, 1937	46	54
May 15, 1937	42	58
May 24, 1937	41	59
June 7, 1937	40	60

Source: Cantril, pp. 149–150.

[a]Actual question in February and March: "Are you in favor of President Roosevelt's proposal regarding the Supreme Court?" (AIPO)

[b]Actual question in April, May, and June: "Should Congress pass the President's plan (proposal) to enlarge the Supreme Court?" (AIPO)

thought that "some kind of change is necessary regarding the Supreme Court." On the two occasions when the question was asked, majorities agreed with the president that some change was necessary.[42] One might expect, then, that Roosevelt's plan would be accepted. However, as Table VI.4 shows, support was not widespread for the executive proposal. At no time during the controversy did a majority accept the specific proposal offered by the president. It appears that neither major rule-value was clearly supported during this conflict.[43] The presidential proposal was rejected. The usual explanation has been that it was denied by Congress because public opinion was opposed. Actually, public opinion was very complex. When given three alternatives in February of 1937, 38 percent thought that Congress should pass the president's plan, 23 percent thought that Congress should modify it, and 39 percent thought it should be defeated.[44]

The present perspective does not view public opinion from the mechanistic "textbook" model. Public attitudes are not viewed as providing precise orders which Congress follows.

42 AIPO, 1937; in Cantril, pp. 149–150.

43 A year later, 57 percent said they had opposed the court packing, 27 percent had favored it, and 14 percent had no opinion (Cantril, p. 151). It should be noted that most of the data do not include "no opinion" responses. Both questions which do show this group find 14 percent without opinion on the subject (Cantril, p. 150).

44 AIPO, 1937; in Cantril, p. 149.

TABLE VI.5 *Judicial value support before and after stress*

Question: Do you think the Supreme Court has recently: (FOR)

	Before value stress April, 1936	*After value stress July, 1937*
Stood in the way of the people	22%	23%
Protected the people	39	43
Neither	6	5
Both	2	4
Express no opinion	31	25

Source: Elmo Roper, *You and Your Leaders: Their Actions and Your Reactions, 1936–1956* (New York: Morrow, 1957), p. 31.

Mass opinion rather takes the form of political culture support which is expressed in a broad form which can be called mood. In the present case the conflict of two accepted values in institutional form led to a feeling of unease and a loss of support for both values. In this type of conflict the value in the defensive position probably has an advantage. The variety of responses (total rejection through modification) tend to act as opposition to the specific proposal made. Although the president's proposal was defeated, did he really lose? The Court was not reorganized but it did accept his New Deal program. Indeed, most political scientists believe that the president won.[45] Yet the "court packing" proposal was the first in which the "conservative coalition" emerged in Congress. Many presidents (including Roosevelt) would question whether the executive did not lose.[46]

Whichever value gained relative to the other, it appears that both suffered some loss of support during the controversy. On the other hand, the Court appeared to return to its previous level of support soon after the controversy ended. Table VI.5 suggests that by July support had returned to the level of the previous April (though the relatively low level of support at that time suggests that some value loss accompanied

[45] William H. Young, *Ogg and Ray's Introduction to American Government*, 12th ed. (New York: Appleton-Century-Crofts, 1962), p. 425.

[46] James T. Patterson, *Congressional Conservatism and the New Deal* (Lexington: University of Kentucky Press, 1967).

the New Deal decisions). Undoubtedly, support for the executive also returned to pre-controversy levels. However, Roosevelt himself never regained the level of support he had in 1936.[47]

By 1949, 60 percent had a favorable view of the Court. Shifts of support for the Court had stabilized to the point where as many were supporting it as were drawing away.[48] During the 1950's, however, the Court again became active in opinion controversy. Opposition to the first controversial decisions, involving school desegregation, were limited to the South. Outside the South, most agreed with the decision.[49] By 1964, however, opinion had changed and most people were looking for some position between desegregation and segregation.[50] The Court's integration position originally had been a positive support for the judicial-rule value. By 1964, although it was not actively opposed, public opinion on the issue did not provide major value support. Other positions taken by the Court in the fifties also appear to have been quite unpopular. Indeed, three studies of the judiciary hold that the *Barenblatt* decision (allowing congressional investigations of wider scope than implied by the earlier *Watkins* decision) and the *Uphaus* decision (limiting the exclusion of states in security cases required by the earlier *Sweezy* decision) were decisions made in response to public opinion.[51] Unfortunately, direct data is not available for these cases. Considering the amount of support for legislative and federalism values within the political culture, however, the conclusions of these authorities may be justified. It may actually be that the earlier unpopular decisions limiting Congress and the states were revised at least partially due to a clash of tradition values where the judicial was the more weakly held value. Toward the end of

[47] See Table VI.16 and the related discussion.

[48] Kenneth M. Dolbeare, "The Public Views the Supreme Court," in Herbert Jacob, ed., *Law, Politics, and the Federal Courts* (Boston: Little, Brown, 1967), p. 202.

[49] *Gallup Opinion Index* (August, 1967), p. 18.

[50] SRC, 1964; in ICPR.

[51] Charles H. Sheldon, "Public Opinion and High Courts: Communist Party Cases in Four Constitutional Systems," *Western Political Quarterly* (June, 1967), pp. 341–360; Walter Murphy, *Congress and the Court* (Chicago: University of Chicago Press, 1962), pp. 246; and C. Herman Pritchett, *Congress Versus the Supreme Court* (Minneapolis: University of Minnesota Press, 1961), p. 132.

TABLE VI.6 *Religious vs. judicial values: school prayer decision*

	1963 (AIPO)[a]	*1964 (SRC)*[b]	*1966 (SRC)*	*1968 (SRC)*
Agree with Court decision	24%	15%	13%	14%
Disagree with Court	70	74	75	81
No opinion, depends, etc.	6	11	12	5

Source: ICPR; *Polls* (Spring, 1965), p. 50.

[a]Actual question: "The United States Supreme Court has ruled that no state or local government may require the reading of the Lord's Prayer or Bible verses in public schools. What are your views on this?"

[b]Actual question for 1964, 1966, 1968: "Some people think the public schools should be allowed to start each day with a prayer. Others feel that religion does not belong in the public schools but should be taken care of by the family and the church. Have you been interested enough in this to favor one side or the other? (If yes), which do you think?"

this decade, there were five times as many persons decreasing support for the Court compared to those who were increasing their support.[52]

It is when judicial values clash directly with the value of religion, however, that the judicial value comes under most stress. Fortunately, direct data is available for this important value clash. When the values of judicial rule and religion conflict, Table VI.6 indicates that the religion value is much more highly preferred. More than seven out of ten of those questioned disagreed with the court decision on prayer in public school. One empirical study found that of the spontaneous mentions of the prayer decision, nine out of ten were unfavorable.[53] Although the religion value seems to have been a central stress to the judicial value in the 1950's and 1960's, other decisions of the Court in that period also tended to be unpopular. Criminal cases, such as *Esposito* and *Miranda,* led a great majority of Americans to believe that courts are not strict enough in applying rules to criminals.[54]

52 Dolbeare, p. 202.

53 Walter F. Murphy and Joseph Tanenhaus, "Public Opinion and the Supreme Court: The Goldwater Campaign," *Public Opinion Quarterly* (Spring, 1968), p. 36. Yet this did not necessarily lead to a rejection of the judicial value; 44 percent of those who prefer prayer in schools said the Court decision was acceptable. See Kenneth Dolbeare, "The Supreme Court and the States," in T. Becker, ed., *The Impact of Supreme Court Decisions* (New York: Oxford University Press, 1969), p. 211.

54 AIPO, 1968, 1969; in *Gallup Opinion Index* (March, 1969), p. 12. In 1968, 63 percent thought the courts were not dealing harshly enough with

TABLE VI.7 *Rating of the Supreme Court*

Question: In general, what kind of a rating would you give the Supreme Court? (AIPO)

	1963	*1967*	*1968*	*1969*
Excellent	10%	15%	8%	8%
Good	33	30	28	25
Fair	26	29	32	31
Poor	15	17	21	23
No opinion	16	9	11	13

Source: AIPO, in *Polls* (Spring, 1965), p. 50; *Gallup Opinion Index* (August, 1967), p. 15; *ibid.* (August, 1968), p. 25; and *ibid.* (July, 1969), p. 14.

There is some evidence to suggest that the tradition value conflicts in which the Court has been involved since the 1930's may actually be harming its position as a cultural value. In 1946 almost as many respondents thought that the Supreme Court decides questions on the basis of politics as thought that it decided on more impersonal grounds.[55] In 1967, although 47 percent believed that the Court was impartial, almost one-third thought that it actually favored one group more than another (see Table IV.24). Most people undoubtedly still support the judicial value but the recent mood has tended to be more and more unfavorable. Those who are changing their support are doing so negatively. A 1957 Gallup study found that only 20 percent of the people had changed their attitude towards the Court in recent years; however, three-quarters of these changes were in an unfavorable direction.[56] Table VI.7 shows the rating of the judicial value between 1963 and 1969. As early as 1963 support is not cultural. In that year, a plurality of 41 percent thought that the Supreme Court was doing only a fair or poor job. This negative and neutral support increases in 1967, and by 1968 a majority actually gives this lower rating to this value. During the period 1963 and 1969, positive support decreases 10 percent.

criminals. By 1969, this opinion was held by 75 percent of the population. Other values were also affected; wiretapping was approved by 45 percent and disapproved by 46 percent in 1969. See *ibid.* (September, 1969), p. 14.

55 AIPO, 1946; in Cantril, p. 151.

56 AIPO, 1957; in *Washington Post and Times Herald* (July 28, 1957), p. E5.

The loss of judicial value support during this period of value conflict may be an indication of permanent weakening of this tradition value. The conflicts, after all, have involved major political culture values — the executive, legislative, federalism, and religion values. The present perspective suggests, however, that this is a temporary loss due to a long series of unpopular decisions.[57] Yet the theoretical position does leave open the possibility that frequent value clashes within the political culture could lead to loss of support even for a tradition value.[58]

In this section three examples of value conflict have been considered. In each situation when two values of the liberal tradition conflict, both values lose support. Presumably stress is resolved in favor of the more strongly held value. Although there is certainly not enough information to provide a hierarchy of these values, some patterns can be noted. The strength of the religion value in each case study is worth notice. A second observation of some importance is that the values appear to be able to return to their consensus level of acceptance after a value clash. Third, over a period of time with repeated value clashes, a particular value could possibly lose its consensus support. Finally, these preliminary observations on value clashes suggest that the idea of shaping derived from the system model and the concept of culture has a utility in trying to explain opinion and value data which, without these conceptual insights, appear only as aberrations.

ENVIRONMENTAL STRESS

In the second general stress situation related to values, the political culture is subject to direct threat from the environment. The stimuli can come from an infinite variety of sources in the intra- or extra-societal environments. Because political culture acts as a value system, most environmental stress is easily tempered and shaped by the culture. This shaping func-

57 Murphy, p. 268. Loss of support is probably at the authority level — i.e., opposed to its then-present membership.

58 See Samuel Krislov, *The Supreme Court in the Political Process* (New York: Macmillan, 1965), p. 155. Any pattern may also be subject to cleavage differences. Judicial support differs by party and region on some decisions. See Dolbeare, pp. 206–213; and Kenneth Dolbeare and Philip E. Hammond, "The Political Party Basis of Attitude Toward the Supreme Court." *Public Opinion Quarterly* (Spring, 1968), p. 24.

tion will be considered in some detail below. At this point in the study, analysis will focus upon powerful stress stimuli which affect the values of the American political culture. Data is fortunately available for three important stress situations: the Great Depression of the 1930's and two war-related stimuli for the 1940's and 1950's.

Environmental Stress from Economic Depression. During the period of time covered by the present study the Great Depression of the 1930's must be considered, in terms of perception or actuality, as a major stress producer. The value of property has been especially subject to the strong environmental forces of this major environmental stimulus. The widespread impression was that the economic system had failed and that it needed substantial reform. Government ownership of property was one proposed solution. Yet member support for the value of property was probably strong even under these threatening circumstances. Little government ownership was actually accomplished although governmental regulation did increase in this period with substantial popular support.[59] Yet as early as 1936 there is evidence of majority support for the property value.[60]

Table IV.54 has shown data for ten areas of support for the property value in the 1930's and 1940's. Little data is available showing support for government ownership through the whole period of the depression.[61] Even the little support which exists for government ownership seems to recede as opinion is measured further and further from the stimulus of the depression. The same pattern is evident for the data of Table VI.8 regarding governmental ownership of the electric power industry. Although the early data is somewhat unreliable because of question wording, the pattern seems evident. In 1940 a simple plurality supported private ownership of electric power prop-

[59] Cantril, pp. 694–696. However, most Americans appear to be satisfied with existing amounts of regulation rather than requesting more. This was true even in 1936. See Cantril, p. 344, question 1.

[60] AIPO, 1936; in *ibid.*, p. 694.

[61] The form of the response categories allowed explains what appears to be substantial support for government ownership of banks. Notice that the 1942 question indicates support for regulation but very little support for actual ownership. This was probably the case through the Depression (see Table IV.54).

TABLE VI.8 *Depression stimulus vs. property*

Question: Do you prefer public (governmental ownership) or private ownership of the electric power industry? (AIPO)

Date of poll	*Private ownership*	*Government ownership*	*Don't know no opinion, etc.*
January 25, 1937[a]	29%	58%	13%
January 18, 1938[a]	34	54	12
January 31, 1940	49	31	20
June 12, 1945	50	29	21
September 6, 1945	61	27	12
December 31, 1946	64	28	8

Source: Cantril, p. 694.

[a]The accuracy of these two polls is highly suspect because they used "public" rather than "governmental" (see note 201 in Chapter IV). Notice that the first poll to use the word "governmental" is also the one to show a marked change in opinion distribution. A majority probably did not support *governmental* ownership in 1937 or 1938; see question 2 in Cantril, p. 694.

erty. Each subsequent poll shows an increase in those who would rather have private ownership until more than six out of ten prefer this alternative in 1946. Table VI.8 shows a rather gross dichotomous breakdown of attitudes toward government ownership of electric power facilities. In 1939, *Fortune* allowed respondents to give somewhat more detailed replies. When asked if the "government should produce and sell all, some, or none of the nation's electric power," 20 percent said all, 24 percent some, and 42 percent none with the others undecided.[62] In this period, stress still appears strong against consensus support for the property value as expressed in responses to the power issue. Yet as Table VI.8 shows, value support tended to increase as the depression stimulus becomes more remote in time. By the Second World War, support for property is near the operational consensus level.

The nature of property value support during the depression period may be gauged from a famous political conflict between the property value and depression stimuli. In 1933 the Tennessee Valley Authority was created as the first in a series of regional public projects over the vigorous objections of those who maintained that this was a serious threat to the property

[62] FOR, 1939; in *ibid.*, p. 695.

value.[63] Questions as to which side had more popular support were hotly debated. The most notable aspect of public opinion on this issue, however, was the lack of public information regarding TVA. In 1945, for example, 40 percent had not heard of TVA and an additional 19 percent defined it incorrectly—a total of 59 percent unfamiliar with the project.[64] During the controversy, in 1938, 70 percent had not followed arguments on the TVA proposal.[65] In 1937 and again in 1938, most people had no opinion on whether there should be more TVA's. On the other hand, in these surveys, pluralities of 38 percent and 28 percent, respectively, of those with an opinion did approve of additional TVA's.[66] Pluralities also approved of TVA itself.[67]

The degree of commitment to further TVA's can be judged by two other surveys. When asked a general question in 1945 concerning the merits of building a Missouri Valley Authority, 27 percent of those familiar with the concept favored the plan and 5 percent opposed it.[68] However, when asked to extend the concept to the Passamaquoddy project in Maine for a specified sum, 73 percent of those who had heard of the project felt the money should not be spent.[69] Again, mass opinion is not mechanistically simple. It may again be proper to consider these opinions as political culture mood. Most people were unfamiliar with the concept of TVA. The depression stimulus created a demand, however, for the authorities to take some action to reduce the stress. While the stress was high, authorities were less restricted by value support for property. As the stress receded property support apparently became more restrictive.[70]

Support for the property value certainly was lessened by depression stimuli strength. On the other hand even during the period of greatest stimuli impact, property support was probably high. Yet under the stress of the Great Depression,

[63] Young, pp. 559–562.
[64] AIPO, 1945; in Cantril, p. 698.
[65] AIPO, 1938; in *ibid.*
[66] AIPO, 1937, 1938; in *ibid.*, p. 696.
[67] AIPO, 1945; in *ibid.*, p. 698.
[68] *Ibid.* The remainder were unsure.
[69] AIPO, 1939; in *ibid.*, p. 697.
[70] Of course, the TVA concept was not extended to other regions.

TABLE VI.9 *War stimuli vs. property*

Question: Do you think that in wartime the federal government should take over all business and industry? (AIPO)

	1937	*1939*
Yes	42%	43%
No	41	40
Qualified, don't know, etc.	17	17

Source: Cantril, p. 345.

property support weakened. Data indicate that both property and achievement [71] values were weakened during this period. But by the early 1940's support for both values seemed on the increase.[72] However, at that time another powerful stimulus appeared which again stressed the property value.

Environmental Stress from War. The complex of Second World War stimuli was the second of the stronger environmental stresses to affect the political system during this period. Once again the property value is under the stress. As can be seen from the section on property in Chapter IV (especially Tables IV.52 and IV.54), support for the value was substantial throughout the war. However, there is evidence that the stress of war does have an effect upon property support. For example, eight out of ten of those interviewed in February, 1943, thought that profits should be limited for the duration of the war.[73] Although Table IV.52 suggests that very few Americans would prefer the government to own and operate business, Table VI.9 suggests that a slight plurality would actually allow government ownership under the stress of war. Other data also suggest that more of the public was willing to accept regula-

[71] FOR, 1937; in *ibid.*, p. 829.

[72] For example, by 1940, 22 percent would like the government to take over all utilities; 55 percent would not like it; and 23 percent did not know. See FOR, 1940; in *ibid.*, p. 982, and Table IV.54. The major sociological study of the period concluded that the depression changed some aspects but "basically the texture of Middletown's culture has not changed." See Robert S. Lynd and Helen Lynd, *Middletown in Transition* (New York: Harcourt, Brace, 1937), p. 489.

[73] FOR, 1943; in Cantril, p. 686.

tion of business during the war.[74] Yet even under the extreme stress of war, almost half of those with an opinion (Table VI.9) opposed government control of productive property. As the other tables show, in most questions asked regarding governmental ownership of property during the war, support for the property value is much higher.[75] Again it is noted that when the value of property is subjected to stress, support is lowered. Opinion takes the form, not of specific instructions, but of a general mood which allows authorities substantial leeway while stress is strong. For the property value this situation apparently existed during both the depression and the war. Support was lessened but value restrictions remained. When the stress was removed support returned to the consensus level.[76]

The liberal tradition value of liberty has also been strongly influenced by the stress of war stimuli. During World War II, most Americans agreed that the various government restrictions on personal liberty were warranted. Rationing of food and other materials was accepted by the people throughout the war.[77] That rationing was accepted only as a temporary measure under the impact of war stimuli can be seen by noting that the mood changed immediately after the fighting ended.[78] During the war, a majority also believed that the government was correct in holding back all information about military losses in Pearl Harbor for the first week after the attack.[79] In 1943 over seven out of ten people believed the Navy Department was right in delaying the full report on Pearl Harbor for a year after the investigation was completed.[80]

More direct data regarding support for freedom of speech under the effects of war stimuli can be derived from a further specification of the data of Table IV.48. These data, in a form appropriate for present concerns, are shown in Table VI.10. Support for the value is shown to be considerably reduced under the stress of war stimuli. After the war was over, two-

74 *Ibid.*, pp. 344–350.
75 See Tables IV.54 and IV.56.
76 See Tables IV.55, IV.56, and IV.57.
77 *Ibid.*, pp. 729–734.
78 AIPO, 1945; in *ibid.*, p. 734.
79 NORC, 1941; in *ibid.*, p. 1144.
80 AIPO, 1943; in *ibid.*

TABLE VI.10 *War stimuli vs. free speech*

	Support liberty value	*Oppose liberty value*	*Don't know*
Prewar (February, 1940)	49%	44%	7%
Early war stimuli (January, 1941)	46	52	2
Early war stimuli (June, 1941)	44	53	3
"In peacetime" (November, 1943)	63	34	3
Postwar (November, 1945)	64	32	4

Source: Table IV.48.

thirds of the population supported the value. Even in 1943, during the war, when people were asked about free speech *after the war,* they overwhelmingly supported the value. However, in the threatening period when Europe was at war and especially in the year prior to American entry, people were much less willing to support freedom of speech under any circumstances. To some extent the different wording used in 1943 and 1945 could account for some of the increase. Yet the change in mood appears substantial. War stimuli appear to explain the change.[81]

People also seem willing to sacrifice liberty in the face of *potential* threats. Table VI.11 shows that the majority support for free speech falls markedly (although not below the majority level) when people perceive that the community is endangered.[82] Another type of freedom, the freedom to strike, shows a similar pattern under the stimuli of war. On the eve of war, majorities believed that strikes were acceptable in non-

[81] It is also interesting that when stress again increased against free speech during the period of controversy over Communist influence in government in the 1950's, support was reduced but not to the levels reached during the war. In 1953, 53 percent supported the liberty value; see Herbert H. Hyman and Paul B. Sheatsley, "Trends in Public Opinion of Civil Liberties," *Journal of Social Issues* (1953), p. 15. For a general review of data on freedom of speech, see Hazel Erskine, "The Polls: Freedom of Speech," *Public Opinion Quarterly* (Fall, 1970), pp. 483–496.

[82] As this statement suggests, war stimuli may have a relationship to the identity system and may perhaps be treated as intracultural stress. However, the strength of the environmental stress leads us to emphasize this aspect of the conflict situation.

TABLE VI.11 *National defense vs. free speech*

Question: Here are some measures which have been suggested to improve our national defense. Should persons who oppose our present form of government be forbidden to express their opinions in public or should they be free to express such opinions?

	July, 1940 (OPOR)	*November, 1940*[a] *(AIPO)*
Should be forbidden	38%	30%
Should be free	50	57
Qualified	9	7
Don't know	3	6

Source: Cantril, pp. 244–245.
[a]In 1940, the wording was slightly altered. For the exact wording, consult the source.

defense industries; but strikes in defense industries were opposed. Once the war began, people began to oppose all strikes and the number of people who opposed defense strikes increased about 10 percent. After the war, most were again willing to accept the right to strike.[83] In each case where war stimuli were present, value support fell and when stimuli receded, support returned.

Environmental Stress from Cold War. The third stimulus is the stress associated with the perceived threat from "cold war." One aspect of this stress, perception of threats to regime survival by internal subversion, has been well documented for the 1950's by the Stouffer study. Perceived threats to internal security, however, have been major limits to the freedom value at least since the period prior to the Second World War. In 1937, 71 percent of those questioned would not allow "Nazis who are American citizens to wear uniforms and parade in this country." [84] Table VI.12 suggests that substantial numbers of Americans would take drastic action against Nazi or Communist adherents – especially during the war years. In

[83] See Cantril, pp. 819–820, qq. 31, 32, 34 (FOR, 1941); p. 818 (OPOR, 1940); p. 822 (NORC, 1942, 1943); p. 818 (FOR, 1942; AIPO, 1942); p. 827 (FOR, 1946). This right is also limited in other sensitive fields; see *Polls,* No. 4 (1968), p. 82.

[84] AIPO, 1937; in Cantril, p. 266.

TABLE VI.12 *What action should be taken against Nazis and Communists?*

	Nazis	*Communists*			
	1940 (FOR)	*1940 (FOR)*	*1941 (FOR)*	*1946 (AIPO)*	*1946 (AIPO)*
Drastic action: capital punishment, deportation, put in jail, etc.	46%	38%	30%	24%	36%
Regulatory action: watch them, keep them out of politics and agitation, etc.	41	49	54	27	23
Other	3	3	5	5	–
Nothing	–	–	–	20	16
Don't know	10	9	12	24	25

Source: Data for 1940, *Public Opinion Quarterly* (December, 1940), p. 710; data for 1941 and 1946, Cantril, pp. 130–131.

1937, 61 percent of those questioned said it would be a good idea to close places which print Communist literature.[85] In nine polls between 1943 and 1964, large pluralities would not allow members of the Communist party the right to speak on the radio.[86] Substantial majorities would not allow Communists to hold public office, or a sensitive job in industry. Culture level majorities approved of registering Communists and of making party membership illegal.[87]

Intolerance of Communists is probably due not only to environmental stress but also to perceptions of threats to sense of national identity. More importantly, as Stouffer has documented, Americans basically think of Communists as anti-religious.[88] Communism as a stimulus, thus, is perceived as a specific threat to at least three major values which are strongly supported within the political culture. Indeed, Communism appears to be perceived by Americans as a total threat to the maintenance of the regime and to all the values of the political culture.[89] A threat of this magnitude is not accorded the

85 AIPO, 1937; in *ibid.*, p. 416.

86 NORC, in *Opinion News* (August 6, 1946), p. 1; Hyman and Sheatsley, p. 8; and Erskine, pp. 488–489.

87 Hyman and Sheatsley, pp. 8–9, 12.

88 Stouffer, p. 186.

89 See the data in Cantril, pp. 130–132. Parsons believes that Americans view Communists as representing a total politicization of the social system; see Talcott Parsons, *Politics and Social Structure* (New York: Free Press, 1969), p. 454.

TABLE VI.13 *Internal security vs. free speech*

	Fascists (AIPO, 1938)[a]	*Communists (AIPO, 1938)*	*Communists (AIPO, 1954)*[b]
Allow speech	34%	35%	27%
Not allow speech	55	56	67
No opinion	11	9	6

Source: data for 1938, Cantril, p. 244; data for 1954, Stouffer, p. 41.

[a]Actual question for 1938: "Do you believe in it [free speech] to the extent of allowing Communists (Fascists) to hold meetings and express their views in this community?"

[b]Actual question: "Suppose an admitted Communist wants to make a speech in your community, should he be allowed to speak or not?"

usual liberty prerogatives of the tradition. This is dramatically expressed in Table VI.13. Most Americans would not allow freedom of speech for Fascists and Communists.[90] Interestingly, the population disapproves just as strongly of allowing people to speak against religion.[91] Yet these reductions in support do not apply to all nonconformists. A culture majority would allow anti-property Socialists to speak [92] and seven out of ten would allow a person whose loyalty has been questioned but who swears he was never a Communist to speak.[93] But even in the cases where majorities would allow anti-tradition value nonconformists to speak, fewer people would allow the speech than would support the general principle of free speech.

Fortunately, the Stouffer study is detailed enough to allow a somewhat closer look at this environmental stress situation. The study was done in 1954 specifically to study the cold war stress against the liberty value. It found that, in the 1950's, the force of this stress reduced support for the value. But the study went beyond this observation. Stouffer found, more precisely,

90 Stouffer (ch. 3) has shown that community leaders are much more willing to allow nonconformists to speak. However, the same general phenomenon is noticeable even if to a lesser extent. Fewer leaders will support the value of free speech as they perceive more basic values to be threatened.

91 *Ibid.*, p. 33.

92 *Ibid.*, p. 29; also see *Public Opinion Quarterly* (Summer, 1946), p. 248.

93 Stouffer, p. 37; also see the position taken as to whether the accused should be able to hold various jobs (p. 38). In 1956 the Survey Research Center found similar results; see *1956 Election Codebook* (Ann Arbor: Inter-University Consortium for Political Research, 1956).

that the greater the perception of the stress, the greater was the reduction of support for the tradition value.[94] However, support was withdrawn below cultural levels only for Communists (and for Nazis in the earlier period), not for other nonconformists.[95] Yet the relationship is apparent. The hypothesis expectation is confirmed by Stouffer. Value support is reduced to the *degree* that stress is perceived.

This study also provides data on the nature of the opinion process. Mass opinion does not appear to meet the expectations of the textbook model. Opinion is not specific. It rather appears quite evanescent and unfocused.[96] The stress situation, once again, merely provides a mood to which the authorities may respond. The authority response may take many forms, however. Political system members will accept authority decisions on how to reduce stress if support for political culture is high and congruent with the regime. During the 1950's stress period, 51 percent of the American public were willing to accept the recommendations of the president or the director of the Federal Bureau of Investigation on how to deal with Communists.[97] The public mood was for stress reduction, but the public would allow the authorities to impose the specific remedy. In sum, during the 1950's stress acted against the tradition value of liberty and reduced its support. When the stress was removed (partially by the authorities), support increased. However, even during the stress situation support was low only for the two groups which threatened the value.[98] Support for the value of liberty, in the more general sense (and for other nonconformists in particular), remained high even during the period of stress.

Values Under Stress. The idea of systemic value stress, then, helps to explain data that the textbook model of democracy can describe only as deviant. By looking at the McClosky or the Prothro-Grigg data from this perspective, one is impressed not by the rejection of the absolutist norms but by the small

[94] Stouffer, p. 208.
[95] Support is reduced for the others, however.
[96] Stouffer, ch. 3.
[97] *Ibid.*, p. 230. Sixty-eight percent of the leaders were so willing.
[98] *Ibid.*, pp. 31, 37–39.

area of deviance under stress from the values of the political culture and the eventual return to consensual support.

Stress temporarily reduces support for values within the political culture. The analysis has suggested, however, that the values themselves are not actually rejected, but that values can be modified if the stimuli are strong enough. The loss of support for values is temporary and even then the value does shape. This phenomenon, interestingly, does not seem to be restricted to the United States. Other democratic nations have also restricted beliefs within their value constellations when exposed to extreme stress such as war. In March of 1945, a plurality of Frenchmen approved of the government suspension of the newspaper *France-Soir* because it had appeared without approval of the military censor.[99] During World War II, only 30 percent of the people of Great Britain thought the government should not be able to suppress opinions against the national interest.[100] On two occasions during the war most Swedes believed that Nazis and Communists should be restricted.[101]

Value conflict has been explored here in a general and rather unsystematic sense. Even with this crude approach, however, certain insights have been gained. Liberal tradition values do seem to shape in a broad sense. This shaping appears expressed as mood which provides Lockean value parameters to governmental action. However, the mood is subject to stress. During stress, support for values is weakened. Weak stimuli can be ignored or highly shaped by the political culture. Environmental stimuli which can reinforce certain values within the liberal tradition value constellation can affect other values more than simple environmental forces. Yet, even under substantial stress, values appear highly resistant to change in their composition. It appears that the liberal tradition values remain within the political culture even under very strong stimuli stress. Values are modified somewhat while under stress but seem to return to the status of a culture value when the stress is removed. The benefits derived from the

99 FIPO, 1945; in Cantril, p. 418.

100 BIPO, 1941; in *ibid.*, pp. 417–418.

101 Swedish Gallup Institute, 1942, 1943; in *ibid.*, p. 870.

system model and the culture perspective, as opposed to the basic stimulus-response analogy, can be simply expressed. By the use of a very few intracultural and environmental stress situations, this study has been able to explain reasonably a great amount of attitudinal data regarding tradition values which the simpler model could explain only as deviant.

SHAPING INFLUENCE OF AMERICAN POLITICAL CULTURE

My major explanatory hypothesis holds that political culture tends to maintain its value composition as it shapes the stimuli which act on its value screen. Unfortunately, since this perspective is relatively new and since my own resources are limited, only a few tests of the hypothesis can be presented. The preceding section has rather unsystematically investigated some value conflicts where political culture values are assumed to shape. Much data can be explained with this assumption. It is now necessary, however, to proceed to a more rigorous test of the shaping hypothesis. Fortuitously, enough data is available for one test of each political culture subsystem. (The quality and amount of data vary widely, however – with the developed subsystems having the better data.) The tests below should be considered the critical tests of the explanatory shaping hypothesis. Each will involve some scientific controls, and some of the tests approach laboratory experiments. Rather than being forced to draw inferences from loosely related opinion data, as was required in the previous section, more stringent design and focus will allow a much more rigorous test of the shaping hypothesis. To sustain this hypothesis, political culture must be shown to shape stimuli in the direction of the liberal tradition values.

IDENTITY SHAPING

Identification with the nation should be an important shaping influence. The hypothesis holds that favorable linkage with the idea of the nation-state should lead to greater acceptance of that attitude; unfavorable linkage should lead to screening rejection. The strength of an identity system will depend upon the amount of attachment to measures of the nation. The data of Chapter III (especially Tables III.2, III.3, and III.4) suggest that vertical identity should have an especially strong shaping quality in the United States.

Some analysts have suggested that terms relating to the nation have an especially-strong shaping influence for Americans.[102] The very terms *American* or *un-American* are apparently used as indications of value approval in ways unknown in other countries.[103] Data do suggest that identity shaping is especially strong in the United States, but it is probably a universal phenomenon.[104] In a Gallup international survey taken in 1946, over 80 percent of the Danes thought Denmark was the best country; while in Finland, 60 percent thought their own country was the best – with another 13 percent choosing nearby Sweden (and 12 percent the United States). In the same year 74 percent of the French thought French nationality best and 80 percent would not want to leave their country. Even so, Americans seem to have an especially strong identity screen – 96 percent of them would not move to another country.[105]

If the shaping hypothesis is correct, one would predict that the introduction of national images should shape perceptions on issues. In other words, without the concept of identity shaping one might expect that values would be treated neutrally – according to their objective merits. The concept of shaping leads one to predict that affinity with the nation would influence perceptions of presumably universal values. Hyman and Sheatsley report on a test of shaping done on

[102] Key, pp. 42–44.

[103] Richard Weaver, *The Ethics of Rhetoric* (Chicago: Regnery, 1965), pp. 218–220. Also see Daniel J. Boorstin, *The Genius of American Politics* (Chicago: University of Chicago Press, 1953), p. 22. Kohn suggests that this is a defense against European attitudes of superiority toward Americans. He believes the attitude is changing in response to European changes (Kohn, p. 59 and ch. 2).

[104] Williams, p. 457.

[105] *Public Opinion Quarterly* (Winter, 1946–1947), p. 629; and Cantril, p. 508. Not surprisingly, American children identify strongly and early with their country. See Robert D. Hess and Judith V. Torney, *The Development of Basic Attitudes and Values Toward Government and Citizenship During the Elementary School Years,* Part I, Cooperative Research Proposal No. 1078 (U.S. Office of Education, 1965), p. 81. Ninety-five percent of the teachers sampled in that study thought "America is the best country in the world" (ibid., p. 80).

attitudes toward the right of countries to impose censorship. In 1945, 69 percent of those Americans questioned believed that American foreign correspondents should be free to report upon everything they wished to see abroad.[106] On the other hand, only a bare majority thought that Russian reporters should be able to report on anything in this country they wished to see.[107] That is, when national identity is evident, it shapes the perceptions that influence the public mood toward press freedom.

Fortunately an even more direct test was performed for the questions of press freedom. An experimental situation was arranged so that the question concerning freedom for American reporters was asked first of a carefully matched group consisting of one half of the sample. The other half of the sample was asked the question concerning Russian reporters first. This allowed a research situation with an experimental and a control group.

The shaping effect was quite evident in the test. Hyman and Sheatsley report:

> It is interesting that when the question of American reporters is experimentally asked *first* and the public is overwhelmingly on record in favor of "freedom", about 70 percent subsequently approve of granting freedom to Russian reporters. But when queried first about allowing *Russian* reporters freedom over here, only about 35 percent express approval.[108]

When identity with the native country is in the shaping position, extension of a valued belief even to a perceived enemy is accepted. When this culturally shaping value is removed as the relevant controlling value-screen, the perception of danger from the Russians outweighs the objective view of the value itself. The screening effect of identity is quite apparent in this experiment. By asking about *American* freedom

[106] Herbert H. Hyman and Paul B. Sheatsley, "The Current Status of American Public Opinion," in Daniel Katz *et al.*, eds., *Public Opinion and Propaganda* (New York: Holt, Rinehart and Winston, 1954), p. 44; also see *Public Opinion Quarterly* (Summer, 1945), p. 256; and *ibid.* (Summer, 1947), p. 280.

[107] Hyman and Sheatsley, "The Current Status of American Public Opinion"; also see *Public Opinion Quarterly* (Winter, 1947–1948), p. 641.

[108] Hyman and Sheatsley, "The Current Status of American Public Opinion," p. 44.

of the press first, freedom is more surely established as an accepted value. Once legitimated by national identity, the individual applies the value to the more "dangerous" situation. Without identity screening the threat perceived as posed by the Russians takes precedence over the other attitude. When the identity value is allowed to screen, shaping takes place.

SYMBOL SHAPING

Democracy is a concept which presumably has a universal meaning. At least this has been assumed to be so. This interpretation can be attributed to the pervasive acceptance of the textbook model of democracy. The cultural approach, on the other hand, suggests that shaping might influence the meaning of the term. It suggests the possibility that the mass of people view such a universally accepted term as *democracy* as a symbol. A widely accepted and emotional term such as democracy would seem to be well suited to become a symbol of the nation. Indeed, it appears that the term is extensively used to describe disparate types of governments, and is even used as a grand classification scheme to separate national friends and enemies.

Table VI.14 shows that for the countries where data is available (France is the only country which deviates greatly),[109] democracy tends to become a symbol of the native political regime. Most people find that there is democracy within their own country. Many fewer believe it exists in other countries. As the national identity data in Chapter II would lead one to expect, this is especially true in the United States. Most Americans not only define their own country as a democracy but they seem very reluctant to call other countries democratic. Shaping is noticeable not only through symbolic identification with the home country but also through historic cultural ties between countries. For example, the second highest percentage of Americans identify England with democracy. The historic connections between Holland and England are also expressed by the data in the table.

[109] Even for France the phenomenon seems to be present. A much higher proportion of Frenchmen believe France is a democracy than does any other country. It should also be noted that the poll was taken just a few years after a series of events that could only have weakened the French political culture: defeat in a major war, occupation by a foreign power, the Vichy collaboration, etc.

TABLE VI.14 *Symbol screening*

Question: Do you think there is democracy in the following countries? (AIPO, 1947)

	Percentage of the people in these countries saying there is democracy in:				
Democracy in	*U.S.*	*France*	*Sweden*	*Holland*	*Norway*
U.S.	88%	74%	85%	59%	74%
England	48	71	86	68	69
France	32	65	59	47	53
Holland	28	–	–	67	–
Argentina	12	–	–	–	–
Poland	9	16	25	–	14
Russia	5	24	14	11	11
Yugoslavia	4	15	–	–	–
Spain	4	2	3	3	1
Sweden	–	–	95	–	–
Norway	–	–	–	–	70
Finland	–	–	65	–	–
Belgium	–	–	–	51	–
Switzerland	–	82	–	–	–
Greece	–	6	–	–	–

Source: *Public Opinion Quarterly* (Summer, 1947), pp. 282–283.

There is, then, some evidence of symbol shaping. The literature previously noted on symbolism suggests that because of the crucial role they play in understanding and communication, all symbols shape perceptions. One would not expect less for a symbol like democracy which has been so central to twentieth century politics. If adequate data were available (say, for example, the legitimization which could be performed by the use of Lincoln symbols), I have little doubt that shaping could be more properly demonstrated for the symbol system. Even the limited data which is available suggests some cultural shaping influence.

LIMITED TERMS AND RULE SHAPING

Fortunately, attitudinal data is more available for the developed culture subsystems. As has been noted above, one of the rule components of the liberal tradition is the belief in limited executive rule. Indeed, it may be said to be one of the characteristic descriptions of the tradition. After all, the *Second Treatise* was written after the Glorious Revolution, partially to justify parliamentary supremacy over the executive. One of the ways this limited rule has been expressed in the liberal tradition is through the idea of limited terms of office.[110] It is on the basis of this value that Washington began the practice of limiting the number of terms a president should serve. Jefferson refused to serve more than two terms specifically because of the liberal principle.[111] The tradition was in turn accepted by Madison and established by Jackson who surely could have been reelected. Although this tradition was assailed by presidents Grant and Theodore Roosevelt, it was not until Franklin Roosevelt that the value was brought under substantial stress. In 1940, Roosevelt became the first president to be elected to more than two terms.

Data available on public attitudes during the two-term controversy allow a rather detailed case study of the political culture shaping process. The case describes how a tradition value under severe stress maintains itself and how it may shape output. The data available allow the use of a laboratory-like situation — as well as the use of other data within a more contrived framework. Although the approach used has certain methodological limitations for some of the data, its usefulness is suggested by the fact that all results tend toward the same conclusion.

All data are related through the system analogy. The value of limited terms is momentarily conceived as the only value within the political culture system. All other forces are considered environmental. To further reduce the complexity of the real world, only two events are considered relevant environmental stimuli — the depression and World War II. Both of these are conceived as creating stress upon the value system to reduce support for the two-term limit value. Of course, this is a very simplified model but it creates the basis for a focused scientific test of the hypothesis.

Table VI.15 shows the distribution of opinion regarding a constitutional amendment providing for a two-term limitation for the president from the depression through the postwar

110 See Table IV.37.
111 Young, p. 339.

TABLE VI.15 *Two-term limitation for president*

Question: Would you favor a constitutional amendment prohibiting any president to run for a third term? (AIPO)[a]

	Favor	*Oppose*	*Don't know*
1937	45%	46%	9%
1938	44	47	9
1939	37	52	11
1940	37	53	10
1943	43	50	7
1944	58	37	5
1945	55	40	5
1946	52	40	8
1947	53	37	10

Source: Cantril, p. 649.

[a]Question wording varied substantially; the source should be consulted for exact wording. The wording consistent with 1940 is used for 1943. The NORC data is used for 1944 as closest to the earlier wording; in 1945, the July data is used because it includes the "don't know" response. For 1947, see *Public Opinion Quarterly* (Summer, 1947), p. 304.

years. This may tentatively be used as a crude operationalization of support for the two-term limit value. No data on this value exist prior to the late 1930's. Although it is dangerous to use elite attitudes to express popular ones, it probably can be accepted that the predominant attitude prior to 1930 was against third terms. With the introduction of the depression stimuli in the 1930's, support for the two-term limit is reduced. The first data of Table VI.15 show that only 45 percent support the value by 1937. As the second stimuli, related to the war, are introduced, support for the tradition value is further reduced to 37 percent. Majorities were actually opposed to an amendment to prohibit third terms between 1939 and 1943. As the perceived danger from the war receded, however, mood changed to favor the tradition value. As early as 1944 majorities began to support a Constitutional amendment to limit a president to two terms of office.

This hypothesis can also be investigated by reference to data collected by Elmo Roper.[112] Depression stimuli can be gauged

112 Elmo Roper, *You and Your Leaders: Their Actions and Your Reactions, 1936–1956* (New York: Morrow, 1957), pp. 31–32.

in somewhat more detail by measuring attitudes toward a third term before and after Roosevelt's attempt to "pack" the Supreme Court to ease acceptance of his New Deal depression program. Table VI.16 shows attitudes toward third terms as affected by depression stimuli before and after the effects of the attack on the judicial value. The data show that when the two values of judiciary and limited rule combine, the shaping effect is substantial. Before the "court packing" attempt, a plurality favored a third term for Roosevelt. After the influence of the judicial value had been felt, a majority were opposed. However, since Table VI.15 shows that support for executive limitation fell again in 1938, one could interpret its temporary support not as a return to culture support but as the effect of support for the judicial value. When the Court crisis had receded in memory, the depression stimuli again effectively weighted support against the liberal tradition value of limited terms.

Table VI.15 shows opinion on whether a constitutional amendment should be passed to prohibit third terms. It shows that the value of limited terms was not accepted before World War II. However, it is possible to believe that an amendment should not be passed and still support the tradition value of limited terms. Roper presents data, which he believes to be representative, which is more closely related to politically relevant behavioral action. The focus is directly upon whether an individual would vote for Roosevelt for a third term in

TABLE VI.16 *Third-term attitudes, depression stimuli, and "court packing"*

	Depression stimuli	
	No judicial support January, 1937	*With judicial support October, 1937*
Negative support for limited terms (pro FDR)	48%	40%
Positive support for limited terms (con FDR)	44	52
Neutral support (no opinion)	8	8

Source: Roper, pp. 31–32. All favorable and conditionally favorable responses are combined, as are simple anti-Roosevelt attitudes and anti–third term for Roosevelt attitudes.

TABLE VI.17 *War stimuli effect upon two-term limit for Roosevelt*

Attitudes toward third term	*Prewar value support*[a]	*Invasion of Poland September, 1939*[b]	*Invasion of Denmark, Norway, and France April–June, 1940*[c]
Vote third term for FDR	33%	43%	52%
Not vote third term for FDR	55	47	39
Others, DK	12	10	9

Source: Derived from Roper, pp. 38, 44.
[a]Poll, May, 1939.
[b]Poll, March, 1940.
[c]Poll, September, 1940.

1940.[113] Roper's data suggest that the value of limited terms was supported by 1939 – at least as applied to Roosevelt. Table VI.17 shows that at that time a substantial 55 percent of the people said they would *not* vote for Roosevelt again in 1940. Four months later, however, after the invasion of Poland, those opposed to Roosevelt fell below a majority and those inclined to vote for him increased ten percentage points. However, a plurality were still opposed to him. After the invasion of Denmark, Norway, and, finally, the fall of France – only then were a majority willing to vote to keep Roosevelt in office for a third term. It seems that only the extreme stress of the war could finally force a majority to agree to vote for Roosevelt against the liberal tradition value of limited terms.

The shaping effect of the value can also be confirmed by Gallup's data. As early as 1937, Gallup shows that only four out of ten say that they would vote for Roosevelt for a third term. As war approached, however, Roosevelt became steadily more popular, until by September, 1939 (the beginning of the war in Europe), he became the choice of 46 percent of those with an opinion. This support remained stable until after the poll released April 1, 1940. On April 9, 1940, German troops entered Denmark, and in a month they broke through to France. Within a few days, on May 16, 1940, Gallup released his next poll. It showed that 57 percent of those with an

113 *Ibid.*, pp. 38–47.

TABLE VI.18 *Support for a fourth term, with and without war stimuli*

Question: If the war is over (still going on) and Roosevelt runs for a fourth term next year, do you think you will vote for or against him? (AIPO)

	If war is going on			*If war is over*		
	Vote for	*Against*	*Other*	*Vote for*	*Against*	*Other*
January, 1943	51%	37%	12%	39%	50%	11%
February, 1943	53	36	11	37	50	13
April, 1943	65	28	7	33	59	8

Source: Cantril, p. 653; and Roper, pp. 53–54.

opinion would now vote for Roosevelt.[114] In only one month Roosevelt's following had increased, by ten percentage points! He was never to lose this advantage as war reports became more ominous.[115]

The most convincing evidence of the influence of war stimuli upon an underlying tradition support for limited terms is provided by opinion data on a fourth term for Roosevelt. On three separate occasions people were asked if they supported Roosevelt for a fourth term, first, if the war was still on and, then, if the war were over.[116] In each sample the *same people* were asked the two questions, thus giving a good approximation of a laboratory test. Table VI.18 reports the rather dramatic results. A majority would vote for Roosevelt if the war was still going on. If the war stimuli are removed, however, in each of the three cases the voting decision is shaped by the tradition value. A majority would support the tradition value by refusing to vote for Roosevelt. This information is confirmed by another question in October, 1944, which asked whom the individual would support in November if Germany surrendered. Under these circumstances, 48 percent said they would support Dewey, 45 percent Roosevelt

114 AIPO, in Cantril, pp. 648–649.

115 *Ibid.*, p. 602.

116 A fourth test is also shown in Cantril but the "don't know" responses are not reported. The results, however, are similar to those reported below; see Cantril, p. 653.

and 7 percent expressed no opinion.[117] Again, the removal of the stimuli reduced stress on the value. More importantly, the liberal tradition shaping hypothesis is sustained in three laboratory-like tests. When severe stress from the war is reduced, the tradition is supported once again as the two-term value shapes the voting decision.

Support for executive limitation as the war stimuli lessened in intensity is also confirmed by the data of Table VI.15. By 1944 a majority were in favor of a constitutional amendment to limit the number of terms that a president could serve. In 1947, the year the tenure limitation was passed by Congress, almost six out of ten of those with an opinion believed that presidents should be restricted to two terms of office.[118] The shaping function of the value when under stress was, thus, very different from its shaping function free of the stress. This is most dramatically shown in the tests reported in Table VI.18. When the extreme stress was eased the value returned to a position of strength within the political culture. Interestingly, in this case where stimuli had overcome a tradition value, an attempt was made to be sure that the value would be maintained in the future even under substantial stress: the value was institutionalized as an amendment to the Constitution. Once institutionalized, the value became even more widely supported. By 1959, although 58 percent said they would like to vote for Eisenhower for a third term in 1960, 64 percent supported the tradition by refusing to change the Twenty-second Amendment to allow him to run.[119]

117 Roper, p. 103.

118 Opinion on this issue is highly related to party preference during the controversy. About seven out of ten Republicans favored limitation of terms through the whole period. A majority of Democrats never favored the value while Roosevelt was president, although 47 percent of those with an opinion did by 1945; see Cantril, pp. 649–650. By 1959 a consensus support was also given by Democrats; see American Institute of Public Opinion news release of March 1, 1959.

119 American Institute of Public Opinion news release of March 1, 1959. An intensive study done in one city found that 66 percent supported the value. Interestingly, this study found that the reasons given for supporting the Amendment were compatible with the tradition. Most of those who supported the value did so because they feared the president could become too powerful; see Roberta S. Sigel and David J. Butler, "The Public and the No Third Term Tradition," *Midwest Journal of Political Science* (February, 1964), pp. 39–54.

PREJUDICE AND BELIEF SHAPING

The liberal tradition values of equality and religion should lead Americans to a belief in the political and moral equality of all men. Public opinion data in the United States does not seem to support this expectation – especially in its application to the Negro. Indeed, one authority on the study of values has suggested that this intolerance is related to a basic reality concerning American values. Francis L. K. Hsu maintains that an excessive self-reliance caused by a belief in a total social and economic equality is the "core value" in explaining the American character.[120] Hsu believes that Americans have expanded the liberal tradition belief in moral and political equality to include belief in economic and social equality. Everyone is considered to be equal in every way and everyone is expected to be highly rationalistic and individualistic. This belief system demands that everyone should be totally self-reliant. Hsu maintains that the impossibility of achieving total self-reliance leads to insecurity which is manifested in intolerance and racial prejudice.[121]

Racial Prejudice. The data of Chapter IV suggest that Hsu's major premise is incorrect. The data suggest that Americans believe in moral and political equality but not Hsu's type of social equality.[122] On the other hand, substantial evidence is available that Americans are racially prejudiced.[123] In 1942, 62 percent of a sample said it would make a difference "if a Negro with just as much income and education as you have,

120 Francis L. K. Hsu, "American Core Value and National Character," in Francis L. K. Hsu, ed., *Psychological Anthropology* (Homewood, Ill.: Dorsey Press, 1961), p. 217.

121 *Ibid.*, pp. 219 ff.

122 See especially Robert E. Lane, *Political Ideology* (New York: Free Press, 1962), ch. 4. Even a work which maintains that equality is the central American value conceives of a limited equality; see C. Peter Mcgrath, Elmer E. Cornwall, Jr., and Jay Goodman, *The American Democracy* (New York: Macmillan, 1969), p. 15.

123 The most influential is Gunnar Myrdal's *An American Dilemma.* Intolerance is not new to the American system. Even before the system was completely established, British loyalists were subject to repressive measures; see Ralph Volney Harlow, *The United States* (New York: Holt, 1949), p. 126.

moved into your block." [124] Forty-eight percent would approve *laws* prohibiting interracial marriages.[125] Myrdal has suggested that there is a hierarchy of relationships which whites would allow.[126] A 1942 survey indicates that this may be so. While 92 percent of a national cross-section of high school students would not marry Negroes and 78 percent would choose a Negro last for a roomate, only 21 percent would not work with Negroes.[127] This study also found very limited approval of the most intimate social relations between the races – especially marriage. About nine out of ten Protestants, Catholics, and Jews would not consider marrying Negroes.[128] In 1968, Gallup found that about seven out of ten Americans did not approve of interracial marriage.[129]

Attitudes regarding the less intimate area of jobs are only slightly more compatible with the tradition. In 1944 and 1946 the population divided with half saying that whites should have *preference* in obtaining jobs.[130] In 1944, 51 percent said it would be all right "if a Negro with the same training had the same kind of job as you at the place where you worked"; but a substantial minority (43 percent) would not like it even if the Negro were better trained.[131] As recently as 1942, about 30 percent would refuse to work on a job with a Negro in an equal position.[132] The desire for social separation is also seen in attitudes toward segregation. The earliest attitude polls available, taken during the period of the race riots in World War II, show substantial support for segregation. In 1942, a

124 NORC, 1942; in Cantril, p. 989. In 1966, 52 percent would be upset if Negroes moved into their neighborhood; see William Brink and Louis Harris, *Black and White* (New York: Simon and Schuster, 1967), p. 109.

125 *Gallup Opinion Index* (August, 1967), p. 16.

126 Myrdal, ch. 3, part 4; also see Donald R. Matthews and James W. Prothro, *Negroes and the New Southern Politics* (New York: Harcourt, Brace and World, 1966), pp. 335–339.

127 FOR, 1942; in Cantril, p. 477.

128 *Ibid.*, p. 431. It should also be noted that 25 percent of the Catholics would not marry Protestants, 20 percent of the Protestants would not marry Catholics, and about half of each would not marry Jews. Over a quarter of the Jews would not marry the Christians (*ibid*).

129 *Gallup Opinion Index* (November, 1968), p. 12.

130 AIPO, 1944, 1946; in *Public Opinion Quarterly* (Winter, 1946–1947), p. 623.

131 NORC, 1944; in Cantril, p. 510.

132 FOR, 1942; in *ibid.*, p. 477.

majority believed in segregation of the armed forces and public transportation; 69 percent believed in separate facilities in restaurants and 84 percent were in favor of separate sections in towns.[133] Two-thirds believed that Negroes and whites should go to different schools.[134] By the end of the war, support for segregation had decreased somewhat. In 1948, although 42 percent supported segregation in public transportation, 49 percent did not.[135]

Evidence is also available that racial prejudice is not directed solely against Negroes. During World War II the public believed in rather severe limitations even on *loyal* Japanese Americans.[136] Forty percent would not give them an equal chance for jobs after the war was over.[137] Jews have also been subject to discrimination, though to a lesser degree. Only 12 percent of the American population said they would support an anti-Jewish campaign in 1938 and on two occasions in 1939; but a further 17 percent would be sympathetic to it.[138] Most disapproved of the treatment given Jews in Germany (even before the concentration camp tortures were widely known), but the country was evenly divided as to whether Jews had too much power in the United States.[139]

Perception and Value Stress. The early empirical data, thus, suggest that over this period of time intolerance does seem to be a characteristic of opinion in the United States. This does not mean, however, that it is a central value of the American culture, as Hsu maintains. This author attempts to make the single value of intolerance explain all American social behavior. However, single-variable explanations are rarely satisfactory. Perhaps if intolerance is related to our model, a more complete explanation can be given. As was stated above, the tradition values of religion and equality would lead one to expect acceptance of moral and religious equality. However, the data suggest otherwise. One of the possible explanations

133 AIPO, NORC, 1942; in *ibid.*, p. 988.
134 NORC, 1942; in *ibid.*, p. 508.
135 AIPO, 1948; in *Public Opinion Quarterly* (Fall, 1948), p. 535.
136 Cantril, pp. 380–381.
137 NORC, 1944; in *ibid.*, p. 381.
138 AIPO, 1938, 1939; in *ibid.*, p. 382.
139 OPOR, 1942; NORC, 1943, 1944, 1945; in *ibid.*, p. 383.

TABLE VI.19 *Perceptions of Negro intelligence*

	1942[a]	*1944*	*1946*	*1968*[b]
Negroes as intelligent	42%	44%	53%	58%
Not as intelligent	48	48	40	14
Don't know	10	8	7	28

Source: NORC, 1942, 1944, 1946; in Cantril, p. 988; and CBS News, *White and Negro Attitudes Towards Race Related Issues and Activities* (Princeton: Opinion Research Corporation, 1968), p. 2. Also see *Polls* (Winter, 1967), p. 77.

[a]Actual question for 1942, 1944, 1946: "In general, do you think Negroes are as intelligent as white people – that is, can they learn things just as well if they are given the same education?" (NORC)

[b]Actual question: "When black babies and white babies are examined right after they are born, the white babies usually have more natural intelligence than the black babies. From what you know or have heard, would you disagree with that statement?" (Opinion Research Corporation)

which has been mentioned is that of intracultural stress. Although there is support for the values of political equality and religion, perhaps certain other values are thought to conflict with these values when applied to blacks. Data suggest that many Americans do evaluate the race problem from the perspective of certain other values of the tradition. Much interracial stress seems related to the value of property. Many perceive the Negro to be a threat to their job. About four out of ten whites believe that improvement of the black job situation will threaten their own income potential.[140] Many others perceive the Negro as not willing to achieve by utilizing the value of education to improve himself.[141] Large proportions of the population also feel that blacks are basically less intelligent than whites. It would seem that these white perceptions must be changed if the values of religion and equality are to influence behavior.

Table VI.19 suggests that these perceptions have been changing. More have been willing to accept Negroes as basically equal in intelligence between 1942 and 1968. Whereas almost half the population thought the Negro was not as intelligent in the 1940's, only 14 percent did in 1968.[142] People not

140 NORC, 1944; in Cantril, p. 510.

141 NORC, 1944; in *ibid.*, pp. 508–509.

142 On the general decrease in prejudice (especially against Negroes and Jews) see Bruno Bettelheim and Morris Janowitz, *Social Change and Prejudice* (New York: Free Press, 1964; earlier edition 1950), ch. 1.

only seem more willing to see intelligence equalities but an increasing percentage of Americans believe that Negroes are discriminated against. Although the questions used vary widely, in the 1940's substantial majorities believed that Negroes were being treated equally.[143] The major Brink and Harris study in the 1960's, however, indicates that six in ten whites believe that blacks are being discriminated against.[144] Once whites accept this position, it would seem more likely that perceptions of racial difference in value application would be noted and changed. Yet poor perception of difference remains. In 1968, 49 percent of whites said that Negroes have not made progress because they do not work hard enough [145] – that is, because they do not accept the achievement value. In comparative surveys taken in 1963 and 1965, Louis Harris found that lack of ambition among Negroes was criticized by more of those interviewed (about 6 out of ten) than any of eight other perceptions of Negro characteristics.[146] The data on Negro beliefs in Chapter V suggest that whites misperceive the values supported by Negroes. Blacks actually support all of the tradition values. It is white misperceptions of this kind that lead to stress on culture support for the equality values.

Although a summary of recent attitudes is difficult, it appears that more recent opinion is somewhat more compatible with the hypothesis. Increasing percentages of whites apparently perceive that blacks are more like themselves. In other words, there is some evidence that Myrdal's "culturally shared myths" are losing some of their power to deflect tradition values from application to Negroes (see Table VI.19). There are some areas, however, where the tradition values have been historically strong enough to overcome racial prejudice. The earliest available data, from 1942, show that more than six out of ten believed that Negroes were as patriotic as whites, and over seven out of ten said blacks should have equal opportunity in the armed forces.[147] The Ku Klux Klan

[143] Cantril, pp. 508, 989; *Public Opinion Quarterly* (Fall, 1944), p. 442.

[144] Brink and Harris, p. 125.

[145] CBS News, *White and Negro Attitudes Towards Race Related Issues and Activities* (Princeton: Opinion Research Corporation, 1968), p. 2.

[146] Harris, 1963, 1965; in *Polls* (Winter, 1967), p. 77.

[147] NORC, 1942; in Cantril, p. 988. Also see Brink and Harris, ch. 8.

TABLE VI.20 *Political equality for Negroes*

Question: If your party nominated a generally well qualified man for president, and he happened to be a Negro, would you vote for him? (AIPO)

	1958	*1963*	*1965*	*1967*	*1969*
Yes	38%	47%	59%	54%	67%
No	53	45	34	40	23
No opinion	9	8	7	6	10

Source: *Gallup Opinion Index* (April, 1969), p. 6.

has been disapproved for the last two decades at least.[148] In 1943, 77 percent of the population believed that no one should make speeches attacking any race in the country.[149] As early as 1939, when Mrs. Franklin Roosevelt resigned her membership in a group because the group would not allow a Negro singer to perform in one of its buildings, two-thirds of the nation approved her action.[150] Eighty-seven percent believed in 1942 that blacks should receive equal pay for equal work.[151]

The most recent data appear to show less liberal tradition value stress from racial prejudice. A slight majority favored the *Brown vs. Board of Education* decision when it was issued in 1954.[152] In 1964, about six out of ten generally approved of civil rights laws,[153] and in 1966 72 percent said they approved of court action to secure equal education. Voting rights legislation was supported by 91 percent,[154] and Negro rights to fair jury trial and nonsegregated use of public transportation by 87 percent each.[155] Table VI.20 shows that Americans have been increasingly willing to vote for a Negro for presi-

148 AIPO, 1946; in Cantril, p. 393; also see Brink and Harris, p. 129; and *Gallup Opinion Index* (August, 1970), p. 11.

149 NORC, 1943; in Cantril, p. 245.

150 AIPO, 1939; Cantril, p. 988.

151 NORC, 1942; in Cantril, p. 509.

152 *Gallup Opinion Index* (August, 1967), p. 18.

153 Lloyd A. Free and Hadley Cantril, *The Political Beliefs of Americans* (New Brunswick, N.J.: Rutgers University Press, 1967), p. 196.

154 Brink and Harris, p. 131; three-quarters would keep their children in school if a court ordered desegregation (SRC, 1964; ICPR).

155 Brink and Harris, p. 127. In 1968, more than two-thirds said that Negroes should be able to live anywhere they can afford (SRC, 1968; ICPR). In 1968, the "in between" increase to 50 percent.

TABLE VI.21 *Hierarchy of prejudice*

Question: I want to read off a list of things about coming in contact with Negroes. For each I wish you would tell me if you personally would or would not object to: (Harris)

Object to contact with Negro	*Percent objecting* 1963	1965
Working on the job	17%	10%
Sitting on the bus	20	13
Sitting next to at lunch counter	20	17
Sitting next to at the movies	23	20
For own children at school	23	20
Using the same restroom	23	25
Trying on the same clothes at a store	32	31
Having a Negro family next door	51	37
Having own child bring to supper	41	41
Close friend marrying	84	85
Own teen-age daughter dating	90	92

Source: *Polls* (Winter, 1967), p. 78.

dent since 1958. A majority were so willing since 1965. Public opinion, however, is by no means settled on the question of liberal tradition rights for minorities. A 1964 question was symptomatic of the ambivalent attitudes felt by the public. When asked if they would favor desegregation, strict segregation, or something in between, 31 percent favored the first, 22 percent the second and 44 percent, the plurality, supported something in between integration and segregation.[156]

Current data suggest that prejudice exists in certain areas but does not exist in others. Two Harris polls provide current empirical substantiation for Myrdal's thesis of an assimilation hierarchy.[157] Table VI.21 shows that few white Americans object to most forms of social contact with blacks. It is

156 SRC, 1964; in ICPR. Interestingly, Negro attitudes on the question are not uniform either. The Matthews and Prothro study found that, in the South, Negro attitudes were 66 percent for desegregation, 15 percent for strict segregation, and 15 percent for something in between (SRC, 1961; in ICPR). In 1968, the "in between" increased to 50 percent.

157 Integration attitudes also appear to have a hierarchical character: 89 percent would not object to sending their children to a school with few Negroes; yet 32 percent would object where half were Negro, and 55 percent would object where more than half were Negro; see *Gallup Opinion Index* (September, 1969), pp. 5–7.

only the closest social relationships, such as courtship and marriage, which are rejected by most whites. In the United States at the present, stress from prejudice is evident but it presently seems restricted to the most intimate contact situations. In these areas, then, whites still do not fully apply to blacks the tradition values of religion and equality.

An American Dilemma: A Test. Although racial prejudice appears currently to be restricted to certain relationships, intolerance does seem to be a recurrent phenomenon in the American public opinion system; but is it part of the political culture? Hsu suggests that intolerance is endemic within Western Christian culture and, indeed, is related to the core value of the nation's character.[158] However, intolerance cannot meet our criteria as a culture value. There is neither a constant substantial difference between the majority and minority positions nor even a consistant majority over time in favor of racial prejudice. In some instances there is support for and in others opposition to intolerance. Attitudes are too volatile to reach the status of a culture value as defined in this study. Intolerance may be better conceived as an environmental stimulus, perhaps related to the personality system of the intrasocietal environment as the result of the insecurity of the human condition. Racial prejudice seems too universal a phenomenon to be related peculiarly to any particular political system.[159]

If intolerance is in the personality system it does not mean that it is unrelated to particular cultures in any systematic manner. It does mean, however, that it should exist rather widely across most complex cultures. A recent survey of thirteen North and South American, and western European countries by Gallup International suggests that, at least among these countries, prejudices exist in all of them. In disapproval of racial intermarriage, the United States ranks first in intolerance. In regard to religious intermarriage, however, it is relatively tolerant. In intolerance toward intermarriage between Jews and non-Jews it ranks fourth, and toward intermarriage

158 Hsu, p. 225.

159 Tamatsu Shibatani and Kian M. Kwan, *Ethnic Stratification* (New York: Macmillan, 1965), esp. p. 572. Also see Williams, p. 457.

between Protestants and Catholics it ranks eighth.[160] The Spearman's rank-order correlation between race prejudice and religious prejudice is only .143, and between race and Jew/Non-Jew prejudice only .336. National prejudice seems to be dependent upon which type of group is the minority in the population. If there is a conspicuous societal religious minority, there is religious prejudice. If there is a conspicuous racial minority, there is racial prejudice. Yet our data is from the West and Hsu suggests that intolerance is only characteristic of Western culture and not of those cultures which teach the individual "to respect authority and external barriers." [161] Yet India has its untouchables; Malays and Filipinos restrict the "overseas Chinese"; Japanese look down on the Ainus, the Ryukyu Islanders, and the Koreans; even the Chinese have their concept of Han, or pure Chinese, and outsider.

While the question is obviously not authoratively settled, these considerations do allow us to assume that racial prejudice is an environmental stumulus. In terms of our model, race prejudice is an environmental stress upon the liberal tradition political culture. If the hypothesis is correct the tradition should shape the stimulus. Frank R. Westie has performed a test which can be applied to the current situation.[162] In 1957 a sample of 103 individuals from Indianapolis were subjected to a laboratory-like test of their support for tradition values and their opinions on the application of these values to Negroes. As we would expect, column 1 of Table VI.22 shows that the tradition beliefs are almost unanimously accepted by the sample (especially the first seven questions which are more closely related to the tradition as it has been defined here). Under such a powerful stimulus as racial prejudice, support for the value falls, as expected, when the stimulus is allowed to act upon the value (columns 2 and 3).

Fortunately, Westie was interested in what I have called the shaping effect of values. The present hypothesis holds that, other things being equal, in the long run the tradition

160 *Gallup Opinion Index* (November, 1968), pp. 12–14. For Brazilian attitudes toward Japanese, see Cantril, p. 478.

161 Hsu, p. 223.

162 Frank R. Westie, "The American Dilemma: An Empirical Test," *American Sociological Review* (August, 1965), pp. 527–538.

TABLE VI.22 *The effect of prejudice stimuli upon tradition values*

Tradition value support		*Value under stress*		*Loss of support from stress*
1. Everyone in America should have equal opportunities to get ahead.		1. I would be willing to have a Negro as my supervisor in my place of work.		
Agree	98%	Agree	60%	38%
Undecided	0	Undecided	2	
Disagree	2	Disagree	38	
2. All people should be treated as equals in the eyes of the law.		2. If I went on trial I would not mind having Negroes on the jury.		
Agree	98	Agree	76	22
Undecided	0	Undecided	5	
Disagree	2	Disagree	19	
3. People should help each other in time of need.		3. If a Negro's home burned down, I would be willing to take his family into my home for a night.		
Agree	99	Agree	64	35
Undecided	1	Undecided	6	
Disagree	0	Disagree	30	
4. Children should have equal educational opportunities.		4. I would not mind having Negro children attend the same school my children go to.		
Agree	98	Agree	79	19
Undecided	1	Undecided	2	
Disagree	1	Disagree	19	
5. Everyone should have equal right to hold public office.		5. I believe that I would be willing to have a Negro represent me in the Congress of the U.S.		
Agree	91	Agree	71	20
Undecided	1	Undecided	6	
Disagree	8	Disagree	23	
6. Each person should be judged according to his own individual worth.		6. I would not mind if my children were taught by a Negro school teacher.		
Agree	97	Agree	67	30
Undecided	1	Undecided	8	
Disagree	2	Disagree	35	

TABLE VI.22 cont.

Tradition value support		*Value under stress*		*Loss of support from stress*
7. I believe in the principle of brotherhood among men.		7. I would be willing to invite Negroes to a dinner party in my home.		
Agree	94	Agree	29	65
Undecided	5	Undecided	4	
Disagree	1	Disagree	67	
8. Public facilities should be equally available to everyone.		8. I would be willing to stay at a hotel that accommodated Negroes as well as whites.		
Agree	83	Agree	61	22
Undecided	4	Undecided	4	
Disagree	14	Disagree	35	
9. Under our democratic system people should be allowed to live where they please if they can afford it.		9. I would be willing to have a Negro family live next door to me.		
Agree	60	Agree	35	25
Undecided	6	Undecided	2	
Disagree	34	Disagree	63	
10. I believe that all public recreational facilities should be available to all people at all times.		10. I don't think I would mind if Negro children were to swim in the same pool as my children.		
Agree	63	Agree	38	25
Undecided	6	Undecided	8	
Disagree	31	Disagree	54	

Source: Frank R. Westie, "The American Dilemma: An Empirical Test," *American Sociological Review* (August, 1965), pp. 531–532.

values of equality will shape the prejudice stimulus in the direction of the tradition values. The Westie experiment was performed explicitly to test the similar Myrdal thesis that the "American Creed" will eventually shape white perceptions so that the equality values of the creed are applied to blacks. To test the shaping influence of the values, Westie first tried to find out what proportion would not apply the values to Negroes. He found that the proportion differed in specific situations but that *approximately six in ten applied the values* and between 20 and 40 percent did not.

From those who *would not apply the values,* Westie attempted to determine if they realized that they did not apply the values; and if they did realize it how they would resolve the conflict. Would they reject the tradition values by refusing to apply them to blacks; or would the subjects support the liberal tradition values and apply them to Negroes? The test would measure whether prejudice or the liberal tradition value was held more strongly.

Westie's experiment constructed a situation where the interviewers recorded both the *spontaneous* realizations of conflict noted by those interviewed, and the responses to several *probes* made by the interviewer. At the first stage, no interviewer action was taken at all. In the second stage the respondent was told to compare his tradition value statement to the more specific application of it without telling him that it could be contradictory. Then three explicit probes were made. The first asked simply, "Any comment?" At the second level the interviewee was asked, "Do you see any contradiction?" If the respondent recognized that there was a contradiction he was then asked, "Any explanation?"[163] Of all of the remarks made, 43 percent were totally spontaneous and an additional 17 percent were made at the second stage without explicit probing. Therefore, about *six out of ten realized that there was a dilemma* without positive probing.[164] Shaping can, thus, be done by about 60 percent of the population without the laboratory artificiality of focused probing; 30 percent more were found able to do it with very little probing.[165]

More importantly for the shaping hypothesis, when the subjects chose between the statements embodying the liberal tradition value of equality and the specific statements affected by prejudice stimuli, a large majority of adjustments were made in the direction of the culture value. As Table VI.23 shows, *more than eight out of ten resolutions of attitude conflict were made in favor of the liberal tradition value of equality. Only 18 percent of the changes were not shaped in the direction of*

[163] For more of the details see Westie, pp. 529–530.
[164] *Ibid.*, p. 534, Table 4.
[165] *Ibid.*

TABLE VI.23 *Shaping affect of the liberal tradition belief in equality*

	Adjustment of liberal tradition beliefs	*Adjustment of specific, stimuli related beliefs*
Percent of adjustments made (N = 1,327)	18%	82%

Source: Westie, p. 537.

the political culture. The Westie experiment thus represents a dramatic example of the strength of the shaping effect of the political culture. Of course, it does not relate directly to behavior. It does suggest at least the possibility that behavior can be affected – if attitudes do affect behavior. In any event, the shaping influence does appear substantial within the value system. When value conflict exists and adjustments are made between the environmental stimuli of race prejudice and the liberal tradition value of equality, it tends to be the liberal tradition value which is maintained.

As Westie suggests, this

> is of some social significance, though it contradicts neither Myrdal's nor widespread sociological expectations. Myrdal predicts that in the long run the general tenets of the American Creed will win out over the contradictory valuations defining American race relations. We had thought (after pre-tests) that we might find that many people would alter the general to fit the specific. Actually, 81.6 percent of the "explanations" were adjustments of specific valuations. This suggests that Myrdal's optimism is not unjustified and lends little support to the fear that people might react to the dilemma by renouncing their allegiance to the democratic tenets of the American Creed.[166]

Westie also finds that conflict between general and specific valuations is less likely for "legal, political and educational relationships."[167] These are the values that have here been identified with the concept of political culture. It may be that the values most related to the liberal tradition political culture have even more shaping power than Westie found. But the shaping evident in the Westie experiment itself is ex-

[166] *Ibid.*, p. 538.
[167] *Ibid.*, p. 537.

tremely important. In over 80 percent of the situations tested, the liberal tradition values shaped the stimuli – a rather strong support for the shaping hypothesis.

A very large amount of public opinion data has been presented as empirical substantiation of the maintenance and shaping hypotheses. In the first part of the empirical analysis, tradition values were assumed to shape. Data were explained from this perspective. Once the assumption was shown to explain the data, it became reasonable to proceed to more demanding tests of the shaping hypothesis.

The culture system perspective seemed much more explanatory of data than the major competing perspective. From the simple, textbook perspective, much opinion data must be classified as aberrant. For example, from this view a decrease from 89 percent support for free speech in general to 63 percent for free speech for those who do not know what they are talking about is a very substantial antidemocratic deviation.[168] The perspective of political culture, on the other hand, emphasizes the shaping effect of the value. More than six out of ten still support free speech for a person who "doesn't know what he is talking about." The culture commitment shaped the second response more toward the free speech value. How else can one explain the still substantial support for the value? Shaping seems the best explanation. This perspective seemed the best explanation for all the data presented in the intracultural and environmental stress sections. Without the culture system assumptions, too much data must remain unexplainable.

Once the possibility of shaping was established, more scientific tests were performed. A suggestive laboratory-type test was shown for the identity system. Less rigorous data were presented for the symbol system. These two tests provide evidence of political culture value shaping. The other two culture subsystems were tested with more scientific controls. Political culture support for the two-term tradition was shown to maintain itself through periods of severe stress. Even during these periods if the stress is experimentally removed, the value

168 McClosky, Table 2, p. 366; Table 3, p. 367.

shapes political attitudes. It is even possible that the violation of the value under stress eventually led to the regime institutionalization of the norm. In a second research situation, support for political and moral equality was shown to shape the strong environmental stimuli of race prejudice. Eighty percent of the changes in attitudinal evaluations were shaped in the direction of the political culture values.

On the basis of these tests, I conclude, in terms of political culture itself, that political culture shapes stimuli to itself as it maintains its value composition. Not all stimuli which stress political systems, however, act upon member political culture. Some avoid its screen and much other is within-system generated. Yet many stimuli do have to interact with the member values. For these, political culture does appear to manifest the characteristics suggested in the earlier theoretical sections. It does appear to be able to screen stimuli which will be processed by the political system.

CHAPTER VII

Maintenance of a Liberal Tradition Regime

STRUCTURAL ASPECTS OF REGIME MAINTENANCE

It has been the purpose of the preceding chapters to describe a value system in the United States which could be empirically designated as a widely supported political culture, to show that these values have been maintained over a substantial period of time and that these member values shape stimuli which act upon them. Yet our most general interest has not been on the maintenance and shaping of values per se but on the influence of values in maintaining a political regime.

A regime is viewed as composed of both structural and value elements. Regime structures are the "formal and informal patterns in which power is distributed and organized," [1] while regime values are the principles, goals, and norms congruent with these structures. If the political culture values are congruent with the regime values, maintenance of the culture is maintenance of the regime. For the United States this is assumed to be the case. But structure can change irrespective of regime value support. Domination by foreign power, elite coups d'état, revolution without widespread mass support, etc., can change a regime. These are important determinants of regime maintenance and should not be ignored. Yet drastic

[1] David Easton, *A Systems Analysis of Political Life* (New York: Wiley, 1965), p. 193.

structure change without value change has not been of great importance for the American political system.

Nonetheless, some formal, regime structures seem to have important independent effects upon regime maintenance. Easton mentions six major structural ways of reducing stress which are positively related to regime maintenance.[2] Federalism, popular representation in legislatures, political party competition, overlapping membership in groups, the use of a written constitution, and formalized adjudication procedures to depoliticize issues — all these are structural means which can serve to mitigate stress against the regime. Each of these political institutions is a central power structure in the United States. Each is a major factor in its regime maintenance. Each is also consistent with the liberal tradition.

Some non-political culture value structures, identified by the political scientist William Mitchell, appear related to regime maintenance. Four socio-psychological structures are mentioned: abundance of natural resources and the consequent feeling that there are enough resources for all; the pursuit of many different goals by different people; social opportunity and mobility; and some repression and displacement of wants to other than political objects. The following more politically related structures are also identified: growth of the institutionalization of conflict; the belief in a common interest; the ability to substitute symbolic for material rewards; the support of decentralized decision-making and elections; moderate levels of conflict; and cooperation among authorities.[3] From the view of the present work, the most interesting aspect of these structures is that they appear characteristic of the United States.[4]

There would, then, appear to be important American structures which are of importance to and aid in regime maintenance. But, once again, a regime is composed of both structure and value elements and the present focus is more on the latter.

[2] Easton, ch. 16, esp. pp. 251–253, 256–257.

[3] William C. Mitchell, *The American Polity* (New York: Free Press, 1962), ch. 9.

[4] *Ibid., passim;* also see Robert A. Dahl, *Pluralist Democracy in the United States: Conflict and Consent* (Chicago: Rand McNally, 1967), *passim.*

More specifically, the present work is interested in the values held by the ordinary member of the political system. Undoubtedly, much value influence is from leaders and elites.[5] Yet for some regimes member values are important in maintaining regimes. To argue this, however, it must be demonstrated that some means exist for ordinary members to transmit their values through structures to influence output. The political system analogy suggests that this can be done in two ways – either through political institutions to authorities or to authorities directly. But (as discussed above) member value influence need not meet overly restrictive tests of rationality. For member values to affect authorities it need only take the form of what has been called mood.[6] For the maintenance and shaping thesis to be sustained, it must only be shown that value *mood* can be transferred to and can have an effect on authorities in the United States.

Yet authorities themselves can have an influence upon regime maintenance. They may take positive actions specifically aimed at maintaining the regime. They can use coercion, they can modify the regime (in some respects to maintain the whole), and they can consciously attempt to inculcate regime values. In the United States, although these means are used to some extent, they are not the only important authoritative influences. It is when authorities react to member demands that they are very related to the regime maintenance process. Authorities appear to be most interested in maintaining their own incumbency. Policies are enacted to some degree in response to demands from members in the hope of support at the next election. If these policies satisfy the members, however, the support generated at the authority level can extend to regime support. This type of regime support appears to be important in the United States.

Because of the generalized, non-directed characteristic of member mood, probably most member demands are not specifically related to very precise positive policies. Values appear to be relayed to the political system more in the nature of

5 Gabriel A. Almond, *The American People and Foreign Policy* (New York: Praeger, 1960), p. 142; and V. O. Key, Jr., *Public Opinion and American Democracy* (New York: Knopf, 1961), p. 537.

6 See pages 41–43, 304–305, 312, 319.

limits to what authorities may do. It would appear that the rationality which exists in voting is most related to this type of member decision.[7] When authorities refuse to take action because the ordinary member would be opposed, they may be said to be acting according to what has been called the "rule of anticipated reactions." [8] Presumably much governmental activity in the United States is based upon this kind of authoritative perception of mood. It may be that Congress's power may be most important through the enforcement of this type of rule in its relation with the more "rational" executive.[9]

Although the law of anticipated reactions undoubtedly operates from sources other than political culture, it would appear that the most effective popular restraints on what can be done by authorities rest within the culture. Could anyone envisage serious consideration of legislation to change the community drastically, to stop the legal observance of Lincoln's birthday, to repeal the Constitution, to eliminate Congress or the presidency, to outlaw free speech, etc. (at least in the absence of an extremely serious threat to survival)? To mention these is to state the strength of political culture mood to restrain actions by authorities. In American politics most no's do not have to be expressed.[10] Political culture allows them to be ignored.

The member can be more specific when he transfers these values to the political system *through institutions.* This process does not require the member to have a specialized focus on the political. Through what has been called the "two-step communication flow" the member with little connection to politics can communicate with those who in turn communicate with authorities. In this manner, the public which is relatively nonattentive to politics may transfer their feeling

[7] In the data which are available, the best explanation of rational voting appears to be based upon the extent to which one disagrees with the way authorities have performed; see, V. O. Key, Jr., *The Responsible Electorate* (Cambridge: Harvard University Press, Belknap Press, 1966), esp. pp. 76, 139–140.

[8] Carl J. Friedrich, *Constitutional Government and Democracy* (Boston: Ginn, 1946; originally published, 1937), pp. 589–591; also see David Truman, *The Governmental Process* (New York: Knopf, 1951), pp. 511–524.

[9] Willmoore Kendall, "The Two Majorities," *Midwest Journal of Political Science* (November, 1960), pp. 325–328.

[10] Friedrich, pp. 590–591; Kendall, *ibid.*

of mood to those who pay attention and to opinion leaders.[11] Among the members more specialized to politics, those in groups relay mass values to interest groups involved in the political process; [12] the active members of the media audience no doubt transfer mood to media elite; local publics transfer mood to local and state governments; [13] and the politically active communicate with party leaders.[14] These political institutions can then relay this mood to authorities. Of course, they may not relay a mood, and they certainly do not precisely relay all feelings of the members. However, they are available to perform the function and the empirical works cited suggest that some transfer takes place in the United States.

Mood can also be *transmitted directly*. A major means of mood transmission takes place through the institutional means of elections. As noted above, specific policy attitudes are too complex to be easily related to elections. However, something as broad as mood appears able to be effective through elections.[15] Mood can also be transmitted between elections. Although transmission is made through most authority structure it appears to be made most often and most effectively through representative legislatures. At least this appears to be so in the United States.[16] Letters are one means of effective communication but only about 17 percent of Americans have ever written their congressmen – although 40 percent of the American attentive public has.[17] Whatever its restrictive aspects, congressmen appear to place great importance on this and other informal communications means.[18] In the United States

11 See notes 23 and 27 in Chapter II.

12 V. O. Key, Jr., *Politics, Parties and Pressure Groups,* 4th ed. (New York: Crowell, 1958), pp. 158–161. Also see Truman, esp. ch. 2.

13 Mitchell, p. 282.

14 John W. Kingdom, *Candidates for Office* (New York: Random House, 1966), pp. 89–95.

15 On this point see especially note 7 and Gerald M. Pomper, *Elections in America* (New York: Dodd, Mead, 1968), esp. chs. 8 and 10.

16 Malcolm E. Jewell and Samuel C. Patterson, *The Legislative Process in the United States* (New York: Random House, 1966), pp. 346–352; and Mitchell, p. 282.

17 Donald J. Devine, *The Attentive Public: Polyarchical Democracy* (Chicago: Rand McNally, 1970), p. 119.

18 Lewis Anthony Dexter, "What Do Congressmen Hear?" in Nelson W. Polsby *et al., Politics and Social Life* (Boston: Houghton Mifflin, 1963), p. 486; Donald R. Matthews, *U.S. Senators and Their World* (New York:

mood can be, and to an extent is, transferred to authorities through elections, letters, and other communications. Indeed, one study suggests that the American Congress may listen too much.[19]

MEMBER VALUES IN REGIME MAINTENANCE

The United States appears to have structures which are important in regime maintenance. Their importance should not be minimized. However, the present emphasis has been upon values. Yet structures and values are highly interrelated. When political culture is supported and when it is congruent with the regime, member support for values can sustain the power structures. When the structures are thus legitimized by member support, the authority occupants of the structure roles can allocate for the society with a reasonable expectation that their outputs will be accepted.

> In political systems, at least in those in which political authorities are not being fundamentally challenged, the capacity of authorities to rule is closely connected to the presence of an ingrained belief, usually transmitted across the generations in the socialization process, that the occupants of the political authority roles have a right to command and the other members of the system a duty to obey. The major source of power for these roles resides in the prevalence of the conviction of their legitimacy.[20]

Without some political culture support it is difficult for a political system to perform the essential function of having its authoritative decisions accepted.[21] It is hard to see how many political systems could act without excessive inefficiencies in the absence of member support for their decisions. Therefore most regimes, no matter how authoritarian, seem to take member values into account – if only to manipulate

---

Random House, Vintage Books, 1960), p. 219; Charles L. Clapp, *The Congressman* (Washington, D.C.: The Brookings Institution, 1963), pp. 73–74; and Jerry Voorhis, *Confessions of a Congressman* (Garden City, N.Y.: Doubleday, 1947), pp. 40–54.

19 John S. Saloma, III, *Congress and the New Politics* (Boston: Little, Brown, 1969), esp. pp. 110, 118–119.

20 Easton, p. 208.

21 *Ibid.*, pp. 281–286, 24–25.

them – so as to assure successful implementation of output. In this way member values have an effect upon what most political systems do. However, in some regimes member value influence is more direct and consequential through its effect upon demands and supports.

DEMANDS AND REGIME MAINTENANCE

The political system is viewed as finite and, thus, able to handle only a limited number of demands.[22] If a system had more demands than it could cope with, it is very unlikely that it could maintain its regime. Every regime, thus, makes provision for the limitation of demands. Probably most wants are never made.[23] They are screened by input structure. The number of members with the skills necessary to submit demands, the capacity of communication channels, collection and comprehension structure quality, and the institutional structure of the regime itself all act as structural limits on the number of demands made upon the system.[24]

Although other structures screen wants, demands are strongly influenced by culture values.[25] Political culture screens many wants which never become demands. Through their value system, members themselves may decide that some wants are inappropriate for admission as demands to be made on the political system. Certain wants are deflected completely and others are modified. Through the screening influence of political culture values, the absolute volume of demands tends to be limited. By limiting the volume of demands, political culture allows the system to concentrate its limited resources to meet the demands which do act on it. If the system can resolve these remaining demands, the likelihood of stress against the regime is reduced.

Even the demands which do act on the system tend to be of a certain type. They tend to be shaped by the political culture. If the political culture is congruent with the regime structure, wants which are not positively related to the values and which are harmful to the power structures tend to be deflected;

22 *Ibid.*, p. 61, 108.
23 *Ibid.*, p. 103; Mitchell, p. 210.
24 Easton, ch. 8–9.
25 *Ibid.*, pp. 95–103; see also Mitchell, p. 286.

others which cannot be deflected tend to be modified so as to be less stressful to the regime; those which are not stressful are allowed; and those supportive of the regime are actually facilitated. If the stimuli are powerful enough, however, potentially stressful demands are made on the political system. Yet a system which has a supported congruent political culture tends to minimize these stressful demands. The fewer the potentially stressful demands, the greater chance for regime maintenance.

In the United States, the regime structure appears congruent with its consensual political culture. Both can be defined as liberal. Both flow from the principles espoused by Locke, and both are based in their details on the norms expressed in *The Federalist Papers.* This political culture would, then, tend to act to maintain its regime by reducing demands – especially stressful ones. On the other hand, some of the values of liberal tradition seem to increase the number of demands which the system must process. Popular rule, the use of many "gatekeepers" (attentive publics, opinion leaders, politicians, legislators, and administrators), and the use of plural political institutions all serve to increase the numbers who can submit demands.[26] The more who are able to submit demands, the greater the potential number of demands. However, increased participation has not led invariably to demand overload. In the Western democratic systems "cultural imperatives have gradually grown up side by side with the slow development of democratic institutions." [27]

In the United States the major form these cultural inhibitors have taken is the liberal tradition. National identity, sense of community, symbolic ideals, institutionalization of rules, and values of liberty, equality, property, and religion have served to limit the number and types of demands made on the regime. This is especially true of the value of limited government as a general value of the American regime. As long as the members believe that most conflict and cleavage disagreement should be settled outside the political system, the number of demands is limited. Through this privatization of wants, the regime has had the capacity to function and thus to allow for

[26] Easton, pp. 88–108.
[27] *Ibid.,* p. 112.

the possibility of its maintenance. As long as the number of demands is limited, and as long as the types of demands acting on the political system are compatible with the power structures, the regime would appear to tend toward maintenance.

DIFFUSE SUPPORT AND REGIME MAINTENANCE

The reduction of the absolute number and the screening of certain types of demands by political culture is an important aspect of the influence of member values upon regime maintenance. However, there is a more direct aspect. The potentially stressful demands which act on the political system, even if not satisfactorily resolved, do not necessarily threaten the maintenance of the regime. These demands must be directed at the regime for the regime to be affected. For stress to threaten the regime – at least in so far as member values are important – support must be reduced for the political culture values themselves. Demands in themselves do not directly threaten the regime.[28]

When evaluating regime maintenance from the member perspective, then, it is support which is crucial. There appear to be three types of mass support that relate to the maintenance of a regime: support for reasons of expediency (including a fear of coercion), support based upon unquestioning tradition, and support based upon a belief in legitimacy.[29] Undoubtedly, the first two types help explain regime maintenance in the United States. In developed national political systems like the American one, however, diffuse support for regime values and the legitimacy generated by this support appear to be most important.[30] It certainly is the most important if a system is going to be able to maintain a regime of the liberal variety for an extended period of time.[31]

Every regime which values some continuity and order is viewed as somewhat dependent upon positive diffuse support from the ordinary members of the system. This diffuse support is determined by the stimuli-culture interaction process. Po-

[28] Easton, p. 58.

[29] *Ibid.*, p. 281.

[30] *Ibid.*, pp. 282, 286, 302–309.

[31] *Ibid.*, p. 293. Also see Taketsugu Tsurutani, "Stability and Instability," *Journal of Politics* (November, 1968), pp. 910–933.

litical culture is the screen through which stimuli to the political system are processed. When stimuli act upon the value screen, if the culture is consensually supported and if the stress is not too severe, the culture will shape stress in the direction of its value composition. If the culture values are congruent with the regime values, the positive support product generated by the culture becomes positive diffuse support for the regime. Through this value screening, a congruent political culture acts to reduce stress on a regime. Once again, however, support for political culture does not guarantee maintenance. Powerful stimuli can change mass support. Then, again, objective circumstance can change regimes without regard for mass feelings. Foreign invasion and elite-based *coups d'état* are two frequent examples. And, of course, political culture values may not be congruent (perhaps they remain from a previous regime overthrown by force) with the actual regime structure. In the latter case support for political culture can lead to regime stress.

In the United States, the power structures of the regime appear to be essentially the same as they were when the regime was instituted in the late eighteenth and early nineteenth centuries. The other major aspect of the regime – its values – has been the focus of the present analysis. Historical materials suggest that the value system which was congruent with the Constitutional regime can be described as the liberal tradition. An extensive analysis of the existing empirical evidence suggests that these values are still the ones supported in the United States as its consensus political culture. It is the argument of the present work that the maintenance of this political culture has helped maintain the regime. Support for a congruent political culture is support for the value aspects of the regime. To the extent member support is critical, then, support for congruent political culture is by definition maintenance of the regime. However, the present argument goes further. It holds that member support is additionally a sustentation of the power structures themselves. It is not assumed that member values are the sole, or even the most important, determinant of regime maintenance.[32] The present thesis does hold that if

[32] Easton, chs. 10, 12.

a political culture is supported and is congruent with its regime values, other things being equal, that regime will be maintained more than one which does not have the protecting screen of political culture. In this sense, American support for the liberal tradition has aided in regime maintenance.

REGIME SUPPORT AND THE LOCKEAN TRADITION

Member support is centrally important for a Lockean regime. The Lockean emphasis upon rationality makes total value acceptance by means of tradition insufficient. Of greater significance, the Lockean emphasis upon consent makes expediency based upon force unacceptable for regime support. Support must, on the contrary, be based upon the belief that the government is legitimate and just.[33] For a Lockean regime, lack of member support is an explicit indicator of illegitimacy and even could provide justification for overthrowing the regime.

Locke, however, does not expect support to be based upon the "textbook model" of constant and total rational questioning. Withdrawal of support does not occur from "every little mismanagement in public affairs. Great mistakes in the ruling part, many wrong and inconvenient laws will be borne by the people without mutiny or murmur." [34] Although Lockean consent is essentially rational, it also involves important aspects of habit. Indeed, Locke seems to describe a process not unlike the one here ascribed to culture.

> People are not so easily got out of their old forms as some are apt to suggest. They are hardly to be prevailed with to amend the acknowledged faults in the frame they have been accustomed to. And if there be any original defects, or adventitious ones introduced by time or corruption, 'tis not an easy thing to get them changed, even when all the world sees there is an opportunity for it. The slowness and adversion in the people to quit their old constitutions has in the many revolutions [that] have been seen in this kingdom . . . still brought us back to our old legislative of kings, lords and commons.[35]

[33] See pages 143–151.

[34] John Locke, *Second Treatise on Civil Government,* in Sir Ernest Barker, ed., *Social Contract* (New York: Oxford University Press, 1962), sec. 225, p. 130.

[35] *Ibid.,* sec. 223, p. 129; sec. 230, p. 133.

But when authorities violate their trust, members can decide to act and they can be justified in overthrowing the regime.[36] A value system which provides for this recourse must place great emphasis on member support if a regime is to maintain itself. On the other hand, the Lockean value system does not encourage regime overthrow. Overthrow may be resorted to only after the values of the regime are seriously assaulted.[37] Indeed, those who lay

> the foundation for overturning the constitution and frame [i.e., the regime] of any just government, he is guilty of the greatest crime I think a man capable of . . . and he who does it is justly to be esteemed the common enemy and pest of mankind, and is to be treated accordingly.[38]

Although a liberal tradition regime would not, then, allow regime dissolution for inconsequential reasons, it appears that a Lockean regime — with its emphasis upon member consent — would require a comparatively high level of value support.

REGIME VALUE SUPPORT IN COMPARATIVE PERSPECTIVE

A comparative perspective again reminds us that the United States is atypical. Regimes may be maintained by various means. Support and coercion are possibilities. Research also suggests that some regimes may be maintained by output manipulation without positive regime support. Authorities may still be able to rule without constant overt coercion or a strong political culture if many members of the system are ambivalent or even if they are opposed to the regime but no alternative can be supported.[39] In this situation authorities rule by sufferance rather than through culture support. For many countries, value consensus does not exist on developed regime values or even, in some cases, upon the basic value component. In these cases, however, the present view is that

[36] *Ibid.*, sec. 222, p. 128.

[37] *Ibid.*, secs. 229–230, p. 133.

[38] *Ibid.*, sec. 230, p. 134.

[39] Daniel Goldrich, *Sons of the Establishment* (Chicago: Rand McNally, 1966), pp. 2–4, 110. As Dahl notes, even those who disagree do not necessarily do anything; see Robert A. Dahl, *Who Governs?* (New Haven: Yale University Press, 1961), p. 332.

TABLE VII.1 *Rule support in Costa Rica (1962) and Panama (1963)*

	In general our system of government and politics is good for the country.		*If a government is doing a good job, it should be allowed to continue in office even if it means postponing elections.*	
	Costa Rica	*Panama*	*Costa Rica*	*Panama*
Agree	91%	40%	29%	54%
Disagree	7	58	71	43
Don't know	1	2	0	3

Source: Daniel Goldrich, *Sons of the Establishment* (Chicago: Rand McNally, 1966), pp. 108, 113.

this tenuous *quid pro quo* support makes a regime less resistant to fundamental change and forces it to look beyond consensual member support for regime maintenance.

Some data regarding regime rule beliefs among elite youth in Costa Rica and Panama (Table VII.1) appear to show the difference between political systems with culture support and those which must rely on other means for maintenance. Costa Rica seems to have political culture support similar to that shown for the United States. The possibility exists that member support performs similar functions in the two countries. Panama, however, presents a very different situation. Majorities actually oppose its rule structure. Panama's regime structure (or at least its nominal one) is opposed but the regime is maintained.

Data available on youth in another South American country show low-level support for community persistence through support for a basic symbol. Colombian support data for the national flag is presented in Table VII.2. Symbol acceptance is much lower than it is in the United States. Table III.9 showed that attachment did not change substantially for American children as they advanced through school. In Colombia, however, the "don't know" category increased substantially. By the IV term only a fragile majority believed the Colombian flag best. Although critical levels of support for national symbols cannot be specified in the present study, Colombia would appear to have weak diffuse support for the nation.

A third study done on Canada, a country often closely associated with the United States as part of an Anglo-American

TABLE VII.2 *Flag symbolism in Colombia*

Question: The Colombian flag is the best flag in the world.

Grade	*Yes*	*Don't know*	*No*
3	86%	13%	1%
4	79	21	0
5	75	24	1
I	83	14	3
II	69	27	4
III	74	20	6
IV	53	42	5

Source: Reid Reading, "Political Socialization in Colombia and the United States," *Midwest Journal of Political Science* (August, 1968), p. 361.

political culture,[40] finds that Canada has a difficult time with national identity. Its historical development, its association with Britain, and its proximity to the United States have created support problems.[41] There are additional identity problems directly related to the existence of both English and French language and culture subsystems within the state.[42] These are most dramatically shown in symbol system attitudes. Table VII.3 presents representative opinions. Again these should be compared to comparable attitudes in the United States. Only on attitudes toward the anthem does opinion approach our culture criteria.

Other comparative support data have been presented throughout the present work. National identity support data for several Western European countries (Tables III.1, III.2, III.3), community trust data for Britain, Germany, Italy, and Mexico (Table III.6), and rule (Table IV.2) and liberty (Table IV.43) support for these same nations have shown a wide range of support for fundamental political values.[43] With such limited data from the above countries, this work cannot presume to make any conclusions. Yet the comparative strength of sup-

40 Gabriel A. Almond, "Comparative Political Systems," *Journal of Politics* (August, 1956), pp. 391–409; and Robert R. Alford, *Party and Society* (Chicago: Rand McNally, 1963), p. 2.

41 Mildred A. Schwartz, *Public Opinion and Canadian Identity* (Berkeley: University of California Press, 1967), ch. 4.

42 *Ibid.*, ch. 5. Also see the articles by G. Horowitz and Erwin C. Hargrove cited in note 60, page 54.

43 Also see Tables IV.39, IV.44, IV.50, IV.56, IV.57, VI.14, Figure III.3, and pages 94, 104, 149, 194, 320, 322, 339–340.

It regulates the number and it shapes the type of demands which are made on the political system. Political culture is also the repository of the diffuse support which will act on the regime. Through political culture the member provides support for the values of the liberal tradition. Since these values are congruent with the American regime structure, support for these values is support for the regime. By thus providing the political system with a protective shaping screen, political culture can become a major means by which the ordinary member can maintain the type of regime he consents to.

THE LIBERAL TRADITION AND DEMOCRACY

Using culture as a shaping aspect of political systems is to some extent incompatible with the "pure" democratic model.[45] Individuals in this model are assumed to discuss rationally and to contest all aspects of politics on the basis of absolute values and arrive at conclusions which can then be translated into policy. For reasons mentioned above, this model is rejected. But, in doing this, is one not forced again to return to the presumed paradox raised earlier? How can a liberal tradition political culture also be a rational, questioning, conflicting, democratic regime?

The paradox is more apparent than real. Although a liberal tradition democracy is not the same as the textbook model, it is not its opposite, either.[46] It does not require total rationality or conflict but it does not exclude them. As a culture system a liberal tradition democracy has *both rational and non-rational aspects.* The acceptance of a political culture does not imply that it is accepted uncritically.[47] It is not passed on by some mysterious force but must use socialization agents who may choose not to pass values on.[48] Many members probably accept values rationally and certainly some rationally reject values. A liberal political culture *may also have conflict* between its

[45] John Dewey, *Freedom and Culture* (New York: Putnam, 1939), esp. pp. 172–173.

[46] See pages 232–240, 294–297.

[47] See Louis Hartz, *The Liberal Tradition in America* (New York: Harcourt, Brace and World, 1955), pp. 10–11.

[48] David Easton and Jack Dennis, *Children in the Political System* (New York: McGraw-Hill, 1969), ch. 2.

values.[49] Indeed, the values themselves may be rejected under sufficient conflict.[50] Value acceptance is merely a tendency. In a liberal tradition culture the possibility for change exists.

More importantly, the textbook model does not explicitly consider the insight of levels at which the political system operates. Although a liberal tradition regime has consensus upon regime values, it *could have substantial policy conflict.* When viewing the major sociological periods in recent comparative history, the most severe cleavages appear to be feudal, liberal, and socialistic. From the perspective of these periods, the United States has had a consensus on the values associated with the liberal period. Major active cleavage has not existed over the value of the liberal tradition values vis à vis uniquely feudal or socialistic values.

This does not mean that the United States has not experienced severe stress. It does mean, however, that stress has not existed to a significant degree at certain levels. Stress in America has not been the result of conflict over the worth of the fundamental values but over their precise meaning and their relationship to each other. It appears that there has been relatively severe conflict in interpreting these values in every generation in American history.[51] Indeed, the major conflicts seem to reduce to five — conflict over: the precise extent of federalism; popular rule; the specific role of the government in regard to property and its regulation; the nature of equality as applied to the Negro; and the interpretation of the necessities of national identity as applied to foreign policy.[52]

Each of the major conflicts of American history — manifested in the Alien and Sedition Acts and the Virginia and Kentucky resolutions, the threatened New England secession during the War of 1812, the South Carolina nullification of the tariff of 1828, the slavery extension controversy and the resulting Civil War, the election of 1876 and the readmission of the South to

[49] Robin M. Williams, Jr., *American Society,* 2d ed. rev. (New York: Knopf, 1960), pp. 573–574.

[50] Easton, p. 106.

[51] Robert A. Dahl, "The American Oppositions: Affirmation and Denial," in Robert A. Dahl, ed., *Political Oppositions in Western Democracies* (New Haven: Yale University Press, 1966), p. 50.

[52] *Ibid.,* pp. 48–53. Note that these differences in interpretation are often due to conflicts between liberal tradition values — eg., between federalism and national identity, popular rule and institutionalization of rule, property and altruism, equality and liberty, and national identity and altruism.

the Union, the rise of agricultural populism and urban labor unionism, the Great Depression and the movement toward the welfare state in the 1930's, and present conflicts relating to Negro rights, foreign policy, and the reaction to New Deal–inspired innovations — these have produced substantial stress. When viewed comparatively, these have not been active cleavages over the merits of the values. They have been interpretational differences which have assumed that the values are worthwhile. There has been substantial conflict, but — with the important exception of the Civil War — it has been settled within the tradition consensus on regime values.

Even considering the extensive examples of stress mentioned above, American conflict has seemed moderate when judged against other political systems. It has seemed so because it has been a conflict within a cultural consensus on basic rules and principles which moderate stress. If all accept a framework within which conflict is to take place, and if all agree to settle conflict within this process and in general accord with certain beliefs, substantial stress can exist at the authority level without threatening the community or regime. In the United States, consensual support for the liberal tradition has provided the framework within which serious conflict has been resolved. Even with substantial authority stress, the shaping strength of political culture has tended to maintain the regime.

To require that rational questioning and conflict occur at the more fundamental levels of community and regime appears to be a most stringent requirement of the textbook model of democracy. The textbook model would apparently continuously expose the members of the political system to new orientations on policy and basic rules at the same time. Few systems could persist under these conditions.[53] Some value consensus appears necessary for any political system.[54]

> However free a society may claim to be, it will nevertheless place some limits on the substantive kinds of demands a member may

[53] Easton, pp. 195–196.

[54] R. M. MacIver, *The Web of Government* (New York: Free Press, 1947; revised 1965), p. 4. Even Kaplan, who tries to adopt Parsons through what he calls a "segmental system" less attuned to value integration, uses agreement on certain values as one necessity; see Harold Kaplan, "The Parsonian Image of Social Structure and Its Relevance to Political Science," *Journal of Politics* (November, 1968), pp. 885–909.

> raise because of their socially disruptive potential. Tolerance in a society cannot permit the unrestricted challenge to any and every value if only because of the effort that would be required to meet each challenge and re-assert a belief in its desirability or presumed necessity. Some of the premises of behavior – its basic rules and goals – must be taken for granted if constant turmoil is to be avoided.[55]

An accepted value system would appear to be even more necessary for a regime based upon consent.[56]

As was mentioned above, much of the argument which holds that a political culture consensus is unnecessary and inconsistent with democracy turns on definition. Although a major democratic theorist has said that consensus on fundamentals is unnecessary and undemocratic, he also wrote that one of the essentials of democratic government is the creation of rules which embody the shared values of the members of the political system.[57] The present view is that a consensus on regime values in no way violates democratic norms other than the overly rationalistic textbook model norms.

EVALUATION

To what extent have our hypotheses been sustained? A great amount of data suggests that the principles and norms of the liberal tradition are the supported values of the American political culture. This does not mean that all the values of the political culture are represented by the liberal tradition. There are undoubtedly others.[58] However, the Lockean liberal

55 Easton, p. 106; also see Mitchell, p. 285.

56 A critic of the American regime notes that "only a non-directive solution to the problem of order would be appropriate to a collectivity organized in accord with a liberal pluralist vision." Daryl Baskin, "American Pluralism: Theory, Practice, and Ideology," *Journal of Politics* (February, 1970), p. 91.

57 Carl Joachim Friedrich, *Man and His Government* (New York: McGraw-Hill, 1963), p. 667; also see pp. 146 ff., 238 ff., 344 ff., 614 ff. Indeed, Friedrich's central concept of authority (as opposed to power) would be imposible without commitment to rule values.

58 For example, Mitchell mentions belief in science, in Protestantism, (p. 212); expectations of honest, effective, and courteous government, opposition to the use of violence, belief in rationality, and belief in positive responsibility to serve the people as political values held by Americans (pp. 270–271). In my opinion these values are not as general or central as the values of the liberal tradition, but they are not always unimportant. Easton and Dennis also identify a more generalized "support for au-

values do appear to represent the core values of the regime, and they probably also represent much of the political culture of the United States. Yet one must also be aware that the test is restricted to the years 1935 to 1970 and that the number and types of questions are limited. One must also note that even the data responses which are available undoubtedly mean something somewhat different to different respondents.[59] It should also be kept in mind that the earliest public opinion polls are less reliable than the current ones and that all polls have some error. Question wording is also a problem in attitude measurement. Beyond these operational problems lies the inherent scientific problem of conceptualization. Conceptualizations are only abstractions of the full complexity of real world values.

Even with these important limitations, however, the data as theoretically organized here seem to confirm the descriptive hypothesis. The summary quality of the liberal tradition as the core of American political culture is impressive. In my opinion, no other systematic thesis is as empirically descriptive. Data are also presented in confirmation of the explanatory hypothesis. The evidence suggests that the political culture values shape in the direction of the culture and that these values tend to be maintained. Since they are here viewed as congruent with the regime values, the regime also tends to be maintained by the same process. The system analogy and the conceptual properties of culture, as integrated with the available data, appear to provide a reasonable explanation of the role of member values in regime maintenance. In this respect, the "limited term" and "American Dilemma" cases were of especial interest.

However, as any framework orders, so does it limit. Absolute criteria for hypothesis verification are probably impossible of fulfillment for any physical or social experiment.[60] Yet, if

---

thority" value (Easton and Dennis, esp. ch. 5; and Easton, pp. 457 ff.). A third study mentions such values as prudent, visible, moderate and legal rules. These also appear important but not as central as the clearly Lockean values. See Paul A. Smith, "The Games of Community Politics," *Midwest Journal of Political Science* (February, 1965), pp. 37–60.

59 Easton, p. 298.

60 Karl R. Popper, *The Logic of Scientific Discovery* (New York: Harper and Row, 1965), ch. 4.

comparability is a reasonable test,[61] I will argue that the liberal tradition hypothesis is the best descriptive statement of a culture value system for the United States and the concepts of political culture maintenance and shaping relatively well explain the member's role in maintaining that regime.

In very recent years, American regime values have seemed to be under more stress than in the past. To some extent this can be explained as characteristic of every age. Each age believes that its values are severely threatened. Again, to some extent, the present stress can be seen as an attempt to fulfill some of the values of the tradition (at the expense, to be sure of some of the others) in a more radical manner — rather than as a total rejection.[62] Today's serious stress, however, does not seem to have (at least as yet) affected the support of the ordinary member. The challenges to the regime seem to be mainly restricted to elites and their lower elite followings on the extreme left and right.[63] Of course, mass support is no guarantee that a regime will survive. History suggests that member support is not among the most important determinants for most regimes. It would seem that a liberal tradition regime, allowing as it does for relatively free political organization, would be especially prone to elite imposed regime change.

Social science analysis cannot predict what will happen in the real world. It can only show forces which are acting and, then, only under very restrictive analytical assumptions. These assumptions always include the caveat of *ceteris paribus.* In the present analysis, I believe that the data show (to the extent that they can) that, at the present, there is consensual support for the liberal tradition values of the American regime. As long as support remains at that level, other things being equal, the regime will be maintained. But, of course, other things are not necessarily equal. Political leaders may ignore member values; environmental stress may make political or other

61 Milton Friedman, *Essays in Positive Economics* (Chicago: University of Chicago Press, 1953), p. 41.

62 As Parsons and White suggest, this may be due to a change in social differentiation (perhaps to a greater concentration in the upper middle class). This is not to say that this phenomenon may not be a potential threat to the regime. It very well may be.

63 See Chapter V, especially the conclusion.

social structure unworkable; elites may organize non–mass-based revolution or *coups d'état;* foreign forces may come to dominate the system; members may come to "rationally" reject the values; or the values may not be transferred to the next generation.

The present study has attempted to show that political culture support by the ordinary member can be a factor in maintaining a regime. For a considerable period of time the regime tested has survived substantial stress. Over two centuries of conflict have not very substantially changed either its values or its regime structure. Whether this will be so for the future is a scientifically unanswerable question.

APPENDIX

1968 Data Test

SECONDARY DATA ANALYSIS presents many problems for the analyst.[1] Not the least of these is the fact that since he is dependent upon others for data collection, the investigator cannot guarantee that the data are equally spread across time. This time dimension is especially important for the present hypothesis, since popular reports suggest that the basic liberal tradition values are no longer supported by important groups – especially youth and blacks.

The nature of the available data make it impossible to completely answer the question of whether the tradition values are still fully supported. However, this appendix is presented as a brief attempt to deal with the thesis using the most current data available within our 1935–1970 time frame. The 1968 Survey Research Center, University of Michigan, election study is the source for the data. Of course, since the study was not conducted to test the present hypothesis, many of the tradition values are not tested and some that are have not been asked in a form most directly related to the thesis. Yet, nine questions somewhat related to Lockean values can be analyzed.

The nine questions relate to the community trust value, support for law, support for government as responsive, support for elections, support for Congress, the federalism value, achievement, and two religion-related values – support for the

[1] See page 69.

Bible and belief that prayers should be allowed in public schools. Again, these questions are not representative of all the tradition values. They are also not the best operationalizations of the value concepts they attempt to measure.[2] For example, the trust question is the same operationalization which has been repeatedly questioned above. This particular question wording appears to be a most conservative indicator of trust support. The better operationalized question would undoubtedly show higher support. The question asking if it is all right to refuse to obey an unjust law if one is willing to go to jail is also a conservative indicator of support for rules: while 61 percent did not think that refusal was proper and thus supported the rule value, 74 percent supported the value in another question in this SRC survey (it asked if the person considered it proper to "stop the government" if all other legal action had failed). The less supported operationalization was purposely used as more rigorous test of the rule value. It should also be noted that the federalism and achievement questions involve conflicts of values which will underestimate the true support for the values.[3]

Even with these limitations Table A.1 shows that all the reported values, with the exception of the trust operationalization, meet the consensual support criteria set in Chapter II. If the better trust operationalization were used (see Table III.5), I believe that it too would easily reach consensual support. Even as operationalized, 55 percent supported the trust value. The table also shows that the *central values* of elections and legislature (Congress) are highly supported. Once again there is substantial support for the religion value. As a whole the most recent data show consensual support for the liberal tradition values.

It was argued in Chapter V, however, that the values should also be supported in the relevant social groupings. Table A.2 reports support in twenty-nine major potential cleavages. As

2 The operational problem is, basically, to find a measure which can concretely represent an abstract concept. See pages 66–76.

3 Because value conflict seriously affects value support (see Chapter VI), the no opinion and qualified responses are excluded for the federalism and achievement questions. This expedient at least allows some idea of value support. The only other recourse was not to use the questions. For the specific question wording, see Tables IV.32 and IV.57.

TABLE A.1 *Support for liberal tradition values, 1968*

	Percent support for value
Community trust	55%
Obey unjust laws	56
Responsive government	65
Elections support	88
Congress support	76
Federalism	63
Achievement	60
Bible	89
School prayer	81

Source: ICPR. For question wording see Tables III.6, IV.5, IV.6, IV.9, IV.32, IV.57, VI.6, and pages 140 and 224.

shown in the overwhelming majority of the tests, the values are supported by a majority of those in the group.

There are two ways to look at Table A.2 in more detail—from the perspective of the values and from the perspective of the groups. It is important to note that 85 percent of the cases where support of values falls below a majority (for trust, unjust laws, federalism, and achievement), it is for values where the operationalization is suspect. One-third of the cases are for the trust question alone. Almost as many fail for the achievement value-conflict situation. Yet the overall impression is one of support for all values. Even the trust operationalization is supported by almost three-quarters of the groups. If the values were better operationalized, I have little doubt that all would be consensually supported by almost every group.

Ten (one-third) of the groups account for all the tests which do not show majority support for the values. Three groups—those with eight, and nine-to-eleven years of education, and the young—do not show majority support for only one value. Two of these tests are for the trust value. Four groups show less than majority support for two values. Democrats show less than majority support for the trust and achievement values, but in both cases support is close to a majority. Those with the lowest income do not support these same two values, but in this case lack of support appears substantial.[4] The apolitical do not show majority support for

[4] Note again that lower achievement support for the low class is conducive to regime maintenance; see page 284, note 143.

trust and Congress. In terms of the conclusion made in Chapter V, however, the surprising finding may be the substantial support given for the other values. Liberals do not show majority support for federalism and achievement. Here especially is the operational problem acute. Indeed to liberals these questions must have seemed highly policy related. Yet support is higher than one might suspect. In evaluating support for these groups, it should be remembered that not supporting two values means that these groups *do* support seven of the nine values reported.

Three of the groups do not show support for four values each — the least educated, Jews, and non-whites. However, note that in each case the weak federalism and achievement value-clash operationalizations are two of the "not supported" values. For the least educated a large percentage of no opinions accounts for the less than majority support for government responsiveness. For Jews the school-prayer question apparently measures not the religion value but an application of it. In some ways the most interesting finding is that Jews and non-whites do not believe in obeying unjust laws. For blacks the identification with the civil rights movement of the 1960's has apparently had an effect. And yet the data reported in Chapters V and VI and especially the support reported in Table A.2 for government responsiveness, elections, and congressional representation leads one to believe that the response to the question on unjust laws is mainly theoretical. In actuality blacks and Jews show as high support for the three major rule values as do the other groups.

Although Table A.2 shows that the young only have less than majority support for one value (and that a plurality — and almost a majority), the idea of a "generation gap" may deserve more detailed consideration. Table A.3 shows support for two major values analyzed by age and education. The particular values were chosen because the religion value might be expected to be less supported by youth — especially college youth — and because the election value is a central measure of the regime. As the table shows, there is almost no evidence of a generation gap. Indeed the young may be slightly more supportive. There is a little more evidence for college-youth nonsupport, but not much. Support is lowest for the college educated 25 to 29 age group for the prayer question, but support

TABLE A.2 *Potential cleavages for American political culture: percentage of support for liberal tradition values, 1968*

	Identity system	*Rule system*					*Belief system*		
Social group	*Community trust*	*Obey unjust laws*	*Responsive government*	*Elections support*	*Congress support*	*Federalism*[a]	*Achievement*[a]	*Bible*	*School prayer*
Political party									
Democrat	47%	54%	65%	88%	74%	52%	48%	92%	84%
Republican	68	59	67	89	78	81	70	90	81
Independent	59	56	65	89	79	65	70	88	76
Apolitical	26	61	58	68	37	77	54	87	88
Political ideology[b]									
Liberal	52	55	67	89	76	45	47	91	79
Conservative	62	62	64	90	78	80	74	91	81
Political public									
Attentive	61	61	69	90	81	67	62	91	81
Non-attentive	53	53	62	86	72	60	59	89	81
Income									
$0–2999	35	54	57	76	63	60	45	93	86
3000–4999	53	50	62	85	67	54	54	90	86
5000–7499	54	55	61	89	76	66	62	92	84
7500 plus	64	59	72	92	84	67	67	89	76
Education									
0–7 years	30	51	48	68	57	49	36	90	89
8 years	37	56	53	82	66	59	57	93	87
9–11 years	46	54	61	90	72	63	58	94	87
12 years	59	64	67	91	81	68	66	92	86
Some college	74	50	78	94	85	63	66	84	67

Sex									
Male	56	58	62	89	76	66	63	87	77
Female	57	54	67	88	76	61	57	92	84
Religion									
Protestant	55	56	64	88	75	67	63	93	83
Catholic	59	58	70	89	78	52	56	91	83
Jew	74	24	61	88	72	47	36	64	41
Other	53	55	67	87	76	66	55	83	74
None	60	55	63	92	76	55	58	59	59
Age									
Young	57	48	72	88	81	61	57	88	74
Middle	58	57	67	91	80	61	60	91	82
Old	51	60	55	79	64	71	62	90	85
Race									
White	59	59	64	88	76	69	66	90	80
Non-white	25	33	69	86	73	19	12	94	92

Source: SRC, ICPR.

[a]No-opinion responses excluded.

[b]Some people are afraid the government in Washington is getting too powerful for the good of the country and the individual person [conservative]. Others feel that the government in Washington is not getting too strong for the good of the country [liberal].

TABLE A.3 *Support for school prayer and elections, by education and age*

	0–7 years education			*8 years education*			*9–11 years education*			
	25–29	*30–59*	*60+*	*25–29*	*30–59*	*60+*	*18–24*	*25–29*	*30–59*	*60+*
Prayer										
Support school prayer	100%	90%	88%	100%	89%	83%	100%	81%	87%	87%
Neutral	0	4	1	0	4	7	0	4	2	4
Anti school prayer	0	6	11	0	7	10	0	15	11	10
Elections										
Support elections	100	83	58	60	88	79	67	87	90	93
Neutral	0	9	25	10	2	3	0	3	3	5
Anti elections	0	8	17	30	10	18	33	10	7	2
N	1	50	81	10	70	69	3	26	140	52

	12 years education				*Some college education*			
	18–24	*25–29*	*30–59*	*60+*	*18–24*	*25–29*	*30–59*	*60+*
Prayer								
Support school prayer	100	83	86	85	85	58	67	83
Neutral	0	3	2	0	0	10	9	6
Anti school prayer	0	14	12	15	15	32	24	11
Elections								
Support elections	100	91	92	84	86	91	95	96
Neutral	0	3	1	6	0	0	0	0
Anti elections	0	6	7	10	14	9	5	4
N	7	65	254	48	13	94	180	46

Source: SRC, ICPR.

is highest for the 18 to 24 age group! The differences shown by age and education for election support are too small to be considered meaningful. Once again the idea of a regime-level generation gap must be rejected.

In summary, although there are cases of non-support, the general impression for 1968 is one of support. Even the three groups which do not show majority support for four values (disregarding for a moment the operational problems), do show majority support for the five other values. No group does not show support for a majority of the values. When one considers the operational problems, however, the more reasonable interpretation is that some groups do not give positive support to one or two of the values of the culture but that, in general, all support the great majority of the values of the tradition.

When considering regime maintenance, moreover, the values most directly related to the regime are supported by all groups. Only one group (the least educated) does not think the government is responsive. Only one group (the apolitical) does not give positive support to the legislative. Not one group does not support elections as effective means to control authorities. Overall, in more than 90 percent of the tests made with the most recent data, the liberal tradition values were supported by a majority of the relevant social groupings of the United States.

Index